# Water Soluble Polymers

Special Issue Editor
**Alexander Penlidis**

MDPI • Basel • Beijing • Wuhan • Barcelona • Belgrade

MDPI

*Special Issue Editor*
Alexander Penlidis
University of Waterloo
Canada

*Editorial Office*
MDPI AG
St. Alban-Anlage 66
Basel, Switzerland

This edition is a reprint of the Special Issue published online in the open access journal *Processes* (ISSN 2227-9717) in 2017 (available at: http://www.mdpi.com/journal/processes/special_issues/soluble_polymers).

For citation purposes, cite each article independently as indicated on the article page online and as indicated below:

Lastname, F.M.; Lastname, F.M. Article title. *Journal Name*. **Year**. *Article number*, page range.

**First Edition 2018**

**ISBN 978-3-03842-695-0 (Pbk)**
**ISBN 978-3-03842-696-7 (PDF)**

# Table of Contents

# About the Special Issue Editor

**Alexander Penlidis**, Professor, Chemical Engineering, University of Waterloo, has thirty seven years of experience in Polymer Engineering and Polymeric Materials, and Mathematical Modelling. He received the 1993 Albright and Wilson Americas Award of the Canadian Society for Chemical Engineering, for distinguished contributions before the age of 40. He is a Fellow of the Chemical Institute of Canada and of the Canadian Academy of Engineering. He has more than 300 publications and has supervised more than 35 Ph.D students. He was the Director of the Institute for Polymer Research (1994–2010) at the University of Waterloo and the founding co-editor of "Polymer Reaction Engineering Journal" (1991–2003).

# Preface to "Water Soluble Polymers"

This book on water soluble polymers (WSP) contains contributions that deal with this extremely popular area of scientific investigation in polymer science and engineering, both in academic and industrial environments. Research and technology on WSP are of current interest not only due to the largely unstudied polymerization kinetics of WSP, but also due to the plethora of technologically innovative applications. Synthetic WSP (copolymers and terpolymers) can modify and improve aqueous solution properties in relation to gelation, thickening, emulsification, stabilization and rheology. Therefore, these polymers find many uses as flocculants and coagulants (for waste water treatment), film-formers, binders, lubricants and coatings, and in enhanced oil recovery (EOR), dewatering, oil-field product and mineral processing, pulp and paper industry (improving papers printing quality), water retention and treatment, and also as biomedical, pharmaceutical and high value cosmetic products.

Hence, the invited contributions cover a wide variety of topics, starting from polymerization kinetics (emphasis on multicomponent systems), clarification of factor effects (for example, ionic strength, pH, monomer concentration, and how they influence important chain characteristics and properties), mathematical modelling, parameter estimation, and process design, and ending with applications (i.e., using the well characterized polymer molecules to deliver specific desirable properties for specific applications (hydrogels, cosmetics, drug release, flocculation, nanotechnology, enhanced oil recovery, polymer flooding, absorbents, crosslinking, and many others)). The contributions integrate experimental and theoretical/ computational studies from both academics and researchers and practitioners from related industry.

This book contains 17 very high quality contributions from author groups that span the globe and represent currently active researchers in the WSP area. The topics are not only current (cutting-edge research) but also of great academic (fundamental phase) and industrial (applied phase) interest. The careful reader will observe the evolution of several common threads (traversing the wide spectrum of polymerization kinetics and modelling, all the way to properties of nano-composites and hydrogels) in the themes of these 17 contributions and the (water soluble) materials they describe.

**Alexander Penlidis**
*Special Issue Editor*

*processes*

MDPI

*Editorial*

# Special Issue: Water Soluble Polymers

Alexander Penlidis

Department of Chemical Engineering, Institute for Polymer Research (IPR), University of Waterloo, Waterloo, ON N2L 3G1, Canada; penlidis@uwaterloo.ca; Tel.: +1-519-888-4567 (ext. 36634)

Academic Editor: Michael Henson
Received: 14 June 2017; Accepted: 14 June 2017; Published: 17 June 2017

This Special Issue (SI) of *Processes* on water soluble polymers (WSP), and the associated Special Issue reprint, contain papers that deal with this extremely popular area of scientific investigation in polymer science and engineering, both in academic and industrial environments. Research and technology on WSP are of current interest not only due to the largely unstudied polymerization kinetics of WSP, but also due to the plethora of technologically innovative applications. Synthetic water soluble copolymers (and terpolymers) can modify and improve aqueous solution properties in relation to gelation, thickening, emulsification, stabilization and rheology. Therefore, these polymers find many uses as flocculants and coagulants (for waste water treatment), film-formers, binders, lubricants and coatings, and in enhanced oil recovery (EOR), dewatering, oil-field product and mineral processing, pulp and paper industry (improving paper's printing quality), water retention and treatment, and also as biomedical, pharmaceutical and high value cosmetic products.

Hence, we have invited papers on a wide variety of topics, starting from polymerization kinetics (emphasis on multicomponent systems), clarification of factor effects (for example, ionic strength, pH, monomer concentration, and how they influence important chain characteristics and properties), mathematical modelling and parameter estimation, and process design, and ending with applications (i.e., using the well characterized polymer molecules to deliver specific desirable properties for specific applications (hydrogels, cosmetics, drug release, flocculation, nanotechnology, enhanced oil recovery, polymer flooding, absorbents, crosslinking, and many others)). We have been particularly interested in receiving manuscripts that integrate experimental and theoretical/computational studies, as well as contributions from industry. We have thus invited not only academics but also researchers and practitioners from related industry to submit manuscripts for this important Special Issue (SI) of *Processes*.

This SI has already published 12 very high quality papers. The author groups clearly span the globe and represent currently active researchers in the WSP area. The topics are not only current (cutting-edge research) but also of great academic (fundamental phase) and industrial (applied phase) interest. The papers are cited below, with brief comments for each paper concerning the main topic and contributions of the paper. The careful reader will observe the evolution of several common threads (traversing the wide spectrum of polymerization kinetics and modelling all the way to properties of nano-composites and hydrogels) in the themes of these 12 papers and the (water soluble) materials they describe.

(1)   Maric, M.; et al. Poly(methacrylic acid-*ran*-2-vinylpyridine) Statistical Copolymer and Derived Dual pH-Temperature Responsive Block Copolymers by Nitroxide-Mediated Polymerization. [1]

The very first submitted and accepted paper of the Special Issue on water soluble polymers deals with nitroxide-mediated polymerization using NHS-BlocBuilder. The main contributions (in fundamental polymer science and chemistry) are twofold: (a) To produce copolymers with tunable water solubility; and (b) To synthesize block copolymers with pH-temperature responsive properties.

(2)   Scott, A.; et al. AMPS/AAm/AAc Terpolymerization: Experimental Verification of the EVM Framework for Ternary Reactivity Ratio Estimation. [2]

The paper describes terpolymerization kinetics of an acrylamide/acrylic acid terpolymer that finds uses in enhanced oil recovery. The main contribution is an experimental demonstration of deriving optimal feed compositions for the design of experiments that lead to optimal reactivity ratio parameter estimation under the EVM framework, which in turn lead to reliable model predictions for terpolymer composition over the full conversion range.

(3)   Tsai, B.; et al. Poly(Poly(Ethylene Glycol) Methyl Ether Methacrylate) Grafted Chitosan for Dye Removal from Water. [3]

This paper's subject is chitosan grafting to produce adsorbing materials for textile dye removal from water (water pollution and textile material production). The adsorbing materials are produced via nitroxide-mediated polymerization 'grafting to' approach. The main contribution, after detailed grafting polymerization chemistry and adsorption studies, is that grafted chitosan is much more effective (by about 30%) than its parent chitosan.

(4)   Rintoul, I. Kinetic control of aqueous polymerization using radicals generated in different spin states. [4]

This paper provides an analysis of experimental conditions required to develop (and potentially exploit) magnetic field (MF) effects in the free radical polymerization of water-soluble polymers. Electron spin states (configuration), MF intensity, and solution viscosity are varied and evaluated for the solution polymerization of acrylamide. It is found that MF effects are significant in photoinitiated polymerizations, specifically at low MF intensities and in viscous reaction media. MF effects are absent in thermally initiated polymerizations (regardless of MF intensity).

(5)   Wu, A.; et al. Simultaneous Monitoring of the Effects of Multiple Ionic Strengths on Properties of Copolymeric Polyelectrolytes during Their Synthesis. [5]

The paper describes an automated online monitoring system with multiple light scattering and viscosity detectors (ACOMP). Results demonstrate the capabilities of ACOMP with the acrylamide/styrene sulfonate copolymerization system for a series of ionic strength levels.

(6)   Hughes, A.; et al. Biodegradable and Biocompatible PDLLA-PEG1k-PDLLA Diacrylate Macromers: Synthesis, Characterisation and Preparation of Soluble Hyperbranched Polymers and Crosslinked Hydrogels. [6]

This paper describes many aspects behind the chemistry and the art of preparing soluble hyperbranched polymers and cross-linked hydrogels. Starting from the ring opening polymerization of D,L-lactide, the target is biodegradable hydrogels with tailored swelling properties for potential applications in regenerative medicine.

(7)   Emaldi, I.; et al. Kinetics of the Aqueous-Phase Copolymerization of MAA and PEGMA Macromonomer: Influence of Monomer Concentration and Side Chain Length of PEGMA. [7]

In-situ NMR is employed to monitor the copolymerization of fully ionized methacrylic acid and PEGMA macromonomer, investigating the effects of monomer concentration and side chain length of PEGMA. Different trends in estimated reactivity ratio values are demonstrated and explained.

(8)   Achilias, D.; et al. Polymerization Kinetics of Poly(2-Hydroxyethyl Methacrylate) Hydrogels and Nanocomposite Materials. [8]

This paper takes us into the world of bio(nano)materials and potential applications in tissue engineering and contact lenses, by studying poly(HEMA)-based hydrogels. There are two main contributions in the paper: (a) The development of a detailed and improved model for HEMA polymerization kinetics over the full conversion range; and (b) The effects of nano-additives on the rate of (in situ) polymerization (nano-silica vs. nano-montmorillonite) with possible explanations of the observed behaviour.

(9)    Fischer, E.; et al.  Aqueous Free-Radical Polymerization of Non-Ionized and Fully Ionized Methacrylic Acid. [9]

The topic of this paper is the free radical polymerization of non-ionized and fully ionized methacrylic acid, in an effort to shed additional light on the peculiar polymerization kinetic behavior of carboxylic acids in aqueous media. The strength of the paper is the clarity of the model development stages and the suggestion of a novel propagation rate expression that takes into account the effect of electrostatic interactions (fully ionized case).

(10)   Han, W.; et al. Applications of Water-Soluble Polymers in Turbulent Drag Reduction. [10]

Turbulent drag reduction is a complex topic which has been reviewed from several different angles (ranging from rheology/fluid mechanics to polymer concentrations used all the way to mathematical models involved). Hence, what would yet another review paper on the topic possibly accomplish? The authors of the current paper critically review 117 recent references and look at turbulent drag reduction from the angle of the types of water soluble polymers and copolymers employed, both synthetic (see Table 1) and natural (see Table 2), along with potential applications.

(11)   Steinmacher, F.; et al. Design of Cross-Linked Starch Nanocapsules for Enzyme-Triggered Release of Hydrophilic Compounds. [11]

The paper describes the synthesis of cross-linked (X-linked) starch (aqueous-core) nanocapsules (NCs) in inverse mini-emulsion and shows experimental data on the influence of X-linker level on several product variables, including morphology. The main contribution of the paper is the design of both a permeable and impermeable NC shell depending on X-linker level. Impermeable shells are further investigated with respect to release studies.

(12)   Pérez-Salinas, P.; et al. Comparison of Polymer Networks Synthesized by Conventional Free Radical and RAFT Copolymerization Processes in Supercritical Carbon Dioxide. [12]

Although this paper's main topic is RAFT vs. regular free radical copolymerization hydrogels in supercritical $CO_2$, the approach for comparing and evaluating polymer networks is general and hence applicable to other types of (co)polymerization. Based on a rich set of experimental results, the main contributions of the paper are twofold: (a) A quantitative criterion is suggested for assessing the degree of heterogeneity (homogeneity) of a polymer network; and (b) The paper concludes with additional information on antibiotic loading, adsorption and release studies for the investigated hydrogels.

One can locate and read these papers via the following link:

http://www.mdpi.com/journal/processes/special_issues/soluble_polymers

In order to complement the above 12 excellent contributions, the deadline for this SI has been extended to 30 September 2017. In addition, *Processes* is planning to produce a Special Issue Reprint (SIR) for this successful SI. The SIR will become available online at Amazon after all the papers have been reviewed and published.

We are looking forward to your contribution before 30 September 2017.

Special Issue Editor

Alexander Penlidis

Department of Chemical Engineering

Institute for Polymer Research (IPR)

University of Waterloo

Canada

## References

1. Maric, M.; Zhang, C.; Gromadzki, D. Poly(methacrylic acid-*ran*-2-vinylpyridine) Statistical Copolymer and Derived Dual pH-Temperature Responsive Block Copolymers by Nitroxide-Mediated Polymerization. *Processes* **2017**, *5*, 7. [CrossRef]
2. Scott, A.; Kazemi, N.; Penlidis, A. AMPS/AAm/AAc Terpolymerization: Experimental Verification of the EVM Framework for Ternary Reactivity Ratio Estimation. *Processes* **2017**, *5*, 9. [CrossRef]
3. Tsai, B.; Garcia-Valdez, O.; Champagne, P.; Cunningham, M. Poly(Poly(Ethylene Glycol) Methyl Ether Methacrylate) Grafted Chitosan for Dye Removal from Water. *Processes* **2017**, *5*, 12. [CrossRef]
4. Rintoul, I. Kinetic control of aqueous polymerization using radicals generated in different spin states. *Processes* **2017**, *5*, 15. [CrossRef]
5. Wu, A.; Zhu, Z.; Drenski, M.; Reed, W. Simultaneous Monitoring of the Effects of Multiple Ionic Strengths on Properties of Copolymeric Polyelectrolytes during Their Synthesis. *Processes* **2017**, *5*, 17. [CrossRef]
6. Hughes, A.; Tai, H.; Tochwin, A.; Wang, W. Biodegradable and Biocompatible PDLLA-PEG1k-PDLLA Diacrylate Macromers: Synthesis, Characterisation and Preparation of Soluble Hyperbranched Polymers and Crosslinked Hydrogels. *Processes* **2017**, *5*, 18. [CrossRef]
7. Emaldi, I.; Hamzehlou, S.; Sanchez-Dolado, J.; Leiza, J. Kinetics of the Aqueous-Phase Copolymerization of MAA and PEGMA Macromonomer: Influence of Monomer Concentration and Side Chain Length of PEGMA. *Processes* **2017**, *5*, 19. [CrossRef]
8. Achilias, D.; Siafaka, P. Polymerization Kinetics of Poly(2-Hydroxyethyl Methacrylate) Hydrogels and Nanocomposite Materials. *Processes* **2017**, *5*, 21. [CrossRef]
9. Fischer, E.; Storti, G.; Cuccato, D. Aqueous Free-Radical Polymerization of Non-Ionized and Fully Ionized Methacrylic Acid. *Processes* **2017**, *5*, 23. [CrossRef]
10. Han, W.; Dong, Y.; Choi, H. Applications of Water-Soluble Polymers in Turbulent Drag Reduction. *Processes* **2017**, *5*, 24. [CrossRef]
11. Steinmacher, F.; Baier, G.; Musyanovych, A.; Landfester, K.; Araújo, P.; Sayer, C. Design of Cross-Linked Starch Nanocapsules for Enzyme-Triggered Release of Hydrophilic Compounds. *Processes* **2017**, *5*, 25. [CrossRef]
12. Pérez-Salinas, P.; Jaramillo-Soto, G.; Rosas-Aburto, A.; Vázquez-Torres, H.; Bernad-Bernad, M.; Licea-Claverie, Á.; Vivaldo-Lima, E. Comparison of Polymer Networks Synthesized by Conventional Free Radical and RAFT Copolymerization Processes in Supercritical Carbon Dioxide. *Processes* **2017**, *5*, 26. [CrossRef]

*processes*

MDPI

*Article*

# Poly(methacrylic acid-*ran*-2-vinylpyridine) Statistical Copolymer and Derived Dual pH-Temperature Responsive Block Copolymers by Nitroxide-Mediated Polymerization

**Milan Maric [1,*], Chi Zhang [1,2] and Daniel Gromadzki [1]**

[1]   Department of Chemical Engineering, McGill University, Montreal, QC H3A 0C5, Canada;
      chi.zhang2@mail.mcgill.ca (C.Z.); dgromadzki.pst@gmail.com (D.G.)
[2]   AstenJohnson Inc., 50 Richardson Side Rd, Kanata, ON K2K 1X2, Canada
*    Correspondence: milan.maric@mcgill.ca; Tel.: +1-514-398-4272

Academic Editor: Alexander Penlidis
Received: 19 January 2017; Accepted: 13 February 2017; Published: 21 February 2017

**Abstract:** Nitroxide-mediated polymerization using the succinimidyl ester functional unimolecular alkoxyamine initiator (NHS-BlocBuilder) was used to first copolymerize *tert*-butyl methacrylate/ 2-vinylpyridine (*t*BMA/2VP) with low dispersity (Đ = 1.30–1.41) and controlled growth (linear number average molecular $M_n$ versus conversion, $M_n$ = 3.8–10.4 kg·mol$^{-1}$) across a wide composition of ranges (initial mol fraction 2VP, $f_{2VP,0}$ = 0.10–0.90). The resulting statistical copolymers were first de-protected to give statistical polyampholytic copolymers comprised of methacrylic acid/2VP (MAA/2VP) units. These copolymers exhibited tunable water-solubility due to the different pKas of the acidic MAA and basic 2VP units; being soluble at very low pH < 3 and high pH > 8. One of the *t*BMA/2VP copolymers was used as a macroinitiator for a 4-acryloylmorpholine/4-acryloylpiperidine (4AM/4AP) mixture, to provide a second block with thermo-responsive behavior with tunable cloud point temperature (CPT), depending on the ratio of 4AM:4AP. Dynamic light scattering of the block copolymer at various pHs (3, 7 and 10) as a function of temperature indicated a rapid increase in particle size >2000 nm at 22–27 °C, corresponding to the 4AM/4AP segment's thermos-responsiveness followed by a leveling in particle size to about 500 nm at higher temperatures.

**Keywords:** nitroxide-mediated polymerization; poly(ampholytes); stimuli-responsive polymers

---

## 1. Introduction

The manipulation of properties by copolymerization (e.g., graft, gradient, block, star architectures) has long been applied to impart desirable properties into polymers. One such class that combines properties are poly(ampholytes) or poly(zwitterions), which contain both negative and positive charges on the chain, either on different monomers or within a single monomer unit [1]. Originally, such copolymers were made via statistical free radical copolymerization of the unalike monomers [2–4]. Such pairs included methacrylic acid (MAA)/2-vinyl pyridine (2VP) [2], MAA/2-dimethylamino ethyl methacrylate (DMAEMA) [3,4] and acrylic acid (AA)/DMAEMA [5]. Later, the uncoupling of the charges was desired to make block copolymers, which led to dramatically different properties in solution. Traditionally, living polymerizations such as ionic [6–11] and later group transfer polymerization [12–16] were used to make polymers with active chain ends that would permit the formation of poly(ampholytes) or schizophrenic block copolymers with a controlled sequence length and narrow molecular weight distributions. In the case of charged species, however, this required protecting group chemistry during the synthesis [6,10]. However, there are drawbacks to using living polymerizations such as ionic polymerizations: meticulous air-free transfers;

extensive purification of solvents and monomers; polymerizations cannot be done in aqueous media; and, in some cases, certain monomer types cannot be polymerized in a desired sequence [17–19]. Consequently, the development of controlled radical polymerization (CRP), more succinctly defined by the International Union of Pure and Applied Chemistry (IUPAC) as reversible de-activation radical polymerization (RDRP) [20], approaches many of the features that make conventional radical polymerization so industrially relevant: ability to be done in dispersed aqueous media; and tolerance of a wide variety of monomers and functional groups, all while approaching the degree of control exhibited by truly living polymerizations. Many of these RDRP methods have been popularized, some of which are nitroxide-mediated polymerization (NMP) [21], atom transfer radical polymerization (ATRP) [22,23] and reversible addition fragmentation transfer polymerization (RAFT) [24].

Generally, ATRP and RAFT have surpassed NMP in terms of versatility, since they can polymerize a wide variety of functional monomers, even though NMP was developed first (NMP was originally restricted to styrenic monomers). Consequently, poly(ampholytic) block copolymers have largely been made using ATRP [25–29] and RAFT [30–32]. However, NMP has narrowed the gap considerably with the development of second-generation nitroxide initiators based on 2,2,5-trimethyl-4-phenyl-3-azahexane-3-oxyl nitroxide TIPNO and the so-called SG1 or BlocBuilder family (based on *N-tert*-butyl-*N*-[1-diethylphosphono-(2,2 dimethylpropyl)] nitroxide) [33,34]. This history of NMP is chronicled in detail elsewhere [35]. Tertiary acrylamides [36–40], acrylates [33,34,41–45] and methacrylates (with a small concentration of co-monomer ~1 mol%–10 mol% to help control the activity of the chain ends) [46–56] were polymerized in a controlled manner using NMP with the commercial unimolecular initiator BlocBuilder. In the latter case with methacrylates, the controlling co-monomer was initially a styrenic and later studies showed that similar monomers can act as controllers. Indeed, many groups applied controlling co-monomers that imparted desirable functional characteristics. For example, acrylonitrile (AN) was used as a controller first for methyl methacrylate (MMA) [47] and then for oligo ethylene glycol methacrylate (OEGMA) [57]. For these cases, the AN imparted better water solubility and cells exhibited non-cytotoxic behavior. In another case, 9-vinylphenyl-9H-carbazole (VBK), acting as a controlling co-monomer for OEGMA/diethylene glycol methacrylate (OEGMA/DEGMA) [54] and dimethylaminoethyl methacrylate (DMAEMA) [55], was shown to add temperature modulated fluorescence, while using only 1 mol% in the initial composition to impart sufficient control of the polymerization. When 4-styrene sulfonic acid, sodium salt (SSNa) was used as a controller for sodium methacrylate (MAA-Na) in homogenous aqueous media, it was subsequently used as a dual initiator/surfactant for ab initio dispersed MMA polymerizations in water [58]. We also reported 2-vinylpyridine (2VP) as a controlling co-monomer for DMAEMA [59] and a protected organo-soluble styrene sulfonic acid controller for glycidyl methacrylate [60]. Many cases describe the use of NMP for making thermo-responsive copolymers; however, poly(ampholytes) and associated multi-responsive systems have not been described in many cases. Here, we statistically copolymerized 2VP with *tert*-butyl methacrylate (*t*BMA), which after deprotection of *t*BMA, results in 2VP/methacrylic acid (MAA) ampholytic statistical copolymers. It should be noted that tolerance to functional groups is relative as we used the protected form of methacrylic acid (MAA), *tert*-butyl methacrylate (*t*BMA). The protected form was used in NMP as the organic acid can attack the alkoxyamine, rendering the chain ends inactive. To circumvent this issue, MAA or acrylic acid (AA) NMP was done with the addition of a small amount of free nitroxide, which was essentially sacrificed to keep the chain ends active throughout the polymerization [48,61,62]. The poly(2VP-*stat*-*t*BMA) copolymers were then used as macroinitiators for a tunable, thermo-responsive segment of poly(acryloyl piperidine-*ran*-acryloyl-morpholine) (4AP-*stat*-4AM), which we reported recently [63]. We thus present the synthesis of a dual responsive (pH and temperature) block copolymer where the respective blocks' response could be tuned by its composition (*t*BMA versus 2VP for pH tuning; 4AP versus 4AM for cloud point tuning). The pH sensitivity of the poly(2VP-*stat*-MAA) copolymers and thermo-responsiveness of a poly(2VP-*stat*-*t*BMA)-b-poly(4AP-*stat*-4AP) block copolymer at various pHs are thus the focus of this report.

## 2. Materials and Methods

### 2.1. Materials

The 2-Vinylpyridine (2VP, 97%), *tert*-butyl methacrylate (*t*BMA, 98%), basic alumina (Brockmann, Type I, 150 mesh), and calcium hydride (90%–95% reagent grade) were purchased from Sigma-Aldrich (Oakville, ON, Canada). Tetrahydrofuran (THF, 99.9%), methylene chloride, methanol, dimethylformamide (DMF) diethyl ether, hexane (all certified grade) and pH buffers were obtained from Fisher (Nepean, ON, Canada); deuterated chloroform (CDCl$_3$) was obtained from Cambridge Isotope Laboratories Inc. (Tewksbury, MA, USA). Trifluroacetic acid (TFA, 99.9%) was purchased from Caledon (Georgetown, ON, Canada).

The 2-({*tert*-butyl[1-(diethoxyphosphoryl)-2,2-(dimethylpropyl]amino}oxy)-2-methylpropionic acid, also known as BlocBuilder$^{TM}$ (Scheme 1b, 99%), was obtained from Arkema (King of Prussia, PA, USA). NHS-BlocBuilder (Scheme 1c) was synthesized via coupling of BlocBuilder and N-hydroxysuccinimide following procedures described previously [64]. The 2VP and *t*BMA monomers were purified by passing through a column of basic alumina mixed with 5 wt% calcium hydride; they were stored in a sealed flask under a head of nitrogen in a refrigerator until needed. All other compounds were used as received.

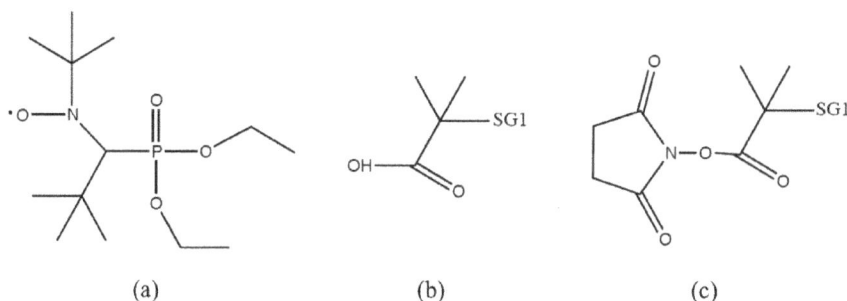

| (a) | (b) | (c) |
|---|---|---|

**Scheme 1.** Structures of (**a**) SG1 nitroxide (**b**) BlocBuilder$^{TM}$ (**c**) N-hydroxy succinimidyl ester-coupled BlocBuilder (NHS-BlocBuilder).

### 2.1.1. Statistical Copolymerization of *tert*-Butyl Methacrylate (*t*BMA) and 2-Vinylpyridine (2VP)

The copolymerizations of *t*BMA and 2VP were performed in a 50 mL three-neck round-bottom glass flask fitted with a reflux condenser, a magnetic stir bar, and a thermal well. BlocBuilder and SG1 (10 mol% relative to BlocBuilder) were dissolved in *t*BMA and 2VP monomers with feed composition ranging from 90 mol% *t*BMA to 10 mol% *t*BMA. Detailed feed solution compositions can be found in Table 1. The solution was then deoxygenated by nitrogen bubbling for 30 min at room temperature prior to heating to 100 °C at a rate of about 8 °C·min$^{-1}$ while maintaining a nitrogen purge. The time when the solution reached 100 °C was taken as the start of the reaction ($t = 0$). Samples were taken periodically to monitor conversion and molecular weight. The final polymer was precipitated in methanol/water ($v/v$ 30/70) for *t*BMA/2VP-90/10 or hexane for the other copolymers, decanted and dried in vacuum at 40 °C overnight. The purified polymer (50% conversion) has number-average molecular weight ($M_n$) = 15.8 kg·mol$^{-1}$ and dispersity (*Đ*) = 1.29. Conversion was determined by gravimetry. Overall conversion was then calculated using the feed composition with respect to *t*BMA ($f_{tBMA,0}$, mol%, *Conv.*$_{ave}$ = $f_{tBMA,0}$ × *Conv.*$_{tBtMA}$ + (1 − $f_{tBMA,0}$) × *Conv.*$_{2VP}$). Molecular weight and *Đ* of the samples were measured by GPC (Waters Breeze) relative to linear PMMA standards (see *Gel*

*Permeation Chromatography* section for full details). Final copolymer composition was determined by [1]H NMR (400 MHz Varian Gemini 2000 spectrometer, CDCl$_3$) using the *tert*-butyl group protons ($\delta$ = 1.0–1.6 ppm) as a marker for the *t*BMA, and the aromatic proton adjacent to the nitrogen atom ($\delta$ = 7.2–7.6 ppm) as a marker for 2VP.

**Table 1.** Experimental conditions for the *tert*-butyl methacrylate/2-vinylpyridine *t*BMA/2VP statistical copolymerizations via nitroxide mediated polymerization (NMP) at 100 °C in bulk.

| Experiment ID [a] | [BlocBuilder] mol·L$^{-1}$ | [SG1] mol·L$^{-1}$ | [*t*BMA] mol·L$^{-1}$ | [2VP] mol·L$^{-1}$ | $f_{tBMA,0}$ [b] mol% | $M_{n,target}$ [c] kg·mol$^{-1}$ |
|---|---|---|---|---|---|---|
| *t*BMA/2VP-90 | 0.036 | 0.004 | 5.733 | 0.631 | 90% | 24.8 |
| *t*BMA/2VP-70 | 0.036 | 0.004 | 4.789 | 2.046 | 70% | 25.1 |
| *t*BMA/2VP-50 | 0.037 | 0.004 | 3.692 | 3.622 | 50% | 24.9 |
| *t*BMA/2VP-30 | 0.038 | 0.004 | 2.399 | 5.629 | 30% | 25.2 |
| *t*BMA/2VP-10 | 0.038 | 0.004 | 0.858 | 7.939 | 10% | 25.5 |

[a] Experimental identification (ID) for copolymers was given by *t*BMA/2VP-X, where X refers to the feed composition with respect to *t*BMA; [b] Feed composition with respect to *t*BMA; [c] Theoretical molecular weight at 100% conversion.

### 2.1.2. Chain Extension of Poly(*tert*-butyl methacrylate-*ran*-2-vinylpyridine)(*t*BMA-*ran*-2VP) Macroinitiator with 4-acryloylmorpholine/4acryloylpiperidine (4AM/4AP) Mixtures

To test the chain end fidelity and to incorporate additional functionality, in this case, thermo-responsiveness, chain extension from a poly(*t*BMA-*ran*-2VP) with a 4-acryloylmorpholine/4-acryloylpiperidine (4AM/4AP) mixture was done. Poly(4AM-*stat*-4AP) copolymers exhibit lower critical solution temperature (LCST) behaviour in aqueous solution that can be tuned by the relative concentrations of 4AM to 4AP in the copolymer [60]. A specific example is given as follows. *t*BMA/2VP-50 was used as a macroinitiator ($\overline{M_n}$ = 4.8 kg·mol$^{-1}$, *Đ* = 1.36, $F_{tBMA}$ = 0.48, see Table 2 for complete characterization). Using the same reactor conditions as described in the section above, 0.1205 g of *t*BMA/2VP-50 was added to 2.9575 g of DMF solvent, 1.0868 g of 4AM and 0.5400 g of 4AP. After purging with nitrogen for 30 min at room temperature, the temperature was increased to a set-point of 120 °C. The time = 0 taken for the chain extension was taken to be when the temperature reached 110 °C. After 75 min, the solution became increasingly viscous, and the sample taken cleanly precipitated the polymer from diethyl ether. The reactor contents were then cooled and the contents were poured into an excess of diethyl ether. The supernatant was decanted and the crude product was re-dissolved in THF and then precipitated again into diethyl ether. The yield of the resulting product was 0.8108 g. Gel permeation chromatography (GPC) revealed some unreacted macroinitiator and the polymer was fractionated again by dissolution in a minimal amount of THF, followed by precipitation slowly with diethyl ether. The polymer was recovered and GPC indicated virtually complete removal of the lower molecular weight impurity. GPC indicated $\overline{M_n}$ = 22.1 kg·mol$^{-1}$, *Đ* = 1.67 with a composition of $F_{tBMA}$ = 0.11, $F_{2VP}$ = 0.11, $F_{4AM}$ = 0.46, $F_{4AP}$ = 0.32 using [1]H NMR (CDCl$_3$ $\delta$ (ppm)): (*t*BMA, (1.0–1.6 ppm, 9H C(CH$_3$)$_3$), 0.9 ppm, -CH$_2$-C(CH$_3$)H on backbone; 2VP, (7.2–7.6 ppm, 4H aromatic); 4AM, $\delta$: (3.5 ppm, 8H, -N-(CH$_2$)$_2$-CH$_2$-CH$_2$-O-); 4AP, $\delta$: (1.5 ppm, 10H, -N-(CH$_2$)$_2$-CH$_2$-CH$_2$-CH$_2$)). Note that the 4AP protons were obscured by the backbone protons and the backbone protons were used to determine the composition. The composition ratio of the *t*BMA/2VP:4AM/4AP blocks corresponds to about 1:4, which is in relatively good agreement with that estimated by the GPC chromatograms. The final yield after fractionation was 0.55 g (70% from the crude polymeric product).

**Table 2.** Molecular weight and composition characterization of tert-butyl methacrylate/2-vinylpyridine (*t*BMA/2VP) statistical copolymers synthesized via nitroxide mediated polymerization (NMP).

| Experiment ID [a] | Reaction Time (min) | Conversion [b] | $F_{tBMA}$ [c] (mol%) | $\overline{M}_n$ [d] (kg·mol$^{-1}$) | $Đ$ [d] |
|---|---|---|---|---|---|
| *t*BMA/2VP-90 | 45 | 31% | 84% | 10.4 | 1.38 |
| *t*BMA/2VP-70 | 202 | 19% | 65% | 7.2 | 1.41 |
| *t*BMA/2VP-50 | 240 | 19% | 48% | 4.8 | 1.36 |
| *t*BMA/2VP-30 | 300 | 18% | 25% | 4.9 | 1.35 |
| *t*BMA/2VP-10 | 301 | 19% | 13% | 3.8 | 1.30 |

[a] Experimental identification (ID) for copolymers was given by *t*BMA/2VP-X, where X refers to the feed composition with respect to tBMA; [b] Monomer conversion was determined gravimetrically; [c] Copolymer composition with respect to *t*BMA was determined by $^1$H NMR; [d] Number-average molecular weight ($M_n$) and dispersity ($Đ$) were determined by gel permeation chromatography (GPC) relative to poly(methyl methacrylate) standards in tetrahydrofuran at 40 °C.

### 2.1.3. Hydrolysis of *tert*-Butyl Groups in the *t*BMA/2VP Statistical Copolymers

The *t*BMA/2VP copolymers were hydrolyzed by trifluoroacetic acid (TFA) to obtain the water-soluble methacrylic acid (MAA)/2VP copolymers. The procedures for the hydrolysis were as follows: The copolymers (~1 g) were dissolved in about 5 mL of methylene chloride. TFA (five times equivalent to the *tert*-butyl group) was then slowly added to the solution. The solution was then stirred at room temperature for up to 24 h or until the MAA/2VP copolymer completely precipitated from the solution. The MAA/2VP copolymer was then rinsed with methylene chloride, re-dissolved in methanol and re-precipitated in diethyl ether, decanted and dried. $^1$H NMR (400 MHz Varian Gemini 2000 spectrometer, CDCl$_3$, Varian, Palo Alto, CA, USA) was used to check the disappearance of the *tert*-butyl protons ($\delta$ = 1.0–1.6 ppm) after hydrolysis.

### 2.1.4. Gel Permeation Chromatography

Molecular weight and $Đ$ of all polymers were characterized by gel permeation chromatography (GPC) (Waters Breeze, Waters Ltd., Mississauga ON, Canada) using THF as the mobile phase at 40 °C in this study. The GPC was equipped with three Waters Styragel HR columns (molecular weight measurement ranges: HR1: $10^2$–5 × $10^3$ g·mol$^{-1}$, HR2: 5 × $10^2$–2 × $10^4$ g·mol$^{-1}$, HR3: 5 × $10^3$–6 × $10^5$ g·mol$^{-1}$) and a guard column. The columns were operated at 40 °C and with a mobile phase flow rate of 0.3 mL·min$^{-1}$ during analysis. The GPC was also equipped with both ultraviolet (UV 2487) and differential refractive index (RI 2410) detectors. The results reported in this paper were obtained from the RI detector. The molecular weight measurements were calibrated relative to linear poly(methyl methacrylate) narrow molecular weight distribution standards.

### 2.1.5. Titration of MAA/2VP Statistical Copolymers

For MAA/2VP-90, MAA/2VP-70 and MAA/2VP-50 copolymers, dissolution was done in 0.1 N sodium hydroxide and then titrated with 1 N hydrochloric acid while the pH of the solutions was monitored with a pH probe calibrated with pH 4, 7, and 10 buffers. For MAA/2VP-30 and MAA/2VP-10 copolymers, dissolution was done in 0.1 N hydrochloric acid and then titrated with 1 N sodium hydroxide while the pH of the solution was monitored. A clear to cloudy transition was observed for all samples as pH changed, and a second cloudy to clear transition was observed for MAA/2VP-90, MAA/2VP-70 and MAA/2VP-50 solutions. Samples were taken at multiple pH values to be further analyzed by dynamic light scattering.

### 2.1.6. Particle Size Measurements of the Aqueous Solutions of MAA/2VP Statistical Copolymers

Dynamic light scattering (DLS) with a Malvern ZetaSizer (Nano-ZS, Malvern, UK) was used to determine the hydrodynamic radius of the MAA/2VP statistical copolymer in aqueous solutions at different pHs taken during titration. The instrument was equipped with a He-Ne laser operating at

633 nm, an avalanche photodiode detector, and a temperature-controlled cell. All clear solutions were filtered using a 0.2 μm pore size filter prior to DLS measurements. All measurements were obtained at 25 °C. For the block copolymer with thermo-responsive segments, 1 wt% solutions were prepared in deionized water and the pH was adjusted with the desired buffer solution (pH = 3, 7, and 10). The samples were filtered through a 0.2 μm pore size filter prior to measurement and then heated in increments of 1.0 °C, allowed to equilibrate for 1 min followed by 10–14 measurements, which were then averaged together to give one value at the corresponding temperature. All DLS measurements were performed at a scattering angle ($\theta$) of 173°. For more accurate measurement of the hydrodynamic radius, $R_h$, the refractive index (RI) is required and it was assumed to be that of PMMA.

## 3. Results

### 3.1. Statistical Copolymerization of tert-Butyl Methacrylate (tBMA) and 2-Vinylpyridine (2VP) Using NHS-BlocBuilder

Table 2 summarizes the polymerization results including reaction time and conversion as well as main characteristics of the statistical copolymers of tBMA/2VP. The polymerization rate decreased exponentially when feed composition of tBMA was decreased from 90% to 70%. This is well known for methacrylate-rich copolymerizations [46–49]. The polymerization kinetics with the various feed compositions is illustrated in Figure 1. The parameters often used to characterize NMP kinetics, $<k_p>$ $<K>$ where $<k_p>$ is the average propagation rate constant and $<K>$ is the average equilibrium constant, were derived from the apparent rate constant from the slopes of the semi-logarithmic kinetic plots shown in Figure 1. The trend is typical for methacrylate/styrenic NMP copolymerizations where a massive increase in $<k_p>$ $<K>$ is witnessed only for low 2VP initial compositions <10 mol% [46–49]. The relatively low Đ of the copolymers (Đ = 1.30 to 1.41) and the linear increases in $M_n$ versus conversion in the range studied suggested that the copolymerizations were relatively well controlled in the conversion range studied (Figure 2).

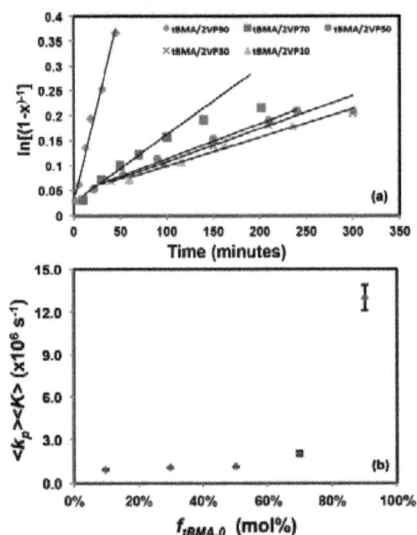

**Figure 1.** (**a**) Kinetic results ($\ln[(1 - x)^{-1}]$ versus time) of the statistical copolymerization of tBMA with 2VP initiated by NHS-BlocBuilder (x = monomer conversion); (**b**) Product of average propagation rate constant $<k_p>$, and propagating radical concentration [P•], $<k_p>$ [P•] (slope of the kinetic plots in (**a**)), versus feed composition with respect to tBMA ($f_{tBMA,0}$); error bars represent the standard deviation of the slopes from (**a**).

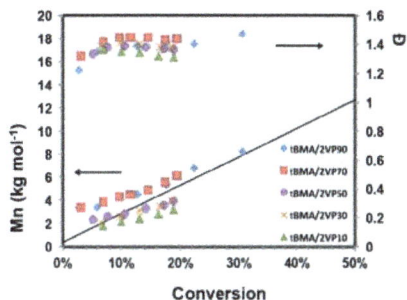

**Figure 2.** Number-average molecular weight ($M_n$) and dispersity (Đ) of the statistical copolymers of *t*BMA and 2VP measured by GPC relative to poly(methyl methacrylate) standards versus conversion; solid line represents the theoretical $M_n$ trend.

### 3.2. pH Sensitivity of the Methacrylic Acid/2-Vinyl Pyridine (MAA/2VP) Copolymers

The *t*BMA/2VP copolymers were hydrolyzed to remove the *tert*-butyl group protecting groups on the *t*BMA units to obtain water-soluble MAA/2VP copolymers. These copolymers were dissolved in aqueous solutions and their particle size was monitored by DLS at various pHs between 1 and 13. Figure 3 below illustrates the results.

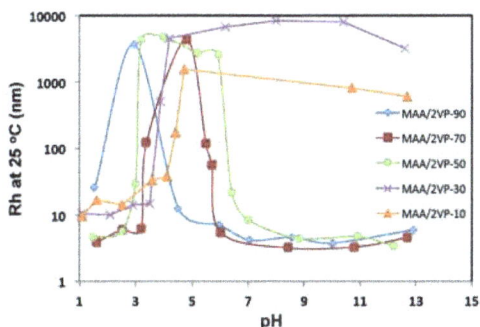

**Figure 3.** Particle size (hydrodynamic radius) of methacrylic acid/2-vinylpyridine (MAA/2VP) copolymers in aqueous solutions at pH ranged from 1 to 14. The sample composition is denoted by MAA/2VP-xx where xx refers to the mol% of MAA units in the copolymer.

### 3.3. Chain Extension of Poly(tert-butyl methacrylate (tBMA)/2-vinylpyridine (2VP)) Macroinitiator with a Mixture of 4-acryloylmorpholine (4AM)/4-acryloylpiperidine (4AP)

To show that a block copolymer consisting of one block of pH-tunable poly(MAA/2VP) could be made with a second block of cloud-point temperature (CPT) tunable poly(4AM/4AP), a chain extension experiment was done with a typical poly(*t*BMA-*stat*-2VP) macroinitiator (*t*BMA/2VP-50) and a batch of 4AM/4AP at a composition that would give a CPT of about 35 °C in a 1 wt% aqueous solution [63]. We understand that the CPT could be altered by the nature of the other segment as it can be moved to lower or higher temperatures (or even extinguished entirely) depending on the composition of the other block. Figure 4 shows the GPC chromatograms before and after chain extension and subsequent fractionation to remove some unreacted macroinitiator. The distribution broadened after chain extension but remained monomodal and $^1$H NMR indicated the presence of the 4AM and 4AP in the block copolymer structure. The poly(*t*BMA/2VP)-*b*-poly(4AM/4AP) block copolymer was sufficiently high in 4AM/4AP content that it was water-soluble without having to de-protect the *t*BMA units.

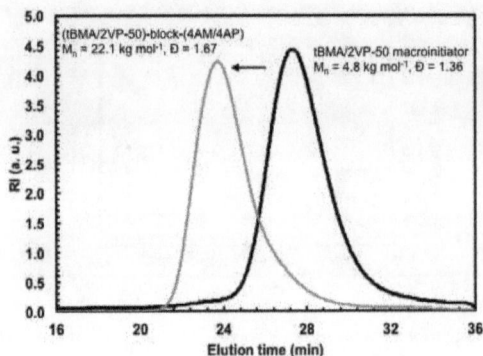

**Figure 4.** GPC chromatograms of the poly(*t*BMA-*stat*-2VP) macroinitiator (*t*BMA/2VP-50) (black line) and the chain-extended product after fractionation of poly(*t*BMA-*stat*-2VP)-*b*-poly(4AM-*stat*-4AP) (grey line).

*3.4. Cloud Point Temperature Measurement of Poly(methacrylic acid-stat-2-vinyl pyridine)-b-poly(4-acryloylmorpholine-stat-4-acryloylpiperidine) (Poly(MAA-stat-2VP)-b-poly(4AM-stat-4AP)) Block Copolymer at Various pH*

The block copolymer described in the previous section was deprotected and confirmation of the removal of the *tert*-butyl group was done with $^{1}$H NMR. The polymer was readily soluble in water at neutral pH. When dissolved in different buffer solutions, the dissolution of 1 wt% solutions was very difficult in pH = 3, easily soluble in pH = 7 and soluble although a bit cloudy in pH = 10 solution. Solutions were shaken vigorously and allowed to sit overnight prior to DLS measurement to ensure that the polymers were solubilized. After passing through a 0.2 μm filter, the solutions were placed in the DLS apparatus and heated to 50 °C.

The DLS results of the three solutions are shown in Figure 5 below. In all cases, a dramatic increase in $R_h$ was observed in all three cases at temperatures 22–27 °C, likely indicating the CPTs due to the 4AM/4AP segment. These CPTs were shifted to lower than what was expected for a 4AM/4AP statistical copolymer with similar composition and solution concentration (35 °C) [60]. At pH = 3, it was expected that the copolymer would be soluble and the $R_h$ matched nearly what it was for the MAA/2VP statistical polymer in solution (<50 nm) that was derived from the macroinitiator of the block copolymer. The CPT was about 24 °C for the solution at pH = 3. At higher temperatures, $R_h$ steadily decreased although remained high after the experiment ~600 nm. Similar profiles were observed for the other solutions. There was a sharp increase in particle size >2000 nm at relatively low temperature (27 °C for the solution at pH = 7 and 22 °C for the solution in pH = 10) and then steady decay in size with terminal values of about 500–600 nm at 50 °C. It was suspected that the decrease in size at higher temperature was due to polymer precipitating out of solution and thus the very large aggregates were not detected as they settled. However, inspection of the samples immediately after removal from the apparatus did not indicate significant settling of polymers. At the pHs studied, it was expected that the MAA/2VP-50 copolymer would be soluble or nearly soluble in all cases (as indicated in Figure 3) and that seems to be reflected in the general trends. However, as noted above, aggregates ~500 nm remained at higher temperatures. It should be noted that such large aggregates may not be useful for drug delivery vehicles, where block copolymer micelles are suggested to be 10–80 nm in size [65]. The larger aggregates observed in the present study may be more amenable to water treatment or enhanced oil recovery (EOR) applications where larger aggregates are more effective in altering the solution viscosity, which is important in EOR [66]. Further, the aggregates seemed to be quite loose, as the solution at pH = 10 was readily filtered through the 0.2 μm filter. Finally, Figure 5 should be cautiously treated as the aggregates are dynamic and it is likely that the profiles will not be very reproducible as there are many fluctuations in size as the temperature increases.

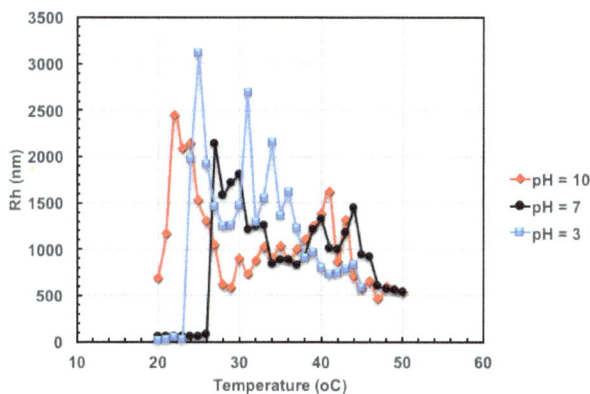

**Figure 5.** Hydrodynamic radius ($R_h$) versus temperature as measured by dynamic light scattering (DLS) for the various 1 wt% aqueous solutions of the poly(MAA-*stat*-2VP)-*b*-poly(4AM-*stat*-4AP) block copolymer in different buffers.

## 4. Discussion

### 4.1. Nitroxide Mediated Polymerization

Various *t*BMA/2VP statistical copolymers with a wide range of initial compositions were synthesized using the succinnimidyl ester functionalized BlocBuilder initiator (NHS-BlocBuilder). Conversions were kept relatively low (19%–31%) to ensure high nitroxide end-group fidelity for subsequent chain extension experiments. The statistical copolymerizations indicated that all copolymerizations had nearly linear $M_n$ versus conversion plots with relatively low Đ = 1.30–1.41. Plots of $<k_p>$ $<K>$ were typical of methacrylic ester/styrenic copolymerizations where a massive increase in this parameter is only witnessed for very rich methacrylate compositions (>90 mol% in the initial mixture). This same trend has been widely reported by Charleux and co-workers and others [45–50]. It should be noted here that additional free nitroxide was not required to control the methacrylate-rich compositions. The same behavior has been seen with other methacrylate-rich compositions [51,58,67] and with a butyl acrylate homopolymerization [64] mediated by NHS-BlocBuilder, which simplifies the formulation, in addition to potentially adding another functional group for coupling other chains [58], although this was not applied here. Furthermore, the *t*BMA/2VP copolymers had final copolymer compositions very close to the initial compositions, suggesting that the copolymerization was polymerizing in essentially a random fashion. Reactivity ratios for this system were not available in the literature. However, there are related systems: 2VP with OEGMA (300 and 1100) had $r_{2VP}$ = 0.99–1.13, $r_{OEGMA300}$ = 0.16–0.25 [68]. Regardless, tuning for this system is relatively predictable.

### 4.2. pH Sensitivity of the Methacrylic Acid/2-Vinyl Pyridine (MAA/2VP) Copolymers

It is known that poly(MAA) is water-soluble at all pH but has increased water solubility at high pH because of the ionization [69]. The reported pKa of PMAA from the literature is 5.4 [70,71]. In contrast, P2VP is only water soluble at low pH and it has a pKa of 4.1 [72,73]. Therefore, it was expected that the copolymers of MAA/2VP will be water soluble at low and high pH but will have decreased water solubility in slightly acidic/basic environments. A similar trend was reported for amphiphilic poly(MAA/diethylamino ethyl methacrylates (DEAEMA))-*b*-poly(methyl methacrylate) diblock terpolymers [9] and with other ampholytic systems [11,74,75]. From Figure 3, at very low pH (pH < 1), all copolymers possessed particle sizes ~10 nm, indicating well-dissolved unimers. As the pH increased, particle size rose sharply to above 1000 nm, indicating aggregation of the water insoluble copolymers. Re-dissolution at higher pH was observed for copolymers that have up to 50% 2VP, whereas copolymers with 70%–90% 2VP remained insoluble.

For the copolymers that re-dissolved at higher pH, the re-dissolution did not happen at the same pH. It appears that as the 2VP content increased in the copolymer, the pH at which re-dissolution occurred shifted to higher values. Additionally, the range of pH where the copolymer was insoluble became wider. This can be explained by the hydrophobicity of P2VP. When the copolymer had a higher 2VP content, more MAA units had to be ionized (and hence the higher pH) for the MAA/2VP copolymer to be sufficiently solubilized.

For the copolymers that had high 2VP compositions >70%, the hydrophobicity of the neutral 2VP units became dominant and the copolymers could not be dissolved even with all MAA units ionized at high pH. However, dissolution of these copolymers due to ionization of 2VP occurred at higher pH, around pH 4, compared to the copolymers with lower 2VP content. As noted above, P2VP has pKa = 4.1 and becomes readily water soluble around pH = 3. The copolymers of MAA/2VP with high 2VP content were water soluble at around pH = 4 (MAA/2VP-10, MAA/2VP-30, MAA-2VP-50) as the result of the hydrophilicity of PMAA, which increased the overall hydrophilicity of the copolymers and allowed the copolymer to be water soluble with a lower degree of 2VP ionization.

### 4.3. Block Copolymers with Both pH and Temperature Sensitivities

To study the dual stimuli-responsive block copolymers where one block has tunable pH sensitivity (via MAA/2VP pH sensitivity) and the other block with tunable CPT (via 4AM/4AP), a candidate MAA/2VP macroinitiator (MAA/2VP-50, 48% MAA content) was chosen to initiate a second batch of 4AM/4AP. The composition of 4AM/4AP was chosen so that a clearly defined CPT ~35 °C could be observed, as we found in our previous study [63]. There was some unreacted macroinitiator but it was readily removed by subsequent fractionation (see Figure 4). After hydrolysis of the tert-butyl protecting groups, the resulting poly(MAA/2VP)-*b*-poly(4AM/4AP) was dissolved as a 1 wt% solution in buffers of three different pHs: 3, 7, and 10. In each case, the pHs were chosen in the range where the macroinitiator was expected to be soluble. However, dissolution was different in each case, as noted in the Results section. Thus, dissolution was carefully done to ensure that the polymers were sufficiently soluble (24 h, and then filtration to remove aggregates). Each solution exhibited the same general trend when they were heated (Figure 5). The CPTs were noticed by the dramatic increase in $R_h$ at 22–27 °C, which is lower than what was expected given the composition of the 4AM/4AP block. A similar drop in CPT was observed for poly(2VP)-*b*-poly(NIPAM) block copolymers in solution, which is accredited to relatively increasing the 2VP content in the block copolymers compared to the poly(NIPAM) segment [76]. This was attributed to the hydrophobicity of the poly(2VP) segment effectively lowering the CPT. In a contrasting case, Schilli et al. examined doubly responsive poly(acrylic acid)-*b*-poly(NIPAM) (PAA-*b*-PNIPAM) block copolymers. With increasing pH from 4.5 to approximately 5–7, the LCST of the PNIPAM increased from 29 to 35 °C [77]. In a system applying similar NMP techniques that was used here, PAA-*b*-poly(diethyl acrylamide) (PAA-*b*-PDEAAm) was made in dispersion and the hydrodynamic radius was measured as a function of pH at different temperatures (15, 25 and 50 °C) [78]. In that study, the aggregate size was greater when the temperature exceeded the cloud point temperature of the PDEAAm block (about 35 °C). Again, the change in pH interacted with the transition temperature and it was higher than that of pure PDEAAm. In the current study, despite the presence of the MAA units, it apparently was not sufficient to counter-balance the effect of the 2VP units. After the initial rise in $R_h$ with temperature, the $R_h$s slowly decreased to about 500–600 nm at 50 °C. This was initially thought to be due to copolymer precipitating that could not be detected by the instrument. However, the solutions did not have any obviously precipitated polymer after removal from the apparatus. Thus, relatively large aggregates ~500 nm were present after the experiment. This suggests that there was a balance between the CPT of the 4AM/4AP segment and the solubility of the MAA/2VP segment. It is likely that the MAA units were sufficiently hydrolyzed to improve the solubility at higher temperature. This system underscores the delicate balances and tunability of properties possible by using dual stimuli-responsive block copolymers consisting of two respective statistical copolymer segments as has been noted in other combinations such as schizophrenic micelles [79] or dual upper critical solution temperature (UCST) and LCST block segments [80]. Future work can add more complex layering by examining the entire span of macroinitiators with different compositions along with second blocks with variable composition to tune the cloud point temperature.

## 5. Conclusions

The versatility of NMP to produce dual stimuli-responsive block copolymers was highlighted where one block consisted of a statistical poly(ampholyte) comprised of methacrylic acid and 2-vinylpyridine units and the other consisted of a statistical thermo-responsive copolymer segment of 4-acryloylmorpholine and 4-acryloylpiperidine. Each block could have its pH-sensitivity and thermo-responsiveness tuned by the relative proportion of co-monomers in each block. The NMP of *t*BMA/2VP mixtures resulted in linear $M_n$ versus conversion plots and relatively narrow molecular weight distributions ($Đ$ = 1.30–1.41) and essentially random microstructure. After hydrolysis of the *t*BMA units, MAA/2VP copolymers exhibited solubility at low pH < 3 and high pH > 8, but formed large aggregates at intermediate pHs. One such *t*BMA/2VP precursor copolymer was used as a mixture to add a thermo-responsive segment consisting of 4AM/4AP units. The macroinitiator was sufficiently active, resulting in the dual responsive block copolymer (after fractionation of a small concentration of inactive macroinitiator). DLS of the block copolymer at various pHs (3, 7 and 10) as a function of temperature indicated a rapid increase in particle size >2000 nm at 22–27 °C, corresponding to the 4AM/4AP segment's thermo-responsiveness followed by a leveling in particle size to about 500 nm at higher temperatures.

**Acknowledgments:** This work was supported by an NSERC Discovery Grant (288125) and Chi Zhang acknowledges scholarship support from NSERC CGS and the MEDA from the Faculty of Engineering, McGill University. Ahmad Khan is thanked for performing some [1]H NMR measurements of the block copolymer.

**Author Contributions:** M.M. and C.Z. conceived and designed the experiments; C.Z. performed the macroinitiator synthesis, characterization and pH DLS experiments while M.M. performed the chain-extension experiments and did the DLS experiments for the dual-responsive block copolymer; C.Z., M.M. and D.G. analyzed the data and wrote the paper.

**Conflicts of Interest:** The authors declare no conflict of interest.

## References

1. Lowe, A.B.; McCormick, C.L. Synthesis and solution properties of zwitterionic polymers. *Chem. Rev.* **2002**, *102*, 4177–4190. [CrossRef] [PubMed]
2. Alfrey, T., Jr.; Morawetz, W. Amphoteric Polyelectrolytes. I. 2-Vinylpyridine—Methacrylic Acid Copolymers 1, 2. *J. Am. Chem. Soc.* **1952**, *74*, 436–438. [CrossRef]
3. Ehrlich, P.; Doty, J. The Solution Behavior of a Polymeric Ampholyte1. *J. Am. Chem. Soc.* **1954**, *76*, 3764–3777. [CrossRef]
4. Tan, B.H.; Ravi, P.; Tan, L.N.; Tam, K.C. Synthesis and aqueous solution properties of sterically stabilized pH-responsive polyampholyte microgels. *J. Colloid Interface Sci.* **2007**, *309*, 453–463. [CrossRef] [PubMed]
5. Alfrey, T.; Pinner, S.H. Preparation and titration of amphoteric polyelectrolytes. *J. Polym. Sci.* **1957**, *23*, 533–547. [CrossRef]
6. Kamachi, M.; Kurihara, M.; Stille, J.K. Synthesis of block polymers for desalination membranes. Preparation of block copolymers of 2-vinylpyridine and methacrylic acid or acrylic acid. *Macromolecules* **1972**, *5*, 161–167. [CrossRef]
7. Kurihara, M.; Kamachi, M.; Stille, J.K. Synthesis of ionic block polymers for desalination membranes. *J. Polym. Sci. Polym. Chem. Ed.* **1973**, *11*, 587–610. [CrossRef]
8. Varoqui, R.; Tran, Q.; Pefferkorn, E. Polycation-polyanion complexes in the linear diblock copolymer of poly(styrene sulfonate)/poly(2-vinylpyridinium) salt. *Macromolecules* **1979**, *12*, 831–835. [CrossRef]

9.  Gotzamanis, G.; Tsitsilianis, C. Stimuli-responsive A-*b*-(B-*co*-C) diblock terpolymers bearing polyampholyte sequences. *Macromol. Rapid. Commun.* **2006**, *27*, 1757–1763. [CrossRef]

10. André, X.; Zhang, M.; Müller, A.H.E. Thermo- and pH-responsive micelles of poly(acrylic acid)-block-poly (*N,N*-diethylacrylamide). *Macromol. Rapid Commun.* **2005**, *26*, 558–563. [CrossRef]

11. Tsitsilianis, C.; Stavrouli, N.; Bocharova, V.; Angelopoulos, S.; Kiriy, A.; Katsampas, I.; Stamm, M. Stimuli responsive associative polyampholytes based on ABCBA pentablock terpolymer architecture. *Polymer* **2008**, *49*, 2996–3006. [CrossRef]

12. Bütün, V.; Billingham, N.C.; Armes, S.P. Unusual aggregation behavior of a novel tertiary amine methacrylate-based diblock copolymer: Formation of micelles and reverse micelles in aqueous solution. *J. Am. Chem. Soc.* **1998**, *120*, 11818–11819. [CrossRef]

13. Webster, O.W. The discovery and commercialization of group transfer polymerization. *J. Polym. Sci. A Polym. Chem.* **2000**, *38*, 2855–2860. [CrossRef]

14. Bütün, V.; Weaver, J.V.M.; Bories-Azeau, X.; Cai, Y.; Armes, S.P. A brief review of 'schizophrenic' block copolymers. *React. Funct. Polym.* **2006**, *66*, 157–165. [CrossRef]

15. Patrickios, C.S.; Hertler, W.R.; Abbott, N.L.; Hatton, T.A. Diblock, Abc triblock, and random methacrylic polyampholytes-synthesis by group-transfer polymerization and solution behavior. *Macromolecules* **1994**, *27*, 930–937. [CrossRef]

16. Chen, W.-Y.; Alexandridis, P.; Su, C.-K.; Patrickios, C.S.; Hertler, W.R.; Hatton, T.A. Effect of block size and sequence on the micellization of ABC triblock methacrylic polyampholytes. *Macromolecules* **1995**, *28*, 8604–8611. [CrossRef]

17. Dobrynin, A.V.; Colby, R.H.; Rubinstein, M. Polyampholytes. *J. Polym. Sci. B Polym. Phys.* **2004**, *42*, 3513–3538. [CrossRef]

18. Kudabergnor, S. Polyampholytes. In *Encyclopedia of Polymer Science and Technology*; Mark, H.F., Ed.; Wiley: New York, NY, USA, 2014; Volume 10, pp. 297–325.

19. Hadjichristidis, N.; Pispas, S.; Floudas, G.A. *Block Copolymers: Synthetic Strategies, Physical Properties and Applications*; Wiley: Hoboken, NJ, USA, 2003.

20. Nicolas, J.; Guillaneuf, Y.; Lefay, C.; Bertin, D.; Gigmes, D.; Charleux, B. Nitroxide-mediated polymerization. *Prog. Polym. Sci.* **2013**, *38*, 63–235. [CrossRef]

21. Georges, M.K.; Veregin, R.P.N.; Kazmaier, P.M.; Hamer, G.K. Narrow molecular weight distribution resins by a free radical polymerization process. *Macromolecules* **1993**, *26*, 2987–2988. [CrossRef]

22. Kato, M.; Kamigaito, M.; Sawamoto, M. Polymerization of methyl methacrylate with the carbon tetrachloride/ dichlorotris-(triphenylphosphine) ruthenium (II)/methylaluminum bis (2, 6-*di-tert*-butylphenoxide) initiating system: Possibility of living radical polymerization. *Macromolecules* **1995**, *28*, 1721–1723. [CrossRef]

23. Wang, J.S.; Matyjaszewski, K. ontrolled/"living" radical polymerization. Atom transfer radical polymerization in the presence of transition-metal complexes. *J. Am. Chem. Soc.* **1995**, *117*, 5614–5615. [CrossRef]

24. Chiefari, J.; Chong, Y.K.; Ercole, F.; Krstina, J.; Jeffery, J.; Le, T.P.T.; Mayadunne, R.T.A.; Meijs, G.F.; Moad, C.L.; Moad, G.; et al. Living free-radical polymerization by reversible addition-fragmentation chain transfer: The RAFT process. *Macromolecules* **1998**, *31*, 5559–5562. [CrossRef]

25. Čadová, E.; Konečný, J.; Kříž, J.; Svitáková, R.; Holler, P.; Genzer, J.; Vlček, P. ATRP of 2-vinylpyridine and *tert*-butyl acrylate mixtures giving precursors of polyampholytes. *J. Polym. Sci. A Polym. Chem.* **2010**, *48*, 735–741. [CrossRef]

26. Jhon, Y.K.; Arifuzzaman, S.; Ozcam, A.E.; Kiserow, D.J.; Genzer, J. Formation of polyampholyte brushes via controlled radical polymerization and their assembly in solution. *Langmuir* **2012**, *28*, 872–882. [CrossRef] [PubMed]

27. Liu, S.; Armes, S.P. Synthesis and aqueous solution behavior of a pH-responsive schizophrenic diblock copolymer. *Langmuir* **2003**, *19*, 4432–4438. [CrossRef]

28. Cai, Y.; Armes, S.P. A zwitterionic ABC triblock copolymer that forms a "trinity" of micellar aggregates in aqueous solution. *Macromolecules* **2004**, *37*, 7116–7122. [CrossRef]

29. Zhang, X.; Ma, J.; Yang, S.; Xu, J. "Schizophrenic" micellization of poly(acrylic acid)-*b*-poly(2-dimethylamino)ethyl methacrylate and responsive behavior of the micelles. *Soft Mater.* **2013**, *11*, 394–402. [CrossRef]

30. Smith, A.E.; Xu, X.; Kirkland-York, S.E.; Savin, D.A.; McCormick, C.L. "Schizophrenic" self-assembly of block copolymers synthesized via aqueous RAFT polymerization: From micelles to vesicles. Paper number 143 in a series on water-soluble polymers. *Macromolecules* **2010**, *43*, 1210–1217. [CrossRef]
31. Du, J.; O'Reilly, R.K. pH-responsive vesicles from a schizophrenic diblock copolymer. *Macromol. Chem. Phys.* **2010**, *211*, 1530–1537. [CrossRef]
32. Savoji, M.T.; Strandman, S.; Zhu, X.X. Switchable vesicles formed by diblock random copolymers with tunable pH-and thermo-responsiveness. *Langmuir* **2013**, *29*, 6823–6832. [CrossRef] [PubMed]
33. Benoit, D.; Grimaldi, S.; Robin, S.; Finet, J.-P.; Tordo, P.; Gnanou, Y. Kinetics and mechanism of controlled free-radical polymerization of styrene and *n*-butyl acrylate in the presence of an acyclic β-phosphonylated nitroxide. *J. Am. Chem. Soc.* **2000**, *122*, 5929–5939. [CrossRef]
34. Benoit, D.; Chaplinski, V.; Braslau, R.; Hawker, C.J. Development of a universal alkoxyamine for "living" free radical polymerizations. *J. Am. Chem. Soc.* **1999**, *121*, 3904–3920. [CrossRef]
35. Moad, G.; Rizzardo, E. The History of Nitroxide Mediated Polymerization. In *Nitroxide Mediated Polymerization: From Fundamentals to Applications in Materials Science*; Gigmes, D., Ed.; Royal Society of Chemistry: London, UK, 2015; pp. 1–44.
36. Magee, C.; Sugihara, Y.; Zetterlund, P.B.; Aldabbagh, F. Chain transfer to solvent in the radical polymerization of structurally diverse acrylamide monomers using straight-chain and branched alcohols as solvents. *Polym. Chem.* **2014**, *5*, 2259–2265. [CrossRef]
37. Götz, H.; Harth, E.; Schiller, S.M.; Frank, C.W.; Knoll, W.; Hawker, C.J. Synthesis of lipo-glycopolymer amphiphiles by nitroxide-mediated living free-radical polymerization. *J. Polym. Sci. A Polym. Chem.* **2002**, *40*, 3379–3391. [CrossRef]
38. Schierholz, K.; Givehchi, M.; Fabre, P.; Nallet, F.; Papon, E.; Guerret, O.; Gnanou, Y. Acrylamide-based amphiphilic block copolymers via nitroxide-mediated radical polymerization. *Macromolecules* **2003**, *36*, 5995–5999. [CrossRef]
39. Phan, T.N.T.; Maiez-Tribut, S.; Pascault, J.-P.; Bonnet, A.; Gerard, P.; Guerret, O.; Bertin, D. Synthesis and characterizations of block copolymer of poly(*n*-butyl acrylate) and gradient poly(methyl methacrylate-*co*-N,N-dimethyl acrylamide) made via nitroxide-mediated controlled radical polymerization. *Macromolecules* **2007**, *40*, 4516–4523. [CrossRef]
40. Delaittre, G.; Rieger, J.; Charleux, B. Nitroxide-mediated living/controlled radical polymerization of N,N-diethylacrylamide. *Macromolecules* **2011**, *44*, 462–470. [CrossRef]
41. Shipman, P.O.; Cui, C.; Lupinska, P.; Lalancette, R.A.; Sheridan, J.B.; Jäkle, F. Nitroxide-mediated controlled free radical polymerization of the chelate monomer 4-styryl-tris(2-pyridyl)borate (StTpyb) and supramolecular assembly via metal complexation. *ACS Macro Lett.* **2013**, *2*, 1056–1060. [CrossRef]
42. Lang, A.S.; Neubig, A.; Sommer, M.; Thalakkat, M. NMRP versus "Click" Chemistry for the Synthesis of Semiconductor Polymers Carrying Pendant Perylene Bisimides. *Macromolecules* **2010**, *43*, 7001–7010. [CrossRef]
43. Deng, L.; Furuta, P.T.; Garon, S.; Li, J.; Kavulak, D.; Thompson, M.E.; Fréchet, J.M.J. Living radical polymerization of bipolar transport materials for highly efficient light emitting diodes. *Chem. Mater.* **2006**, *18*, 386–395. [CrossRef]
44. Lessard, B.; Graffe, A.; Maric, M. Styrene/*tert*-butyl acrylate random copolymers synthesized by nitroxide-mediated polymerization: Effect of free nitroxide on kinetics and copolymer composition. *Macromolecules* **2007**, *40*, 9284–9292. [CrossRef]
45. Lessard, B.; Schmidt, S.C.; Maric, M. Styrene/acrylic acid random copolymers synthesized by nitroxide-mediated polymerization: Effect of free nitroxide on kinetics and copolymer composition. *Macromolecules* **2008**, *41*, 3446–3454. [CrossRef]
46. Charleux, B.; Nicolas, J.; Guerret, O. Theoretical expression of the average activation-deactivation equilibrium constant in controlled/living free-radical copolymerization operating via reversible termination. Application to a strongly improved control in nitroxide-mediated polymerization of methyl methacrylate. *Macromolecules* **2005**, *38*, 5485–5492.
47. Nicolas, J.; Brusseau, S.; Charleux, B. A minimal amount of acrylonitrile turns the nitroxide-mediated polymerization of methyl methacrylate into an almost ideal controlled/living system. *J. Polym. Sci. A Polym. Chem.* **2010**, *48*, 34–47. [CrossRef]

48. Dire, C.; Charleux, B.; Magnet, S.; Couvreur, L. Nitroxide-mediated copolymerization of methacrylic acid and styrene to form amphiphilic diblock copolymers. *Macromolecules* **2007**, *40*, 1897–1903. [CrossRef]
49. Nicolas, J.; Couvreur, P.; Charleux, B. Comblike polymethacrylates with poly(ethylene glycol) side chains via nitroxide-mediated controlled free-radical polymerization. *Macromolecules* **2008**, *41*, 3758–3761. [CrossRef]
50. Lessard, B.; Maric, M. Incorporating glycidyl methacrylate into block copolymers using poly(methacrylate-*ran*-styrene) macroinitiators synthesized by nitroxide-mediated polymerization. *J. Polym. Sci. A Polym. Chem.* **2009**, *47*, 2574–2588. [CrossRef]
51. Lessard, B.; Tervo, C.; De Wahl, S.; Clerveaux, F.J.; Tang, K.K.; Yasmine, S.; Andjelic, S.; D'Alessandro, A.; Maric, M. Poly(*tert*-butyl methacrylate/styrene) macroinitiators as precursors for organo-and water-soluble functional copolymers using nitroxide-mediated controlled radical polymerization. *Macromolecules* **2010**, *43*, 868–878. [CrossRef]
52. Moayeri, A.; Lessard, B.; Maric, M. Nitroxide mediated controlled synthesis of glycidyl methacrylate-rich copolymers enabled by SG1-based alkoxyamines bearing succinimidyl ester groups. *Polym. Chem.* **2011**, *2*, 1121–1129. [CrossRef]
53. Ting, S.R.S.; Min, E.-H.; Escale, P.; Save, M.; Billon, L.; Stenzel, M.H. Lectin recognizable biomaterials synthesized via nitroxide-mediated polymerization of a methacryloyl galactose monomer. *Macromolecules* **2009**, *42*, 9422–9434. [CrossRef]
54. Lessard, B.H.; Ling, E.Y.J.; Maric, M. Fluorescent, thermoresponsive oligo(ethylene glycol) methacrylate/9-(4-vinylbenzyl)-9*H*-carbazole copolymers designed with multiple LCSTs via nitroxide mediated controlled radical polymerization. *Macromolecules* **2012**, *45*, 1879–1891. [CrossRef]
55. Lessard, B.H.; Maric, M. "Smart" poly(dimethylaminoethyl methacrylate-*ran*-9-(4-vinylbenzyl)-9*H*-carbazole) copolymers synthesized by nitroxide mediated radical polymerization. *J. Polym. Sci. A Polym. Chem.* **2011**, *49*, 5270–5283. [CrossRef]
56. Wang, Z.-J.; Maric, M. Synthesis of narrow molecular weight distribution norbornene-lactone functionalized polymers by nitroxide-mediated polymerization: Candidates for 193 nm photoresist materials. *Polymers* **2014**, *6*, 565–577. [CrossRef]
57. Chenal, M.; Mura, S.; Marchal, C.; Gigmes, D.; Charleux, B.; Fattal, E.; Couvreur, P.; Nicolas, J. Facile synthesis of innocuous comb-shaped polymethacrylates with PEG side chains by nitroxide-mediated radical polymerization in hydroalcoholic solutions. *Macromolecules* **2010**, *43*, 9291–9303. [CrossRef]
58. Brusseau, S.; Belleney, J.; Magnet, S.; Couvreur, L.; Charleux, B. Nitroxide-mediated copolymerization of methacrylic acid with sodium 4-styrene sulfonate: Towards new water-soluble macroalkoxyamines for the synthesis of amphiphilic block copolymers and nanoparticles. *Polym. Chem.* **2010**, *1*, 720–729. [CrossRef]
59. Zhang, C.; Maric, M. pH-and temperature-sensitive statistical copolymers poly[2-(dimethylamino) ethyl methacrylate-stat-2-vinylpyridine] with Functional succinimidyl-ester chain ends synthesized by nitroxide-mediated polymerization. *J. Polym. Sci. A Polym. Chem.* **2012**, *50*, 4341–4357. [CrossRef]
60. Consolante, V.; Maric, M. Nitroxide-mediated polymerization of an organo-soluble protected styrene sulfonate: Development of homo and random copolymers. *Macromol. React. Eng.* **2011**, *5*, 575–586. [CrossRef]
61. Couvreur, L.; Lefay, C.; Belleney, J.; Charleux, B.; Guerret, O.; Magnet, S. First nitroxide-mediated controlled free-radical polymerization of acrylic acid. *Macromolecules* **2003**, *36*, 8260–8267. [CrossRef]
62. Couvreur, L.; Charleux, B.; Guerret, O.; Magnet, S. Direct synthesis of controlled poly(styrene-*co*-acrylic acid)s of various compositions by nitroxide-mediated random copolymerization. *Macromol. Chem. Phys.* **2003**, *204*, 2055–2063. [CrossRef]
63. Savelyeva, X.; Li, L.; Bennett, I.; Maric, M. Stimuli-responsive 4-acryloylmorpholine/4-acryloylpiperidine copolymers via nitroxide mediated polymerization. **2016**, Submitted.
64. Vinas, J.; Chagneux, N.; Gigmes, D.; Trimaille, T.; Favier, A.; Bertin, D. SG1-based alkoxyamine bearing a *N*-succinimidyl ester: A versatile tool for advanced polymer synthesis. *Polymer* **2008**, *49*, 3639–3647. [CrossRef]
65. Torchilin, V.P. Structure and design of polymeric surfactant-based drug delivery systems. *J. Controll. Release* **2001**, *73*, 137–172. [CrossRef]
66. Wever, D.A.Z.; Picchioni, F.; Broekhuis, A.A. Polymers for enhanced oil recovery: A paradigm for structure-property relationship in aqueous solution. *Prog. Polym. Sci.* **2011**, *36*, 1558–1628. [CrossRef]
67. Darabi, A.; Shirin-Abadi, A.R.; Jessop, P.G.; Cunningham, M.F. Nitroxide-mediated polymerization of 2-(diethylamino)ethyl methacrylate (DEAEMA) in water. *Macromolecules* **2015**, *48*, 72–80. [CrossRef]

68. Driva, P.; Bexis, P.; Pitsikalis, M. Radical copolymerization of 2-vinyl pyridine and oligo(ethylene glycol) methyl ether methacrylates: Monomer reactivity ratios and thermal properties. *Eur. Polym. J.* **2011**, *47*, 762–771. [CrossRef]
69. Mori, H.; Müller, A.H.E. New polymeric architectures with (meth)acrylic acid segments. *Prog. Polym. Sci.* **2003**, *28*, 1403–1439. [CrossRef]
70. Merle, Y. Synthetic polyampholytes. V: Influence of nearest-neighbor interactions on potentiometric curves. *J. Phys. Chem.* **1987**, *91*, 3092–3098. [CrossRef]
71. Ho, B.S.; Tan, B.H.; Tan, J.P.K.; Tam, K.C. Inverse microemulsion polymerization of sterically stabilized polyampholyte microgels. *Langmuir* **2008**, *24*, 7698–7703. [CrossRef] [PubMed]
72. McParlane, J.; Dupin, D.; Saunders, J.M.; Lally, S.; Armes, S.P.; Saunders, B.R. Dual pH-triggered physical gels prepared from mixed dispersions of oppositely charged pH-responsive microgels. *Soft Matter* **2012**, *8*, 6239–6247. [CrossRef]
73. Dupin, D.; Fujii, S.; Armes, S.P.; Reeve, P.; Baxter, S.M. Efficient synthesis of sterically stabilized pH-responsive microgels of controllable particle diameter by emulsion polymerization. *Langmuir* **2006**, *22*, 3381–3387. [CrossRef] [PubMed]
74. Iatridi, Z.; Mattheolabakis, G.; Avgoustakis, K.; Tsitsilianis, C. Self-assembly and drug delivery studies of pH/thermo-sensitive polyampholytic (A-*co*-B)-*b*-C-*b*-(A-*co*-B) segmented terpolymers. *Soft Matter* **2011**, *7*, 11160–11168. [CrossRef]
75. Gotzamanis, G.; Papadimitriou, K.; Tsitsilianis, C. Design of a C-*b*-(A-*co*-B)-*b*-C telechelic polyampholyte pH-responsive gelator. *Polym. Chem.* **2016**, *7*, 2121–2129. [CrossRef]
76. Corten, C.; Kretschmer, K.; Kuckling, D. Novel multi-responsive P2VP-block-PNIPAAm block copolymers via nitroxide-mediated radical polymerization. *Beilstein J. Organ. Chem.* **2010**, *6*, 756–765. [CrossRef] [PubMed]
77. Schilli, C.M.; Zhang, M.; Rizzardo, E.; Thang, S.H.; Chong, Y.K.; Edwards, K.; Karlsson, G.; Müller, A.H.E. A new double-responsive block copolymer synthesized via RAFT polymerization: poly(*N*-isopropylacrylamide)-block-poly(acrylic acid). *Macromolecules* **2004**, *37*, 7861–7866. [CrossRef]
78. Delaittre, G.; Save, M.; Gaborieau, M.; Castignolles, P.; Rieger, J.; Charleux, B. Synthesis by nitroxide-mediated aqueous dispersion polymerization, characterization, and physical core-crosslinking of pH-and thermoresponsive dynamic diblock copolymer micelles. *Polym. Chem.* **2012**, *3*, 1526–1538. [CrossRef]
79. Liu, S.; Billingham, N.C.; Armes, S.P. A schizophrenic water-soluble diblock copolymer. *Angew. Chem. Int. Ed.* **2001**, *40*, 2328–2331. [CrossRef]
80. Arotcarena, M.; Heise, B.; Ishaya, S.; Laschewsky, A. Switching the inside and the outside of aggregates of water-soluble block copolymers with double thermoresponsivity. *J. Am. Chem. Soc.* **2002**, *124*, 3787–3793. [CrossRef] [PubMed]

*processes*

MDPI

*Article*

# AMPS/AAm/AAc Terpolymerization: Experimental Verification of the EVM Framework for Ternary Reactivity Ratio Estimation

**Alison J. Scott, Niousha Kazemi and Alexander Penlidis ***

Institute for Polymer Research (IPR), Department of Chemical Engineering, University of Waterloo, Waterloo, Ontario, N2L 3G1, Canada; ajscott@uwaterloo.ca (A.J.S.); nkazemi@uwaterloo.ca (N.K.)
* Correspondence: penlidis@uwaterloo.ca; Tel.: +1-519-888-4567 (ext. 36634)

Academic Editor: Michael Henson
Received: 5 January 2017; Accepted: 21 February 2017; Published: 25 February 2017

**Abstract:** The complete error-in-variables-model (EVM) framework, consisting of both design of experiments and parameter estimation stages, is applied to the terpolymerization of 2-acrylamido-2-methylpropane sulfonic acid (AMPS, $M_1$), acrylamide (AAm, $M_2$) and acrylic acid (AAc, $M_3$). This water-soluble terpolymer has potential for applications in enhanced oil recovery, but the associated terpolymerization kinetic characteristics are largely unstudied. In the current paper, EVM is used to design optimal experiments (for the first time in the literature), and reactivity ratios are subsequently estimated based on both low and medium-high conversion data. The results from the medium-high conversion data are more precise than those from the low conversion data, and are therefore used next to predict the terpolymer composition trajectory over the full course of conversion. Good agreement is seen between experimental data and model predictions, which confirms the accuracy of the newly determined ternary reactivity ratios: $r_{12} = 0.66$, $r_{21} = 0.82$, $r_{13} = 0.82$, $r_{31} = 0.61$, $r_{23} = 1.61$, $r_{32} = 0.25$.

**Keywords:** 2-acrylamido-2-methylpropane sulfonic acid; acrylamide; acrylic acid; error-in-variables-model; polymerization kinetics; reactivity ratio estimation; terpolymerization

## 1. Introduction

Water-soluble terpolymers have applications in a wide variety of areas such as enhanced oil recovery (EOR), dewatering, mineral processing and flocculation. Most of these applications rely on the fact that the addition of the polymeric material can alter the rheology of an aqueous medium [1].

Among synthetic water-soluble polymers, polyacrylamide is used as a base in many applications. It is often difficult for one single polymer to meet all of the requirements of an application, so copolymers and terpolymers can be employed to deliver specific properties. One of the most widely used copolymers of acrylamide is the acrylamide/acrylic acid (AAm/AAc) copolymer. The AAm/AAc copolymer can be used in many of the above-mentioned applications, including enhanced oil recovery [2]. However, it has been observed that the AAm/AAc copolymer degrades in hostile environments (typical EOR conditions). Therefore, the addition of a third comonomer resulting in a terpolymer backbone with higher thermal and shear stability has been suggested to overcome this problem. One such comonomer that can enhance the stability of the AAm/AAc copolymer in harsh environments is 2-acrylamido-2-methylpropane sulfonic acid (AMPS). AMPS is a larger monomer molecule compared to AAm and AAc, which, when incorporated in the terpolymer, provides better thermal and shear stability and, as a result, makes the final polymer more suitable for EOR applications [3].

The AMPS/AAm/AAc terpolymer is a largely unstudied system that has only appeared in the literature within the past ten years. Some applications for this new terpolymer have been reported in EOR [3], oil-field drilling [4], superabsorbent hydrogels [5], sludge dewatering [6], and controlled drug-delivery systems [7]. In these few studies, only the final properties of the terpolymer substrate (such as swelling behavior, resistance to temperature and shear stress) have been discussed, but kinetic characteristics of the terpolymerization have not been reported. Since the final application properties of this terpolymer are directly related to its microstructure, it is essential to have a clear understanding of the terpolymerization kinetics.

Given that there are three different possibilities for the terminal monomer (on the growing radical), and three options for the added monomer, nine different propagation steps are possible according to the terminal model:

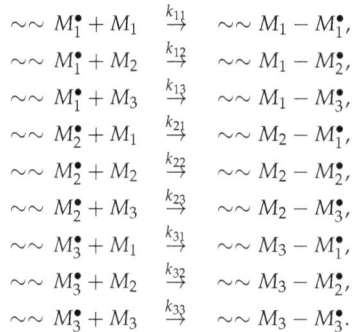

$$\sim\sim M_1^\bullet + M_1 \xrightarrow{k_{11}} \sim\sim M_1 - M_1^\bullet,$$
$$\sim\sim M_1^\bullet + M_2 \xrightarrow{k_{12}} \sim\sim M_1 - M_2^\bullet,$$
$$\sim\sim M_1^\bullet + M_3 \xrightarrow{k_{13}} \sim\sim M_1 - M_3^\bullet,$$
$$\sim\sim M_2^\bullet + M_1 \xrightarrow{k_{21}} \sim\sim M_2 - M_1^\bullet,$$
$$\sim\sim M_2^\bullet + M_2 \xrightarrow{k_{22}} \sim\sim M_2 - M_2^\bullet,$$
$$\sim\sim M_2^\bullet + M_3 \xrightarrow{k_{23}} \sim\sim M_2 - M_3^\bullet,$$
$$\sim\sim M_3^\bullet + M_1 \xrightarrow{k_{31}} \sim\sim M_3 - M_1^\bullet,$$
$$\sim\sim M_3^\bullet + M_2 \xrightarrow{k_{32}} \sim\sim M_3 - M_2^\bullet,$$
$$\sim\sim M_3^\bullet + M_3 \xrightarrow{k_{33}} \sim\sim M_3 - M_3^\bullet.$$

In this series of reactions, $M_i \bullet$ represents a radical species with monomer $i$ at the chain end ($i = 1, 2, 3$). Similarly, $M_j$ represents monomer $j$ that is being added to the chain end ($j = 1, 2, 3$). Each of the nine reactions has a rate constant, $k_{ij}$ (radical $i$ adding monomer $j$).

Six parameters, called monomer reactivity ratios ($r_{ij}$), can be used to describe the potential for homopropagation relative to the potential for cross-propagation.

$$r_{12} = \frac{k_{11}}{k_{12}}, \qquad r_{13} = \frac{k_{11}}{k_{13}}, \qquad r_{23} = \frac{k_{22}}{k_{23}},$$

$$r_{21} = \frac{k_{22}}{k_{21}}, \qquad r_{31} = \frac{k_{33}}{k_{31}}, \qquad r_{32} = \frac{k_{33}}{k_{32}}.$$

Reactivity ratios are crucial to the study of the kinetics of multicomponent polymerization systems. Terpolymerization systems are frequently utilized in industry and possess valuable information for academic research, yet there is a considerable lack of reactivity ratio estimation studies for such systems. This is partially due to the complexity of the terpolymer composition model, the Alfrey–Goldfinger model (Equation (1)). $F_i$ is the instantaneous mole fraction of monomer $i$ incorporated (bound) in the terpolymer, $r_{ij}$ are the monomer reactivity ratios relating $i$ and $j$, and $f_i$ is the corresponding mole fraction of unreacted (free) monomer $i$ (often referred to as the feed mole fraction). Equation (1) relates instantaneous (not cumulative) copolymer composition properties:

$$\frac{F_1}{F_2} = \left(\frac{f_1}{f_2}\right)\left(\frac{f_1/r_{31}r_{21} + f_2/r_{21}r_{32} + f_3/r_{31}r_{23}}{f_1/r_{12}r_{31} + f_2/r_{12}r_{32} + f_3/r_{32}r_{13}}\right)\left(\frac{f_1 + f_2/r_{12} + f_3/r_{13}}{f_2 + f_1/r_{21} + f_3/r_{23}}\right), \qquad (1a)$$

$$\frac{F_1}{F_3} = \left(\frac{f_1}{f_3}\right)\left(\frac{f_1/r_{31}r_{21} + f_2/r_{21}r_{32} + f_3/r_{31}r_{23}}{f_1/r_{13}r_{21} + f_2/r_{23}r_{12} + f_3/r_{13}r_{23}}\right)\left(\frac{f_1 + f_2/r_{12} + f_3/r_{13}}{f_3 + f_1/r_{31} + f_2/r_{32}}\right), \qquad (1b)$$

$$\frac{F_2}{F_3} = \left(\frac{f_2}{f_3}\right)\left(\frac{f_1/r_{12}r_{31} + f_2/r_{12}r_{32} + f_3/r_{32}r_{13}}{f_1/r_{13}r_{21} + f_2/r_{23}r_{12} + f_3/r_{13}r_{23}}\right)\left(\frac{f_2 + f_1/r_{21} + f_3/r_{23}}{f_3 + f_1/r_{31} + f_2/r_{32}}\right). \qquad (1c)$$

However, more importantly, the knowledge gap in terpolymerization kinetics is related to a de facto accepted analogy between copolymerization and terpolymerization mechanisms; researchers often use reactivity ratios obtained for binary pairs (from copolymerization experiments) in terpolymerization models. However, because the error in the binary data tends to propagate into the ternary system, binary reactivity ratios cannot be used to describe ternary systems; at best, this provides an approximation [8]. More importantly, ternary reactivity ratios are never determined using ternary experimental data, and differences in the system make it imprudent to use binary and ternary reactivity ratios interchangeably. Using inappropriate reactivity ratios may affect the model performance for predicting terpolymer composition (and sequence length characteristics, since these also depend on reactivity ratio values) and the determination of other terpolymerization characteristics (such as the azeotropic point). Despite these risks, all studies performed previously have employed binary reactivity ratios directly in terpolymer models. This should be avoided [8].

Problems associated with reactivity ratio estimation and design of experiments for terpolymer systems have largely been resolved using the error-in-variables-model (EVM), which was discussed recently by Kazemi et al. [9] (and will be reviewed briefly in the current paper). Thus, in what follows, the AMPS/AAm/AAc terpolymer is investigated by implementing the EVM framework for accurate determination of ternary reactivity ratios. Parameter estimation and the implementation of the design of experiments strategy are demonstrated, and the reactivity ratio estimates are analyzed in terms of both precision and accuracy. Finally, reactivity ratio values based on optimally selected experiments are suggested. Comparisons between low and medium-high conversion level data are also included to examine the effect of the data set (and its inherent errors) on reactivity ratio estimation results.

## 2. Experimental

### 2.1. Design of Experiments

Optimal design of experiments leads to increased information content while minimizing the number of experiments and obtaining more precise parameter estimates. Under the error-in-variables-model (EVM) framework, one can design experiments that consider error in all variables involved (both independent and dependent) in the process model [9,10]. A brief guide for estimating reliable ternary reactivity ratios from terpolymerization data is provided in Figure 1, and a comprehensive evaluation with detailed explanations has been previously published by Kazemi et al. [8,9]. Additional comments related to the steps of Figure 1 are provided below:

1. Review literature for any information on polymerization kinetics of the system in question.
2. Use literature values (if any) and prior knowledge to find (or guess) reasonable preliminary reactivity ratios for the system; determine whether there are any constraints on the feed compositions.

   2.a. In the absence of any prior information, run three preliminary experiments. Each should have a composition rich in one of the three monomers (e.g., 80% or higher), and any experimental limitations (constraints) in the feasible experimental region should be carefully recorded. Estimate preliminary ternary reactivity ratios.

3. Choose three optimal feed compositions according to the EVM framework. Each recipe should contain 80% of one monomer and an equal amount of the other two (that is, $f_i$ /$f_j$ /$f_k$ : (0.8/0.1/0.1), (0.1/0.8/0.1) and (0.1/0.1/0.8)). If a polymerization recipe containing 80% of one monomer is not achievable (due to possible feed composition constraints), choose lower percentages as necessary.
4. Perform experiments at low conversion ($\leq$ 5%–10%) and/or up to medium-high conversion levels (50%–70%). Collect data for terpolymer composition and corresponding conversion values.
5. Use the EVM parameter estimation methodology for estimating reactivity ratios and construct joint confidence regions (JCRs). Refer to the work by Kazemi et al. [8,10] for a detailed implementation of this method.

6. If satisfied with the precision of the results, move to the next step. If not satisfied, perform independent replicates of the optimal experiments and re-estimate reactivity ratios.

7. Present reactivity ratio estimates and their joint confidence regions.

This methodology is used in the terpolymerization of AMPS/AAm/AAc; details for each step are presented in what follows. For the first time in the literature, the EVM framework for ternary reactivity ratio estimation is verified experimentally.

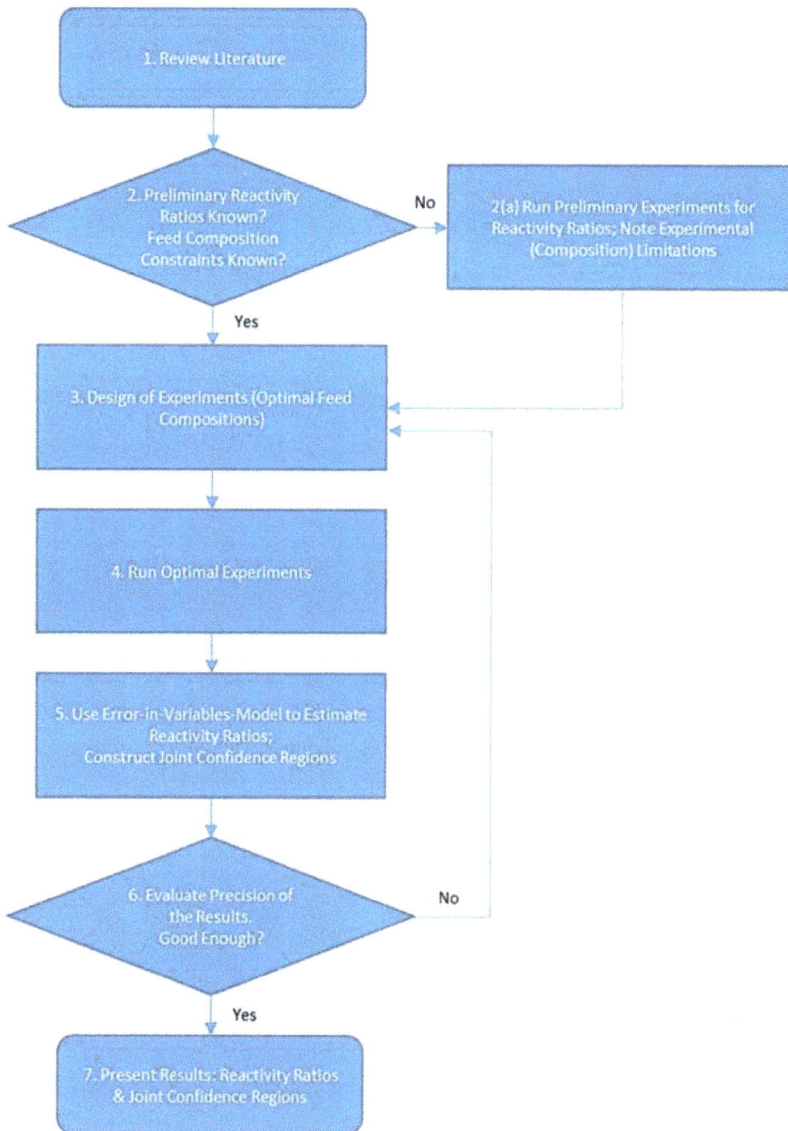

**Figure 1.** Flowchart of the EVM (error-in-variables-model) framework for ternary reactivity ratio estimation.

## 2.2. Reagent Purification

Monomers 2-acrylamido-2-methylpropane sulfonic acid (AMPS; 99%), acrylamide (AAm; electrophoresis grade, 99%), and acrylic acid (AAc; 99%) were purchased from Sigma-Aldrich (Oakville, Ontario, Canada). AAc was purified via vacuum distillation at 30 °C, while AAm and AMPS were used as received. Initiator (4,4'-azo-bis-(4-cyanovaleric acid), ACVA), inhibitor (hydroquinone) and sodium hydroxide were also purchased from Sigma-Aldrich. Sodium chloride from EMD Millipore (Etobicoke, Ontario, Canada) was used as received. Water was Millipore quality (18 MΩ·cm), and acetone and methanol were used as received from suppliers. Nitrogen gas (4.8 grade) purchased from Praxair (Mississauga, Ontario, Canada) was used for degassing solutions.

## 2.3. Polymer Synthesis

Aqueous monomer solutions with a monomer concentration of 1 M and an initiator (ACVA) concentration of 0.004 M (relative to the total solution volume) were prepared. Specific feed compositions (that is, "pre-polymerization" solution compositions) will be discussed in Step 3 of the Results and Discussion section (see Section 3.3).

As demonstrated in a recent study [11], constant pH and ionic strength are extremely important in water-soluble copolymer and terpolymer synthesis. Therefore, solutions were titrated with sodium hydroxide to adjust the pH to approximately 7 ($\pm 0.5$). Similarly, to ensure constant ionic strength (IS) between experiments, sodium chloride was added to each pre-polymerization solution. The highest IS occurs when the solution is rich in AMPS, so sodium chloride was added to all other pre-polymerization solutions to reach that same IS value.

The solutions were then purged with 200 mL/min nitrogen for 2 h. After degassing, aliquots of ~20 mL of solution were transferred to sealed vials using the cannula transfer method [12]. Terpolymerizations were run in a temperature controlled shaker-bath (OLS200; Grant Instruments, Cambridge, UK) at 40 °C and 100 rpm. Vials were removed at selected time intervals, placed in ice and further injected with approximately 1 mL of 0.2 M hydroquinone solution to stop the polymerization. Polymer samples were isolated by precipitating the products in acetone or methanol, filtered (paper filter grade number 41, Whatman; Sigma-Aldrich, Oakville, Ontario, Canada) and vacuum dried for 1 week at 50 °C. All polymerizations were independently replicated.

## 2.4. Polymer Characterization

Conversion of the polymer samples was determined using gravimetry. The mass of the sodium ions was also considered in conversion calculations, as per the recommendation of Riahinezhad et al. [11,13]. Copolymer composition was measured using elemental analysis (CHNS, Vario MICRO Cube, Elementar, Isospark Canada Inc., Montreal, Canada). Calculation of composition did not include H content, as residual water has been known to affect the determined H content [12]. Select samples were independently replicated.

## 3. Results and Discussion

The error-in-variables-model framework, which was outlined in Figure 1, is applied to the terpolymerization of AMPS/AAm/AAc. This is the first time in the literature that the entire framework, from preliminary investigation and design of experiments to parameter estimation, is verified experimentally.

## 3.1. Step 1: Review Literature for Polymerization Kinetics

In recent years, several studies have investigated the AMPS/AAm/AAc terpolymer. These have focused on synthesis, characterization, and potential applications for this terpolymer; none of the studies have included terpolymerization kinetics.

For example, Bao et al. [5] grafted the AMPS/AAm/AAc terpolymer onto sodium carboxymethyl cellulose and montmorillonite (MMT) to create a superabsorbent hydrogel. In this case, physical properties of the synthesized terpolymer (such as degree of swelling, water retention, and morphology) were the focus of the analysis. Similarly, Ma et al. [6] synthesized the AMPS/AAm/AAc terpolymer via UV irradiation for use as a flocculent. While this group did provide more information about their synthesis steps, the overall focus of the paper was applications. Polymer characteristics including intrinsic viscosity, dissolution time and flocculation performance were presented. This particular terpolymer has also been used in drug-delivery applications [7]. The drug-delivery system uses superabsorbent polymer composites, so characteristics such as swelling capacity and drug encapsulation efficiency were studied. While the investigation included release profiles for drug-delivery, it did not discuss details surrounding the polymerization kinetics.

In perhaps the most relevant papers to the current work, Peng et al. [4] and Zaitoun et al. [3] have studied the AMPS/AAm/AAc terpolymer for petrochemical applications. The work by Peng et al. [4] describes the free-radical terpolymerization of AMPS/AAm/AAc and its application as a high-temperature resistant filtration control agent. Zaitoun et al. [3] have investigated the potential to use AMPS/AAm/AAc in enhanced oil recovery (EOR) applications, as the AMPS comonomer is expected to improve shear stability and limit thermal degradation (compared to standard AAm/AAc copolymers). However, in both of these cases, the authors make no mention of polymerization kinetics.

The kinetic characteristics of the terpolymer being synthesized are directly related to its microstructure. Therefore, it is important to have a clear understanding of the terpolymerization kinetics. Since this information is not available in the literature, reliable reactivity ratios for this AMPS/AAm/AAc system will be determined experimentally in what follows.

## 3.2. Step 2: Determine Preliminary Reactivity Ratios and Establish Feed Composition Constraints

Since, to date, there have been no kinetic studies for this particular terpolymerization in the literature, binary values for the associated copolymer pairs were used as preliminary reactivity ratios. These binary values were obtained experimentally, which allowed for the same experimental set-up to be used for both the co- and terpolymerizations.

For the AMPS/AAm and AMPS/AAc copolymerizations, an in-depth study was recently completed by Scott et al. [14]. Different reactivity ratio estimates for these copolymers were published previously, but most of these estimates were subject to numerous sources of error (namely, linear parameter estimation techniques for non-linear parameter estimation with no independent replication). In an attempt to provide the most accurate reactivity ratio estimates possible, Scott et al. [14] used the error-in-variables-model (EVM) technique to design experiments and estimate reactivity ratios for both AMPS/AAm and AMPS/AAc copolymerizations. As an additional advantage, the polymerization conditions (pH, ionic strength, etc.) that were used in these copolymerizations are also used for the terpolymerizations described in the current paper.

Similarly, for the AAm/AAc copolymerization, Riahinezhad et al. [12] conducted a thorough investigation of the reactivity ratios for this copolymerization system. Again, the same polymerization conditions were used for the AAm/AAc binary system and for the terpolymerizations described in the current paper. Therefore, the reactivity ratios arrived at for the above binary systems can confidently be considered as the "best" binary reactivity ratios for the different pairs. These values are summarized in Table 1.

**Table 1.** Binary reactivity ratios for copolymerizations associated with 2-acrylamido-2-methylpropane sulfonic acid (AMPS)[1]/acrylamide (AAm)[2]/acrylic acid (AAc)[3].

| Source | T(°C) | pH | $r_{12}$ | $r_{21}$ | $r_{13}$ | $r_{31}$ | $r_{23}$ | $r_{32}$ |
|--------|-------|-----|----------|----------|----------|----------|----------|----------|
| Scott et al. [14] | 40 | 7 | 0.18 | 0.85 | 0.19 | 0.86 | - | - |
| Riahinezhad et al. [12] | 40 | 7 | - | - | - | - | 1.33 | 0.23 |

One of the advantages associated with EVM is the ability to introduce feed composition constraints on the experimental design. However, since very little kinetic information is available for the AMPS/AAm/AAc terpolymerization, it is difficult to establish whether such constraints exist. In studying the AMPS/AAc copolymer, Scott et al. [14] reported that the polymerization was extremely slow and minimal precipitate formed when the AAc fraction was high in the feed ($f_{AAc,0} = 0.85$). Similarly, Ryles and Neff [1] observed that a preliminary feed composition of $f_{AAc,0} = 0.80$ for the AMPS/AAc copolymer made polymer isolation (and subsequent filtering/precipitation) difficult as a result of phase separation. Therefore, for the AMPS/AAm/AAc terpolymer, the feed composition of AAc was constrained such that $f_{AAc,0} \leq 0.70$.

### 3.3. Step 3: Apply Design of Experiments to Select Optimal Feed Compositions

The basic idea behind design of experiments is to select optimal feed compositions (experimental trials) which minimize variability in the parameter estimates. As mentioned previously, terpolymerization studies often (incorrectly!) use reactivity ratios extracted from binary systems; based on this analogy, a ternary system is treated as three separate binary copolymerizations. This approach is approximate at best, as it propagates the inherent error present in the binary reactivity ratio estimates, overlooks the effect of the interactions between all three monomers on their reactivity towards each other and is not at all reliable for predicting ternary compositions.

An additional problem with terpolymerization studies is the form of the Alfrey–Goldfinger (A–G) model that is typically used to evaluate instantaneous terpolymerization composition (see Equation (1)). Kazemi et al. [15] recently showed that selecting different combinations of ratios of mole fractions in the A-G model (e.g., $F_1/F_2$ and $F_1/F_3$ versus $F_1/F_2$ and $F_2/F_3$) can affect the precision of the reactivity ratio estimates. This work exposed the fact that the model suffers from symmetry issues; final results depend on the arbitrary choice of different combinations of copolymer mole fractions into the parameter estimation scheme. Therefore, in current and future terpolymerization investigations, a recast version of the model (courtesy of Kazemi et al. [8]) should be used. The recast A–G model presents each instantaneous terpolymer mole fraction as a single response (see Equation (2)). While these expressions may seem more complex than the conventional A–G model, this formulation is symmetrical and error structures are not distorted [8]:

$$F_1 = \frac{f_1\left(\frac{f_1}{r_{21}r_{31}} + \frac{f_2}{r_{21}r_{32}} + \frac{f_3}{r_{31}r_{23}}\right)\left(f_1 + \frac{f_2}{r_{12}} + \frac{f_3}{r_{13}}\right)}{f_1\left(\frac{f_1}{r_{21}r_{31}} + \frac{f_2}{r_{21}r_{32}} + \frac{f_3}{r_{31}r_{23}}\right)\left(f_1 + \frac{f_2}{r_{12}} + \frac{f_3}{r_{13}}\right) + f_2\left(\frac{f_1}{r_{21}r_{31}} + \frac{f_2}{r_{12}r_{32}} + \frac{f_3}{r_{13}r_{32}}\right)\left(f_2 + \frac{f_1}{r_{21}} + \frac{f_3}{r_{23}}\right) + f_3\left(\frac{f_1}{r_{13}r_{21}} + \frac{f_2}{r_{23}r_{12}} + \frac{f_3}{r_{13}r_{23}}\right)\left(f_3 + \frac{f_1}{r_{31}} + \frac{f_2}{r_{32}}\right)}, \quad (2a)$$

$$F_2 = \frac{f_2\left(\frac{f_1}{r_{21}r_{31}} + \frac{f_2}{r_{12}r_{32}} + \frac{f_3}{r_{13}r_{32}}\right)\left(f_2 + \frac{f_1}{r_{21}} + \frac{f_3}{r_{23}}\right)}{f_1\left(\frac{f_1}{r_{21}r_{31}} + \frac{f_2}{r_{21}r_{32}} + \frac{f_3}{r_{31}r_{23}}\right)\left(f_1 + \frac{f_2}{r_{12}} + \frac{f_3}{r_{13}}\right) + f_2\left(\frac{f_1}{r_{12}r_{31}} + \frac{f_2}{r_{12}r_{32}} + \frac{f_3}{r_{13}r_{32}}\right)\left(f_2 + \frac{f_1}{r_{21}} + \frac{f_3}{r_{23}}\right) + f_3\left(\frac{f_1}{r_{13}r_{21}} + \frac{f_2}{r_{23}r_{12}} + \frac{f_3}{r_{13}r_{23}}\right)\left(f_3 + \frac{f_1}{r_{31}} + \frac{f_2}{r_{32}}\right)}, \quad (2b)$$

$$F_3 = \frac{f_3\left(\frac{f_1}{r_{13}r_{21}} + \frac{f_2}{r_{23}r_{12}} + \frac{f_3}{r_{13}r_{23}}\right)\left(f_3 + \frac{f_1}{r_{31}} + \frac{f_2}{r_{32}}\right)}{f_1\left(\frac{f_1}{r_{21}r_{31}} + \frac{f_2}{r_{21}r_{32}} + \frac{f_3}{r_{31}r_{23}}\right)\left(f_1 + \frac{f_2}{r_{12}} + \frac{f_3}{r_{13}}\right) + f_2\left(\frac{f_1}{r_{12}r_{31}} + \frac{f_2}{r_{12}r_{32}} + \frac{f_3}{r_{13}r_{32}}\right)\left(f_2 + \frac{f_1}{r_{21}} + \frac{f_3}{r_{23}}\right) + f_3\left(\frac{f_1}{r_{13}r_{21}} + \frac{f_2}{r_{23}r_{12}} + \frac{f_3}{r_{13}r_{23}}\right)\left(f_3 + \frac{f_1}{r_{31}} + \frac{f_2}{r_{32}}\right)}. \quad (2c)$$

With this new information in mind, the goal in this step of the procedure is to apply the EVM design criterion to the recast Alfrey–Goldfinger model so that optimal feed compositions (that can lead to the most reliable reactivity ratios) are selected.

EVM considers error in all terms, so using a design of experiments technique within the EVM context helps to account for the error in both the independent variables (feed compositions) and the dependent variables (terpolymer compositions). Details have been presented previously [8,9], but the key points are briefly revisited below. The EVM design criterion aims to maximize the determinant of the information matrix ($\underline{G}$), which is the inverse of the variance–covariance matrix of the parameters:

$$\underline{G} = \sum_{i=1}^{n} r_i \underline{Z}'_i \left(\underline{B}_i \underline{V} \underline{B}'_i\right)^{-1} \underline{Z}_i, \quad (3)$$

where $r_i$ = number of replicates at the $i^{th}$ trial (out of n trials), $\underline{Z}_i$ = vector of partial derivatives of the model function with respect to the parameters (in this case, the partial derivatives of the recast A–G model (Equation (2)) with respect to the reactivity ratios), $\underline{B}_i$ = vector of partial derivatives of the model function with respect to the variables (again, in this case, the partial derivatives of the recast A-G model (Equation (2)) with respect to the feed ($f$) and terpolymer ($F$) compositions), and $\underline{V}$ = variance–covariance matrix of the variables (which provides information about measurement error and possible correlation of the variables involved).

As explained in previous work by Kazemi et al. [9], three optimal experiments are sufficient to estimate terpolymerization reactivity ratios in this nonlinear model scenario. In the terpolymerization problem, the EVM model consists of three equations (see again Equation (2)) and five variables ($f_1, f_2, F_1, F_2, F_3$); only two of the three feed compositions are independent ($f_3 = 1 - f_1 - f_2$) and the terpolymer compositions are measured independently. Therefore, for the terpolymerization, there are two independent variables (5 variables – 3 equations = 2) and six (6) parameters (reactivity ratios). The number of optimal experiments needed can be calculated by dividing the number of parameters by the number of independent variables (see Bard [16] and Duever et al. [17]); hence, 6 ÷ 2 = 3.

For ternary reactivity ratio estimation, optimal feed compositions are typically located at the corners of the triangular (terpolymerization) composition plot [9]. To help visualize the location of the optimal feed compositions for this system, Figure 2 combines the "suboptimal" regions (shaded areas) as well as the three optimal points located close to or inside these regions. Polymerizations run using these three optimal feed compositions (recipes) will provide sufficient information for reliable reactivity ratio estimation.

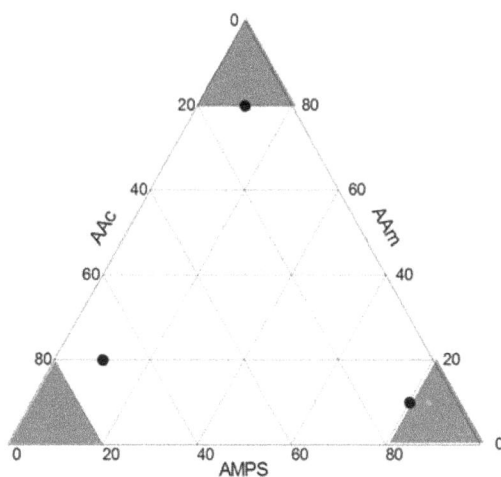

**Figure 2.** Optimal feed regions and compositions for 2-acrylamido-2-methylpropane sulfonic acid (AMPS)/acrylamide (AAm)/acrylic acid (AAc) terpolymerization.

*3.4. Step 4: Perform Experiments; Collect Conversion and Composition Data*

3.4.1. Low Conversion Experiments

The first attempt at estimating the reactivity ratios for the AMPS/AAm/AAc terpolymerization was conducted by analyzing low conversion data, similar to the conventional approaches for estimating reactivity ratios of copolymerizations and terpolymerizations [9,18]. As shown in Table 2, the three feed compositions correspond to the optimal feed compositions of Figure 2 (reflecting also process constraints), and the experimental data were limited to low conversion ($X_w < 0.100$). Conversion values with a "*" indicate results from an independently replicated polymerization; the same classification will be used in Table 3. That is, these replicates were synthesized entirely independently, using freshly made solutions, etc. Conversion was determined again using gravimetry, and samples were independently characterized using elemental analysis, as described in Section 2.4. Note also that experimental data in Tables 2 and 3 are presented in terms of mass conversion ($X_w$), which should not be confused with molar conversion ($X_n$; see, for example, Equations (4) through (6)).

**Table 2.** Experimental data for AMPS/AAm/AAc terpolymerization; low conversion.

| Optimal Feed Composition | $X_w$ | $f_{AMPS,0}$ | $f_{AAm,0}$ | $f_{AAc,0}$ | $\overline{F}_{AMPS}$ | $\overline{F}_{AAm}$ | $\overline{F}_{AAc}$ |
|---|---|---|---|---|---|---|---|
| #1 | *0.066 | 0.8 | 0.1 | 0.1 | 0.761 | 0.135 | 0.104 |
| | *0.071 | 0.8 | 0.1 | 0.1 | 0.778 | 0.137 | 0.085 |
| | 0.083 | 0.8 | 0.1 | 0.1 | 0.783 | 0.136 | 0.081 |
| | 0.091 | 0.8 | 0.1 | 0.1 | 0.750 | 0.146 | 0.104 |
| #2 | 0.036 | 0.1 | 0.8 | 0.1 | 0.123 | 0.826 | 0.052 |
| | *0.043 | 0.1 | 0.8 | 0.1 | 0.122 | 0.824 | 0.054 |
| | 0.071 | 0.1 | 0.8 | 0.1 | 0.107 | 0.836 | 0.057 |
| | *0.079 | 0.1 | 0.8 | 0.1 | 0.115 | 0.817 | 0.069 |
| | *0.088 | 0.1 | 0.8 | 0.1 | 0.110 | 0.823 | 0.067 |
| #3 | *0.021 | 0.1 | 0.2 | 0.7 | 0.123 | 0.374 | 0.503 |
| | 0.025 | 0.1 | 0.2 | 0.7 | 0.121 | 0.382 | 0.496 |
| | 0.030 | 0.1 | 0.2 | 0.7 | 0.122 | 0.384 | 0.494 |
| | *0.034 | 0.1 | 0.2 | 0.7 | 0.118 | 0.368 | 0.514 |
| | *0.040 | 0.1 | 0.2 | 0.7 | 0.117 | 0.372 | 0.511 |
| | 0.056 | 0.1 | 0.2 | 0.7 | 0.125 | 0.377 | 0.498 |
| | 0.059 | 0.1 | 0.2 | 0.7 | 0.124 | 0.374 | 0.502 |
| | 0.064 | 0.1 | 0.2 | 0.7 | 0.121 | 0.368 | 0.511 |
| | *0.087 | 0.1 | 0.2 | 0.7 | 0.114 | 0.344 | 0.542 |

In this and in Table 3, $X_w$ = conversion, $f_{i,0}$ = initial feed composition (monomer $i$) and $\overline{F}_i$ = cumulative terpolymer composition.

Since the conversion level was kept low for these runs, it can be assumed that the composition drift is negligible. Therefore, the instantaneous terpolymer composition model (that is, the recast Alfrey–Goldfinger model, Equation (2)) and the EVM parameter estimation technique were employed [9]. Details regarding the parameter estimation technique, the resulting reactivity ratio estimates, and the corresponding joint confidence regions (JCRs) will be presented in Step 5 (Section 3.5).

### 3.4.2. Medium-High Conversion Experiments

The recast A–G model (Equation (2)) is a significant improvement over Equation (1), but it is only valid for low conversion data sets. In order for copolymer composition drift to be negligible (that is, for the initial feed composition to remain unchanged and for the (measurable) cumulative copolymer composition to be equal to its instantaneous value), experimental data must be collected at very low conversion levels. This restrictive assumption introduces additional sources of error, including significant experimental difficulties.

As an alternative, a cumulative ternary composition model has been considered in order to estimate ternary reactivity ratios using the full conversion trajectory. The cumulative model (essentially the Skeist equation applied to terpolymerization), shown in Equation (4), relates the cumulative terpolymer composition for each monomer ($\overline{F}_i$) to the initial mole fraction of monomer in the feed ($f_{i,0}$), and the corresponding mole fraction of unreacted monomer ($f_i$) and molar conversion ($X_n$):

$$\overline{F}_1 = \frac{f_{1,0} - f_1(1 - X_n)}{X_n}, \tag{4a}$$

$$\overline{F}_2 = \frac{f_{2,0} - f_2(1 - X_n)}{X_n}, \tag{4b}$$

$$\overline{F}_3 = \frac{f_{3,0} - f_3(1 - X_n)}{X_n}. \tag{4c}$$

In this step of the procedure, reactivity ratios for the AMPS/AAm/AAc terpolymerization were estimated using the same optimal feed compositions of Figure 2, but this time running the terpolymerizations to medium-high conversion levels (see Table 3). Since it is no longer valid to assume constant composition (that is, composition drift is no longer negligible), $f_i$ must be evaluated over conversion $X_n$, according to the model in ordinary differential equation form, shown in Equation (5) (where $F_i$ values are calculated using Equation (2)). Given the initial conditions $f_i = f_{i,0}$ at $X_n = 0$, a numerical solution can be used to evaluate terpolymer compositions along the full conversion trajectory:

$$\frac{df_1}{dX_n} = \frac{f_1 - F_1}{1 - X_n}, \tag{5a}$$

$$\frac{df_2}{dX_n} = \frac{f_2 - F_2}{1 - X_n}, \tag{5b}$$

$$\frac{df_3}{dX_n} = \frac{f_3 - F_3}{1 - X_n}. \tag{5c}$$

It is important to note that molar conversion ($X_n$) is used in both Equations (4) and (5), but that mass conversion ($X_w$) is reported in the experimental data tables (see Tables 2 and 3). Molar conversion and mass conversion are related using monomer molecular weights ($MW_i$), as shown in Equation (6):

$$X_n = X_w \frac{MW_1 f_{1,0} + MW_2 f_{2,0} + MW_3 f_{3,0}}{MW_1 \overline{F}_1 + MW_2 \overline{F}_2 + MW_3 \overline{F}_3}. \tag{6}$$

This methodology (using direct numerical integration (DNI) to evaluate the cumulative composition model) has been described previously by Kazemi et al. [8]. The current approach (i.e., integrating the instantaneous terpolymer composition model (Equation (2)) over conversion via Equations (4) and (5) and conducting parameter estimation via EVM simultaneously) is generally preferable for parameter estimation, as it includes all available information from the system (not only at low conversion as per typical approaches), and does not suffer from the limiting assumptions or experimental difficulties associated with low conversion data analysis. As was the case for the low conversion experiments, estimation details along with reactivity ratio estimates and corresponding JCRs will be shown in Step 5 (Section 3.5).

**Table 3.** Experimental data for AMPS/AAm/AAc terpolymerization; medium-high conversion.

| Optimal Feed Composition | $X_w$ | $f_{AMPS,0}$ | $f_{AAm,0}$ | $f_{AAc,0}$ | $\overline{F}_{AMPS}$ | $\overline{F}_{AAm}$ | $\overline{F}_{AAc}$ |
|---|---|---|---|---|---|---|---|
| | *0.066 | 0.8 | 0.1 | 0.1 | 0.761 | 0.135 | 0.104 |
| | *0.071 | 0.8 | 0.1 | 0.1 | 0.778 | 0.137 | 0.085 |
| | 0.083 | 0.8 | 0.1 | 0.1 | 0.783 | 0.136 | 0.081 |
| | 0.091 | 0.8 | 0.1 | 0.1 | 0.750 | 0.146 | 0.104 |
| | *0.143 | 0.8 | 0.1 | 0.1 | 0.771 | 0.122 | 0.106 |
| #1 | 0.190 | 0.8 | 0.1 | 0.1 | 0.753 | 0.146 | 0.101 |
| | *0.208 | 0.8 | 0.1 | 0.1 | 0.765 | 0.134 | 0.101 |
| | 0.382 | 0.8 | 0.1 | 0.1 | 0.757 | 0.138 | 0.105 |
| | *0.469 | 0.8 | 0.1 | 0.1 | 0.780 | 0.114 | 0.105 |
| | 0.512 | 0.8 | 0.1 | 0.1 | 0.760 | 0.132 | 0.107 |
| | 0.733 | 0.8 | 0.1 | 0.1 | 0.769 | 0.120 | 0.112 |
| | 0.836 | 0.8 | 0.1 | 0.1 | 0.775 | 0.108 | 0.116 |

**Table 3.** *Cont.*

| Optimal Feed Composition | $X_w$ | $f_{AMPS,0}$ | $f_{AAm,0}$ | $f_{AAc,0}$ | $\bar{F}_{AMPS}$ | $\bar{F}_{AAm}$ | $\bar{F}_{AAc}$ |
|---|---|---|---|---|---|---|---|
| | 0.036 | 0.1 | 0.8 | 0.1 | 0.123 | 0.826 | 0.052 |
| | *0.043 | 0.1 | 0.8 | 0.1 | 0.122 | 0.824 | 0.054 |
| | 0.071 | 0.1 | 0.8 | 0.1 | 0.107 | 0.836 | 0.057 |
| | *0.079 | 0.1 | 0.8 | 0.1 | 0.115 | 0.817 | 0.069 |
| | *0.088 | 0.1 | 0.8 | 0.1 | 0.110 | 0.823 | 0.067 |
| | 0.114 | 0.1 | 0.8 | 0.1 | 0.110 | 0.834 | 0.056 |
| | 0.138 | 0.1 | 0.8 | 0.1 | 0.111 | 0.828 | 0.061 |
| #2 | *0.183 | 0.1 | 0.8 | 0.1 | 0.106 | 0.825 | 0.069 |
| | *0.186 | 0.1 | 0.8 | 0.1 | 0.109 | 0.824 | 0.068 |
| | 0.227 | 0.1 | 0.8 | 0.1 | 0.110 | 0.827 | 0.063 |
| | 0.360 | 0.1 | 0.8 | 0.1 | 0.109 | 0.814 | 0.077 |
| | *0.370 | 0.1 | 0.8 | 0.1 | 0.105 | 0.824 | 0.072 |
| | *0.447 | 0.1 | 0.8 | 0.1 | 0.110 | 0.806 | 0.084 |
| | 0.496 | 0.1 | 0.8 | 0.1 | 0.117 | 0.797 | 0.086 |
| | 0.525 | 0.1 | 0.8 | 0.1 | 0.109 | 0.812 | 0.078 |
| | 0.544 | 0.1 | 0.8 | 0.1 | 0.112 | 0.808 | 0.080 |
| | *0.021 | 0.1 | 0.2 | 0.7 | 0.123 | 0.374 | 0.503 |
| | 0.025 | 0.1 | 0.2 | 0.7 | 0.121 | 0.382 | 0.496 |
| | 0.030 | 0.1 | 0.2 | 0.7 | 0.122 | 0.384 | 0.494 |
| | *0.034 | 0.1 | 0.2 | 0.7 | 0.118 | 0.368 | 0.514 |
| | *0.040 | 0.1 | 0.2 | 0.7 | 0.117 | 0.372 | 0.511 |
| | 0.056 | 0.1 | 0.2 | 0.7 | 0.125 | 0.377 | 0.498 |
| | 0.059 | 0.1 | 0.2 | 0.7 | 0.124 | 0.374 | 0.502 |
| | 0.064 | 0.1 | 0.2 | 0.7 | 0.121 | 0.368 | 0.511 |
| #3 | *0.087 | 0.1 | 0.2 | 0.7 | 0.114 | 0.344 | 0.542 |
| | 0.103 | 0.1 | 0.2 | 0.7 | 0.121 | 0.359 | 0.520 |
| | *0.125 | 0.1 | 0.2 | 0.7 | 0.112 | 0.329 | 0.559 |
| | 0.139 | 0.1 | 0.2 | 0.7 | 0.126 | 0.353 | 0.521 |
| | *0.151 | 0.1 | 0.2 | 0.7 | 0.115 | 0.349 | 0.536 |
| | 0.191 | 0.1 | 0.2 | 0.7 | 0.123 | 0.337 | 0.540 |
| | *0.199 | 0.1 | 0.2 | 0.7 | 0.116 | 0.348 | 0.536 |
| | 0.259 | 0.1 | 0.2 | 0.7 | 0.119 | 0.341 | 0.540 |
| | *0.260 | 0.1 | 0.2 | 0.7 | 0.117 | 0.342 | 0.541 |
| | *0.282 | 0.1 | 0.2 | 0.7 | 0.116 | 0.342 | 0.542 |

### 3.5. Step 5: Use EVM to Estimate Reactivity Ratios; Construct Joint Confidence Regions

The error-in-variables-model (EVM), described previously for the design of experiments (see Section 3.3), can also be used in the current parameter estimation step. EVM is one of the most powerful non-linear regression approaches available, as it considers all sources of experimental error (both in the independent and dependent variables) [19]. EVM not only forces the experimenter to consider all sources of error, but also provides estimates of the true values of other variables involved in the model along with the parameter estimates. Therefore, it is by far the most statistically correct and comprehensive approach for reactivity ratio estimation [20].

In this step, the EVM approach is used to estimate reactivity ratios for both the low and medium-high conversion data. However, as discussed in Section 3.4, the terpolymerization model differs for each data set: low conversion data are analyzed using the recast instantaneous terpolymerization model (Equation (2)) along with the related low conversion assumptions, whereas the medium-high conversion data are analyzed with the direct numerical integration (DNI) of the instantaneous model, i.e., using the cumulative composition model (see Equations (2), (4) and (5)). The nested-iterative EVM implementation has been described in detail in several previous references (for instance, see references [8–10,17,18,21,22] cited herein), so no further details will be presented.

The terpolymerization reactivity ratio estimates and corresponding JCRs for both data sets (Tables 2 and 3), along with the corresponding binary (copolymerization) reactivity ratios (Table 1) are presented in Figure 3. In all cases, the results show that JCRs from medium-high conversion data are smaller (and therefore more precise) than JCRs from low conversion data. These results are as expected; utilizing all of the experimental information available and eliminating potentially inaccurate assumptions can improve the precision of the point estimates. These results confirm (for the first time) that EVM can successfully be used to analyze directly experimental data from terpolymerizations over the whole conversion range. In addition, the results prove that three optimally designed ternary feed compositions can provide sufficient information to estimate reactivity ratios with very little correlation for this terpolymerization system, as one can realize from the orientation of the JCRs. In addition, in all cases, the literature binary reactivity ratio estimates are located outside the terpolymerization JCRs.

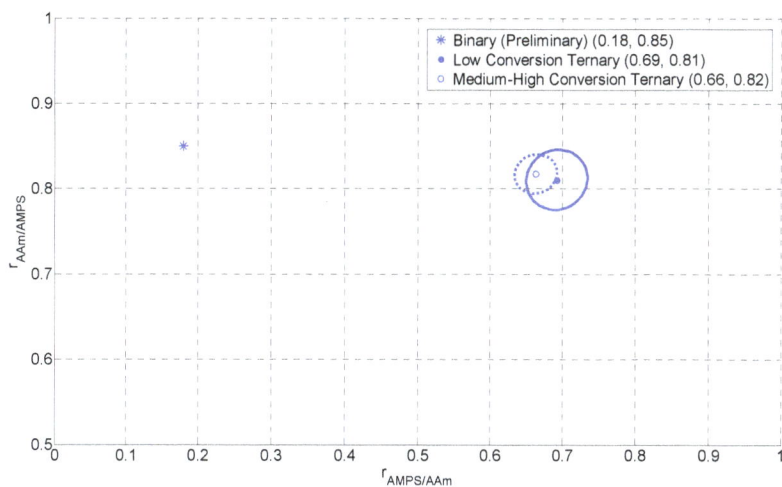

(a)

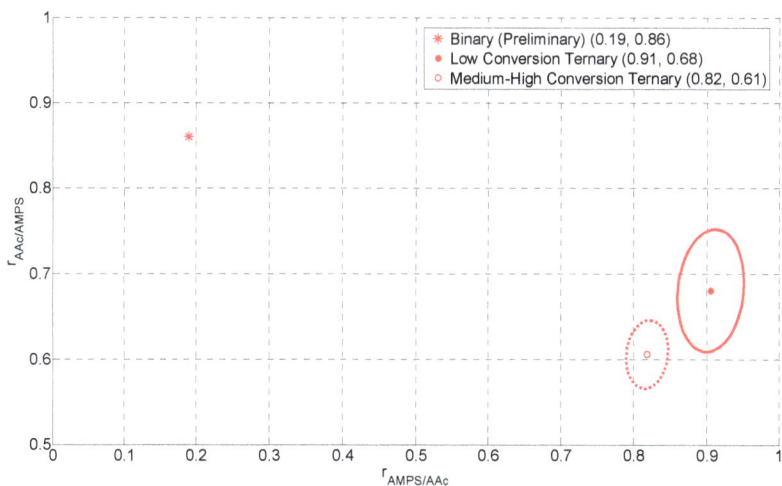

(b)

**Figure 3.** *Cont.*

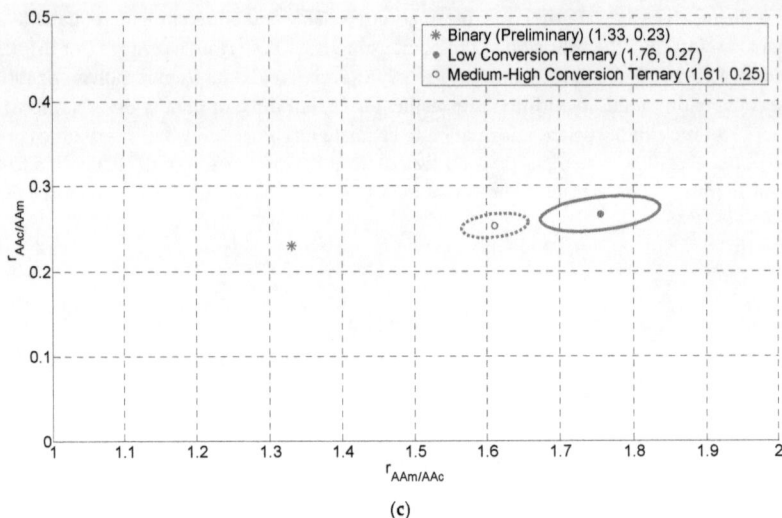

(c)

**Figure 3.** Reactivity ratio estimation results for AMPS/AAm/AAc terpolymerization (with preliminary copolymerization estimates from literature [12,14]).

### 3.6. Step 6: Decide Whether Results Are Precise Enough

In Step 5 (Section 3.5), the reactivity ratio estimates and associated JCRs confirm that the EVM-based experimental design and parameter estimation method for ternary systems work very well with experimental data directly from the AMPS/AAm/AAc terpolymerization. The reliability of the results is first established by examining the size of the JCRs and by noting the lack of correlation between the parameters (see Figure 3).

In the second diagnostic stage, it is important to investigate the accuracy of the reactivity ratios by running additional checks. One of the most common diagnostic checks is to evaluate the behavior/profiles of the cumulative terpolymer composition. Model predictions (using reactivity ratio estimates) for the terpolymer composition over the polymerization trajectory are compared to experimentally measured terpolymer compositions. An acceptable agreement between predicted and experimental results reflects the reliability and accuracy of the reactivity ratios for the terpolymerization system.

Thus, direct numerical integration (DNI) was applied to the recast version of the instantaneous terpolymer composition equation (see Equation (2)) using newly determined reactivity ratio estimates. Since the medium-high conversion data provided more information (and smaller JCRs), the reactivity ratios estimated from the data of Table 3 were used. The predicted cumulative terpolymer composition trajectories versus conversion, as well as the experimental points obtained via elemental analysis, are shown in Figure 4 for all three of the optimally designed feed compositions.

Figure 4 shows that, in all three cases, the predicted terpolymer composition trajectories (from ternary reactivity ratio estimates) capture the experimentally observed behavior satisfactorily. At low conversion, however, there are some minor discrepancies between the model and the experimental results. The noise seen in the experimental data is typical at such low conversions, which confirms the need for higher conversion experiments. Thus, in spite of the natural variation in experimental results, it is possible to conclude that the cumulative terpolymer composition (DNI) model and the EVM-based estimated ternary reactivity ratios can successfully predict the behavior of the system. This is an important diagnostic check for this system (and for terpolymerization studies, in general), as it indicates that employing the terpolymerization composition model and the EVM framework leads to precise and reliable reactivity ratios for the system.

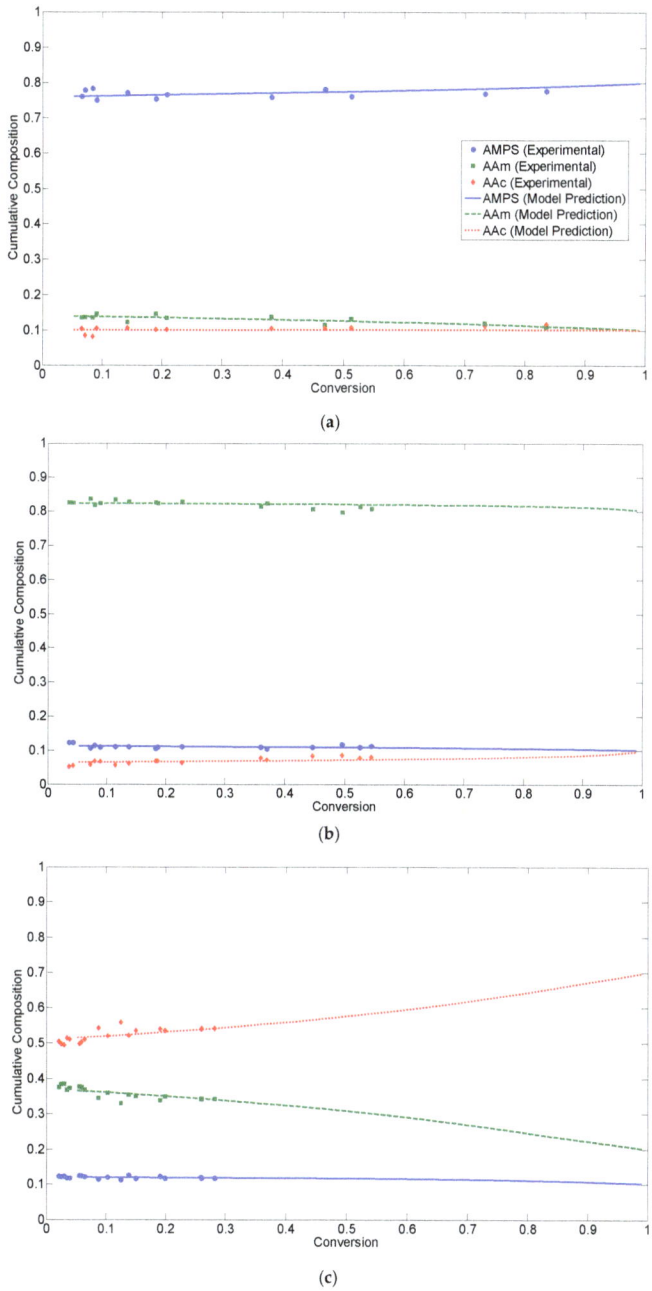

**Figure 4.** Cumulative terpolymer composition, $\overline{F}_i$, for AMPS/AAm/AAc (experimental data and model predictions) for $f_{AMPS,0}/f_{AAm,0}/f_{AAc,0}$ = (**a**) 0.8/0.1/0.1, (**b**) 0.1/0.8/0.1 and (**c**) 0.1/0.2/0.7.

### 3.7. Step 7: Present Reactivity Ratios and JCRs

It was shown in Step 5 (Section 3.5) that using medium-high conversion data provides smaller joint confidence regions (JCRs) for reactivity ratio estimation (compared to conventional low conversion data analysis). In addition, in Step 6 (Section 3.6), the reactivity ratio estimates from medium-high conversion data were successfully used to predict the cumulative terpolymer composition. Therefore, of the reactivity ratio estimates presented in Table 4, the information in the last row (for ternary reactivity ratio estimation of medium-high conversion data) is the most precise. The JCRs associated with these estimates have been presented previously in Figure 3.

**Table 4.** Summary of reactivity ratio estimates for $AMPS^1/AAm^2/AAc^3$ terpolymerization.

| Experimental Data | Conversion | Type | $r_{12}$ | $r_{21}$ | $r_{13}$ | $r_{31}$ | $r_{23}$ | $r_{32}$ |
|---|---|---|---|---|---|---|---|---|
| Literature [12,14] | Medium-High | Binary | 0.18 | 0.85 | 0.19 | 0.86 | 1.33 | 0.23 |
| Optimally Designed Data | Low | Ternary | 0.69 | 0.81 | 0.91 | 0.68 | 1.76 | 0.27 |
| Optimally Designed Data | Medium-High | Ternary | 0.66 | 0.82 | 0.82 | 0.61 | 1.61 | 0.25 |

## 4. Conclusions

We have discussed the effectiveness of the error-in-variables-model (EVM) framework for experimental applications. More specifically, accurate ternary reactivity ratios have been established for the AMPS/AAm/AAc terpolymerization. In a comparison of reactivity ratio estimation results for low conversion and medium-high conversion level data, the point estimates were fairly consistent. However, in terms of precision, the medium-high conversion level data provided much smaller JCRs, which indicates a much higher degree of confidence in the results (compared to the low conversion data results). This represents an improvement, since the collected data at medium-high conversion levels contain more information; in addition, potentially inaccurate assumptions (required for analyses with low conversion level data and instantaneous models) are avoided. For the first time, EVM was successfully applied to experimental terpolymerization data at medium-high conversion levels. The analysis has also shown that sufficient information is available from three optimally designed feed compositions; ternary reactivity ratios were estimated with high precision and very little correlation for the AMPS/AAm/AAc system.

**Acknowledgments:** The authors wish to acknowledge financial support from the Natural Sciences and Engineering Research Council (NSERC) of Canada and the Canada Research Chair (CRC) program. In addition, thanks go to UWW/OMNOVA Solutions, Akron, OH, USA, for special support to A.J.S.

**Author Contributions:** The experimental data collection and reactivity ratio estimation was performed by A.J.S. The EVM framework (for design of experiments and parameter estimation) was based on the Ph.D. thesis by N.K. A.P. supervised the work done by both N.K. and A.J.S.

**Conflicts of Interest:** The authors declare no conflict of interest.

## References

1. Ryles, R.G.; Neff, R.E. Thermally stable acrylic monomer for profile modification applications. In *Water-Soluble Polymers for Petroleum Recovery*; Stahl, G.A., Schulz, D.N., Eds.; Springer Science & Business Media: New York, NY, USA, 1988.
2. Wei, B.; Romero-Zerón, L.; Rodrigue, D. Oil displacement mechanisms of viscoelastic polymers in enhanced oil recovery (EOR): A review. *J. Pet. Explor. Prod. Technol.* **2014**, *4*, 113–121. [CrossRef]
3. Zaitoun, A.; Makakou, P.; Blin, N.; Al-Maamari, R.; Al-Hashmi, A.; Abdel-Goad, M.; Al-Sharji, H. Shear Stability of EOR Polymers. In Proceedings of the Society of Petroleum Engineers International Symposium, The Woodlands, TX, USA, 11–13 April 2011.

4. Peng, B.; Peng, S.; Long, B.; Miao, Y.; Guo, W.Y. Properties of high-temperature-resistant drilling fluids incorporating acrylamide/(acrylic acid)/(2-acrylamido-2-methyl-1-propane sulfonic acid) terpolymer and aluminum citrate as filtration control agents. *J. Vinyl Add. Tech.* **2010**, *16*, 84–89. [CrossRef]
5. Bao, Y.; Ma, J.; Li, N. Synthesis and swelling behaviors of sodium carboxymethyl cellulose-g-poly(AA-co-AM-co-AMPS)/MMT superabsorbent hydrogel. *Carbohydr. Polym.* **2011**, *84*, 76–82. [CrossRef]
6. Ma, J.; Zheng, H.; Tan, M.; Liu, L.; Chen, W.; Guan, Q.; Zheng, X. Synthesis, characterization, and flocculation performance of anionic polyacrylamide P(AM-AA-AMPS). *J. Appl. Polym. Sci.* **2013**, *129*, 1984–1991. [CrossRef]
7. Anirudhan, T.S.; Rejeena, S.R. Biopolymer-based stimuli-sensitive functionalized graft copolymers as controlled drug delivery systems. In *Surface Modification of Biopolymers*; Thakur, V.K., Singha, A.S., Eds.; John Wiley & Sons, Inc.: Hoboken, NJ, USA, 2015; pp. 291–334.
8. Kazemi, N.; Duever, T.A.; Penlidis, A. Demystifying the estimation of reactivity ratios for terpolymerization systems. *AIChE J.* **2014**, *60*, 1752–1766. [CrossRef]
9. Kazemi, N.; Duever, T.A.; Penlidis, A. A powerful estimation scheme with the error-in-variables model for nonlinear cases: Reactivity ratio estimation examples. *Comput. Chem. Eng.* **2013**, *48*, 200–208. [CrossRef]
10. Riahinezhad, M.; McManus, N.T.; Penlidis, A. Effect of monomer concentration and pH on reaction kinetics and copolymer microstructure of acrylamide/acrylic acid copolymer. *Macromol. React. Eng.* **2015**, *9*, 100–113. [CrossRef]
11. Riahinezhad, M.; Kazemi, N.; McManus, N.T.; Penlidis, A. Optimal estimation of reactivity ratios for acrylamide/acrylic acid copolymerization. *J. Polym. Sci. Part A: Polym. Chem.* **2013**, *51*, 4819–4827. [CrossRef]
12. Riahinezhad, M.; Kazemi, N.; McManus, N.T.; Penlidis, A. Effect of ionic strength on the reactivity ratios of acrylamide/acrylic acid (sodium acrylate) copolymerization. *J. Appl. Polym. Sci.* **2014**, *131*, 40949. [CrossRef]
13. Scott, A.J.; Riahinezhad, M.; Penlidis, A. Optimal design for reactivity ratio estimation: A comparison of techniques for AMPS/acrylamide and AMPS/acrylic acid copolymerizations. *Processes.* **2015**, *3*, 749–768. [CrossRef]
14. Kazemi, N. Reactivity Ratio Estimation Aspects in Multicomponent Polymerizations at Low and High Conversion Levels. MASc Thesis, Department of Chemical Engineering, University of Waterloo, Waterloo, ON, Canada, 8 July 2010.
15. Bard, Y. *Nonlinear Parameter Estimation*; Academic Press: New York, NY, USA, 1974.
16. Duever, T.A.; Keeler, S.E.; Reilly, P.M.; Vera, J.; Williams, P. An application of the error-in-variables-model parameter estimation from Van Ness-type vapour-liquid equilibrium experiments. *Chem. Eng. Sci.* **1987**, *42*, 403–412. [CrossRef]
17. Kazemi, N.; Duever, T.A.; Penlidis, A. Reactivity ratio estimation from cumulative copolymer composition data. *Macromol. React. Eng.* **2011**, *5*, 385–403. [CrossRef]
18. Dube, M.A.; Amin Sanayei, R.; Penlidis, A.; O'Driscoll, K.F.; Reilly, P.M. A microcomputer program for estimation of copolymerization reactivity ratios. *J. Polym. Sci. Part A: Polym. Chem.* **1991**, *29*, 703–708. [CrossRef]
19. Polic, A.L.; Duever, T.A.; Penlidis, A. Case studies and literature review on the estimation of copolymerization reactivity ratios. *J. Polym. Sci. Part A: Polym. Chem.* **1998**, *36*, 813–822. [CrossRef]
20. Reilly, P.M.; Patino-Leal, H. A Bayesian study of the error-in-variables model. *Technometrics.* **1981**, *23*, 221–231. [CrossRef]
21. Reilly, P.M.; Reilly, H.V.; Keeler, S.E. Parameter estimation in the error-in-variables model. *J. R. Stat. Soc. Ser. C (Appl. Stat.)* **1993**, *42*, 693–701.
22. Hagiopol, C. *Copolymerization: Toward a Systematic Approach*; Springer: New York, NY, USA, 2012.

_processes_

MDPI

_Article_

# Poly(Poly(Ethylene Glycol) Methyl Ether Methacrylate) Grafted Chitosan for Dye Removal from Water

Bryan Tsai [1], Omar Garcia-Valdez [2], Pascale Champagne [1] and Michael F. Cunningham [2,*]

[1] Department of Civil Engineering, Queen's University, Kingston, ON K7L 3N6, Canada;
0bt12@queensu.ca (B.T.); pascale.champagne@queensu.ca (P.C.)
[2] Department of Chemical Engineering, Queen's University, Kingston, ON K7L 3N6, Canada;
omargv86@gmail.com
* Correspondence: michael.cunningham@queensu.ca; Tel.: +1-613-533-2782

Academic Editor: Alexander Penlidis
Received: 1 February 2017; Accepted: 11 March 2017; Published: 14 March 2017

**Abstract:** As the demand for textile products and synthetic dyes increases with the growing global population, textile dye wastewater is becoming one of the most significant water pollution contributors. Azo dyes represent 70% of dyes used worldwide, and are hence a significant contributor to textile waste. In this work, the removal of a reactive azo dye (Reactive Orange 16) from water by adsorption with chitosan grafted poly(poly(ethylene glycol) methyl ether methacrylate) (CTS-GMA-$g$-PPEGMA) was investigated. The chitosan (CTS) was first functionalized with glycidyl methacrylate and then grafted with poly(poly(ethylene glycol) methyl ether methacrylate) using a nitroxide-mediated polymerization grafting to approach. Equilibrium adsorption experiments were carried out at different initial dye concentrations and were successfully fitted to the Langmuir and Freundlich adsorption isotherm models. Adsorption isotherms showed maximum adsorption capacities of CTS-$g$-GMA-PPEGMA and chitosan of 200 mg/g and 150 mg/g, respectively, while the Langmuir equations estimated 232 mg/g and 194 mg/g, respectively. The fundamental assumptions underlying the Langmuir model may not be applicable for azo dye adsorption, which could explain the difference. The Freundlich isotherm parameters, $n$ and $K$, were determined to be 2.18 and 17.7 for CTS-$g$-GMA-PPEGMA and 0.14 and 2.11 for chitosan, respectively. An "$n$" value between one and ten generally indicates favorable adsorption. The adsorption capacities of a chitosan-PPEGMA 50/50 physical mixture and pure PPEGMA were also investigated, and both exhibited significantly lower adsorption capacities than pure chitosan. In this work, CTS-$g$-GMA-PPEGMA proved to be more effective than its parent chitosan, with a 33% increase in adsorption capacity.

**Keywords:** PEGMA; grafting; nitroxide-mediated polymerization; chitosan; wastewater; dye

## 1. Introduction

As the discharge of environmental pollutants in receiving waters continues to be a serious concern across the globe, it is important to develop effective approaches for their removal. Synthetic dyes found in wastewater effluents are of particular concern due to their high toxicity and low biodegradability in water [1]. They are used widely in the textile, pharmaceutical, cosmetics, and food industries because of their coloring capabilities. In the textile industry, an estimated 10,000 different dyes and pigments are used, and over $7 \times 10^5$ tons of synthetic dyes are annually produced worldwide; up to 200,000 tons of these dyes are discharged to receiving environments every year [2]. With the large quantities of dyes released and their known toxicity to aquatic environments, as well as to human health, the removal of textile dyes has received increasing research attention in recent years. Dyes have been reported to exhibit carcinogenicity, mutagenicity, and resistance to natural degradation [3]. Some of these dyes also have the potential to cause kidney failure and dysfunction of the brain and reproductive and central nervous systems.

Another important aspect of textile wastewater treatment is the de-colouration of dyes. Even in small amounts, dyes can be toxic to aquatic life and lead to changes in salinity and the visible coloration of receiving waters. Additionally, they can also reduce sunlight penetration, which hinders natural photosynthesis and disinfection processes [4]. Unfortunately, these dyes exhibit high stabilities to light, temperature, water, and detergents, allowing them to escape conventional physio-chemical and biological wastewater treatment processes [2,5]. Furthermore, textile industries consume a substantial amount of water in the dyeing process and have been classified as among the most polluting of all industrial sectors. Hence, as the demand for textile products and synthetic dyes increases with the growing global population, textile dye wastewater is becoming a more significant water pollution contributor.

Adsorption is one of the most commonly used approach for the treatment of wastewaters from the textile industry. The applications of most traditional wastewater treatment approaches are often limited by process cost, efficiency, and sludge handling requirements after treatment. Since adsorption processes are highly versatile in terms their applicability for a range of contaminants, their use has been growing in popularity. Additionally, their potential simplicity, economic feasibility, and high level of efficiency make them suitable solutions for the treatment of a number of wastewater contaminants. Adequate, cost-effective adsorbents that could provide sufficient adsorption capacity, adsorption rate, and mechanical strength are in increasing demand. Activated carbon has been reported to be very successful and efficient in adsorbing dyes, but its popularity has decreased due to its high market value and the operational costs associated with its use. Other adsorbents such as silicon polymers and kaolin have also been explored. Natural materials and polymers are often suggested alternatives to activated carbons because of their renewability and availability [3]. These natural materials can be used with or without pretreatment and can often be modified to selectively target particular pollutants under specific conditions. Polymeric adsorbents are demonstrating increasing advantages over activated carbon because of their simple processing and possibility for modifications. Moreover, polymeric adsorbents can be tailored to have reversible adsorption capabilities, meaning the adsorbent could be recovered through separation processes to separate the adsorbent from the dye from for future reuse.

One example of a natural polymer is chitosan (CTS) (Figure 1). Chitosan is a naturally abundant polysaccharide derived from chitin, which is found and harvested from crustacean shells. Chitin is one of the most abundant natural polymers in the world, second to cellulose, therefore making chitosan inexpensive and readily accessible. Among a wide range of applications, chitosan is a natural adsorbent and has been shown to adsorb substantial amounts of heavy metals and organic pollutants due to its high hydrophilicity and presence of amino and hydroxyl functional groups that serve as active sites for adsorption.

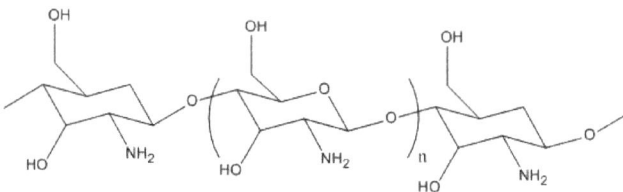

**Figure 1.** Chemical structure of chitosan.

The ability for chitosan to adsorb textile dyes has been well documented over the past decade, showing promising results and expanding its potential applications. Azo dyes are one of many classes of dyes and are characterized by having one or more azo groups ($-N=N-$). Azo dyes represent 70% of dyes produced annually and consist of a wide range of classifications that describes their reactivity with the dyeing materials. Reactive azo dyes, specifically, are the most used due to their bright colors, excellent color fastness, and ease of application [5]. Reactive Orange 16 (RO16) (Figure 2) was selected as a representative reactive azo dye for this work as it is a commonly used textile dye.

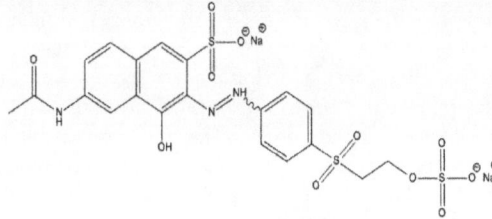

**Figure 2.** Chemical structure of Reactive Orange 16.

Polymer graft modification can introduce desired properties and broaden the field of potential applications for chitosan through the selection of different types of polymer side chains with different functionalities. By grafting and functionalizing additional side chains onto chitosan, the adsorption capacity of the grafted copolymer could potentially be increased. For reactive azo dyes, Singh et al. reported a poly(methyl methacrylate) (PMMA) grafted chitosan with an adsorption capacity for azo dyes of three times that of pure chitosan. In addition, a separate study demonstrated that triphenylphosphine (TPP) crosslinked chitosan improved the adsorption capacity of chitosan by 40% in the removal of reactive azo dyes [5]. These studies would suggest that with tailored polymeric side chains, the adsorption efficiency of chitosan can be improved extensively through functionalization. In this work, the adsorption capacity of glycidyl methacrylate (GMA) functionalized chitosan with grafted poly-(poly(ethylene glycol) methyl ether methacrylate) (PPEGMA) side chains (CTS-*g*-GMA-PPEGMA) was investigated and compared with pure chitosan for the removal of RO16 dye. PPEGMA has a similar structure to PMMA (Figure 3), and would therefore be expected to exhibit similar adsorption properties. PPEGMA also has repeating ester groups, while PMMA only has one, which would suggest that PPEGMA could yield higher adsorption capacities. More importantly, PPEGMA is water soluble, whereas PMMA is water insoluble. Since chitosan is soluble in water under acidic conditions, the synthesis of PPEGMA grafted chitosan could be performed in an aqueous system, in comparison to the synthesis of PMMA grafted chitosan which would require organic solvents, leading to a much greener and cost efficient process resulting from the reduction in organic waste solvents and fewer synthesis steps.

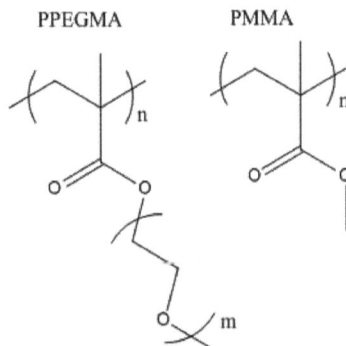

**Figure 3.** Chemical structure of poly-(poly(ethylene glycol) methyl ether methacrylate) (PPEGMA) and poly(methyl methacrylate) (PMMA).

The modification of chitosan to yield application-specific properties can generally take place at the amino ($NH_2$) or primary hydroxyl (OH) functionalities, although the amino functionality is mostly used. However, the natural adsorption capacity of chitosan is largely attributed to the $NH_2$ group. As such, the hydroxyl group was targeted for the modification of chitosan in this work, hence preserving the $NH_2$ functionality and its natural adsorption capacity. The modification of chitosan through polymer grafting was achieved via nitroxide-mediated polymerization and a grafting to approach [6]. The objective was to examine the adsorption capacity of CTS-*g*-GMA-PPEGMA for RO16 relative to pure chitosan in water for textile wastewater treatment applications. The present work describes the synthesis and application of PPEGMA-grafted chitosan for the removal of Reactive Orange 16. Figure 4 illustrates the steps we used in the synthesis of the CTS-*g*-GMA-PPEGMA. Batch adsorption experiments were carried out to develop adsorption isotherms for both chitosan and PPEGMA-grafted chitosan and experimental data were fitted to the Langmuir and Freundlich isotherms for the computation of respective equation parameter constants.

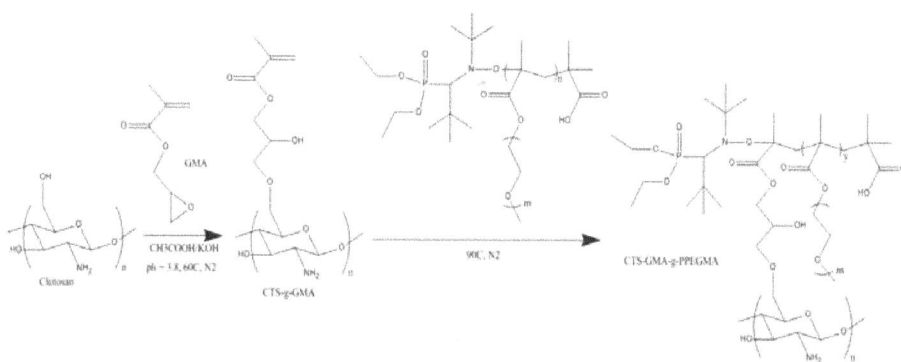

**Figure 4.** Synthesis of CTS-*g*-GMA-PPEGMA. CTS,1 g; 0.4 M acetic acid solution, 100 mL; 0.05 M KOH solution, 5 mL; hydroquinone solution ($9.09 \times 10^{-5}$ mol in 10 mL of $H_2O$); glycidyl methacrylate (GMA) (0.024 mol, 3.53 g, 3.30 mL) was added to the system dropwise. The system was degassed for 30 min under nitrogen atmosphere prior to increasing the temperature to 65 °C for 2 h. pH of the mixture was 3.8.

## 2. Experimental

### 2.1. Materials

Chitosan (CTS, Aldrich, Oakville, ON, Canada, low molecular weight, 85% degree of deacetylation), glycidyl methacrylate (GMA, Aldrich, 97%), acetic acid (Fisher, Waltham, MA, USA, 99.7%), and acetone (ACP, 99.5%) were used as received. Poly(ethylene glycol) methyl ether methacrylate (PEGMA, Aldrich, Mn of 300 g/mol) and styrene (St, Aldrich, 99+%) were passed through aluminum oxide columns (Aldrich, ~150 mesh, 58 A) prior to polymerization. BlocBuilder (2-methyl-2-(*N-tert*-butyl-*N*-(1-diethoxyphosphoryl-2,2-dimethylpropyl)aminoxy)-propionic acid alkoxyamine) (BB, 99%) and SG1 (4-(diethoxyphosphinyl)-2,2,5,5-tetramethyl-3-azahexane-*N*-oxyl) (85%) were supplied by Arkema. Reactive Orange 16 was kindly supplied by the Ramsay research group in the Department of Chemical Engineering at Queen's University (Kingston, ON, Canada).

### 2.2. Instrumentation

[1]H NMR spectroscopy was performed at room temperature on a FT-NMR Bruker Advance 400 MHz spectrometer (Billerica, MA, USA) with a total of 256 scans, using $D_2O$ as the solvent at 5 mg/mL. Thermogravimetric analysis (TGA) curves were recorded on TA Instruments Q500 TGA Analyzer (New Castle, DE, USA) by heating the sample from 75 °C to 600 °C at a rate of 10 °C per

minute. Gel Permeation Chromatography (GPC, Waters, Milford, MA, USA) analyses were performed with a Waters 2690 Separation Module and Waters 410 Differential Refractometer with THF as the eluent. Adsorption experiments were conducted using a WSR Shaker Model 3500 (VWR, Radnor, PA, USA) and Lambda XLS Spectrometer (Perkin Elmer, Waltham, MA, USA).

### 2.3. Synthesis of CTS-g-GMA

The functionalization of chitosan with GMA followed a previously published synthesis approach by García-Valdez et al. [6]. Chitosan (1 g) was first dissolved in 100 mL of acetic acid (0.4 M) in a three neck round bottom flask with 5 mL of 0.05 M potassium hydroxide (KOH) and 10 mL of 9.08 M hydroquinone. The flask was then purged with nitrogen for 30 min and heated to 60 °C in an oil bath. GMA was added drop wise at 60 °C and magnetically stirred for 2 h. CTS-g-GMA was precipitated in acetonitrile, washed with THF and dried for $^1$H NMR and TGA analysis.

### 2.4. Synthesis of Poly(PEGMA) via Nitroxide-Mediated Polymerization

PEGMA monomer (50 g, 53 mL), styrene (1.92 g, 2.14 mL), BlocBuilder (0.70 g), and 0.06 mL of SG1 were mixed in a three neck round bottom flask with magnetic stirring and oxygen purging for 30 min. (A small amount of styrene comonomer provides better control over the PEGMA polymerization). The flask was then heated to 90 °C for one hour. Unreacted monomer was removed through precipitation in diethyl ether. The polymer was dried under vacuum and analyzed by $^1$H NMR and TGA.

### 2.5. Synthesis of CTS-GMA-g-PPEGMA via Grafting to Approach

The modification of CTS-g-GMA was carried out following a previously reported approach developed in our research group [5]. Briefly, CTS-g-GMA (1 g) was dissolved in 100 mL of 0.1 M acetic acid in a three neck round bottom flask. KOH was added to the CTS-g-GMA solution to increase the pH to 5 and mechanically stirred under nitrogen for 30 min before increasing the temperature to 90 °C. PPEGMA (0.5 g) was dissolved in 60 mL of de-ionized (DI) water and was purged of oxygen in an inert nitrogen atmosphere. 20 mL PPEGMA was added every 30 min once the CTS-g-GMA solution reached 90 °C and the reaction was left to react for an additional 1.5 h. CTS-g-GMA-PPEGMA was washed in THF and extracted using a rotary evaporator.

### 2.6. Dye Adsorption Experiments

A 100 ppm dye stock solution (100 mg in 1 L of DI water) was diluted to lower concentrations for the construction of a calibration curve. Concentrations ranged from 0.5 ppm to 80 ppm with a $R^2$ of 0.99. The $\lambda_{max}$ of RO16 dye was measured to be 494 nm. Adsorption experiments were carried out using CTS-g-GMA-PPEGMA, chitosan, chitosan-PPEGMA (50/50) physical mixture, and PPEGMA as adsorbents on an orbital shaker table set at 240 rpm. The experiments were conducted at room temperature for a predetermined equilibrium time of 36 h and neutral pH. Each adsorbent was thoroughly mixed with 20 mL dye solution, and once the samples were shaken for the desired experimental time, the suspensions were filtered through Whatman 0.90 mm filter paper. The aqueous dye samples were diluted, if necessary, and analyzed for dye concentration. Initial dye concentrations ranged from 20 to 1200 ppm with 50 mg of adsorbent and each adsorption batch experiment was conducted in duplicate. The adsorption capacity of each adsorbent from the equilibrium solution was calculated using Equation (1).

$$Q_e = (C_o - C_e) \times \frac{V}{W} \tag{1}$$

where $Q_e$ is the adsorption capacity, describing the mass of dye adsorbed (mg) for each gram of adsorbent (mg/g), $C_o$ is the initial concentration of dye in mg/L, $C_e$ is the equilibrium concentration of dye in solution (mg/L), $V$ (L) is the volume of dye solution and $W$ (g) is the weight of the adsorbent used.

## 3. Results and Discussion

### 3.1. CTS-g-GMA

The $^1$H NMR of CTS-g-GMA (Figure 5) shows several peaks that would distinguish it from pure CTS. The $^1$H NMR for chitosan showed a singlet and a doublet between 3–4 ppm, as expected. The proton neighboring the amino group had an expected chemical shift at 2–3 ppm, while the protons around the primary and secondary alcohol exhibited chemical shifts at 3.5–4 ppm. When compared to the results of CTS-g-GMA, additional signals were observed at 4.3 ppm and between 5.5–6 ppm. The doublet observed at 4.3 ppm was identified represent the protons neighboring the ether (HC-OR) and the alcohol (HC-OH) groups on GMA, which had an expected chemical shift at 3.3–4.3 ppm. Further confirmations of the presence of GMA were two signals observed between 5.5–6 ppm. The vinylic double bond (C=C–H) on GMA was believed to be responsible for these two signals, with expected chemical shifts in the range of 5.5–7.5 ppm.

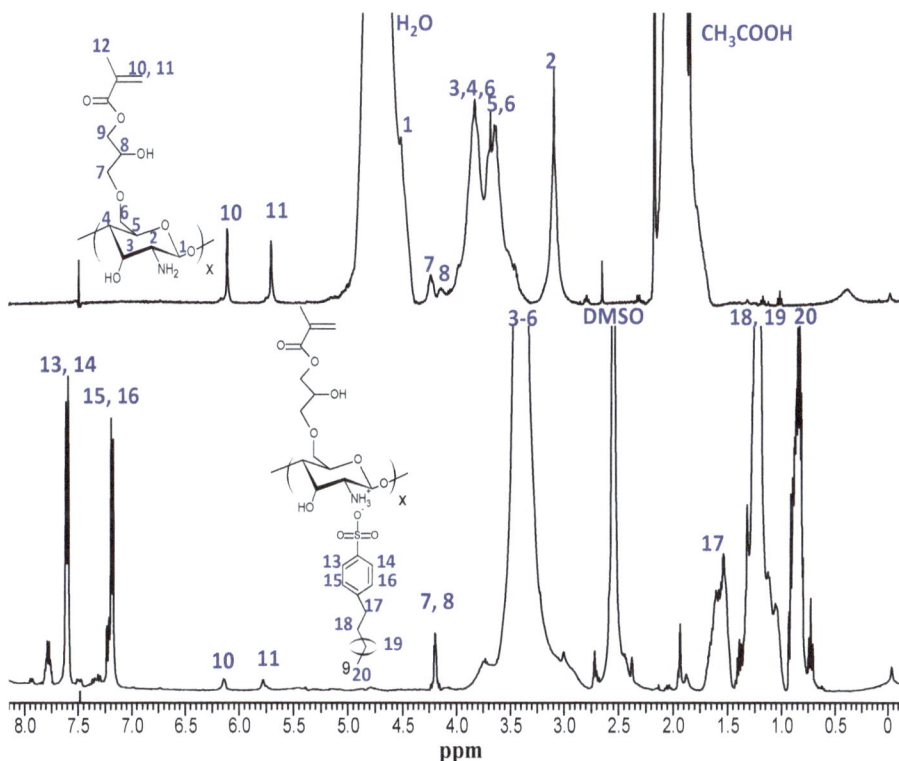

**Figure 5.** $^1$H NMR of CTS-g-GMA (**Top**) and CTS (**Bottom**).

### 3.2. PPEGMA

The TGA curve of PPEGMA (Figure 6) showed a thermal decomposition of the material starting at around 340 °C and ending at 400 °C. GPC results (Figure 7) of PPEGMA showed a relatively narrow peak, which indicates that the PPEGMA polymerization was well controlled. The average molecular weight (Mn) was 11,500 g/mol with a dispersity (Đ) of 1.18.

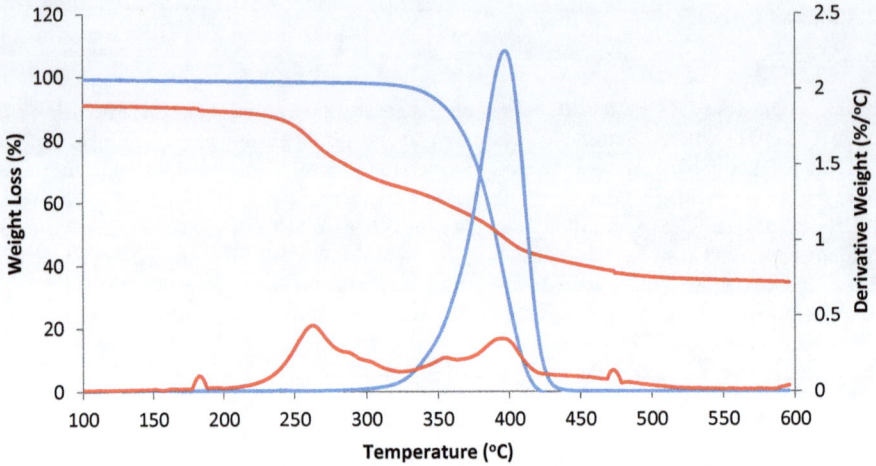

**Figure 6.** Thermogravimetric analysis (TGA) of PPEGMA (blue line) and CTS-*g*-GMA-PPEGMA (red line).

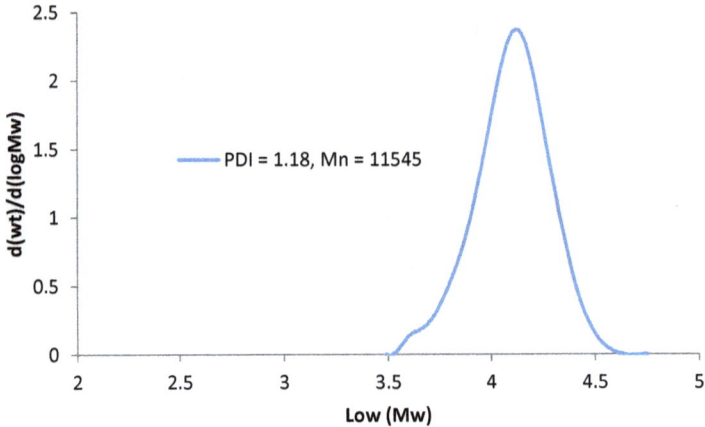

**Figure 7.** Gel Permeation Chromatography (GPC) analysis of the polymerization of PPEGMA via NMP.

## 3.3. CTS-g-GMA-PPEGMA

TGA results shown in Figure 6 characterized the decomposition of the grafted polymer starting at approximately 200 °C and ending at 500 °C. The first weight loss of the material from 200 °C to 340 °C was attributed to the decomposition of the chitosan backbone, while PPEGMA decomposed between 340 and 500 °C. The synthesized CTS-*g*-GMA-PPEGMA was estimated to be approximately 55% CTS and 45% PPEGMA. [1]H NMR (Figure 8) showed multiple defining peaks for CTS-*g*-GMA-PPEGMA. The medium intensity peak at 3.25 ppm represented the methyl groups at the end of PEGMA chains, while the high intensity peaks around 3.5 ppm were the repeating (O–CH$_2$CH$_2$) chains. When compared with the [1]H NMR result for CTS-*g*-GMA, traces of CTS-*g*-GMA could still be seen. More importantly, signals for the double bonds on the GMA chains could still be observed, with significantly reduced intensity, thus indicating that most GMA double bonds were reacted.

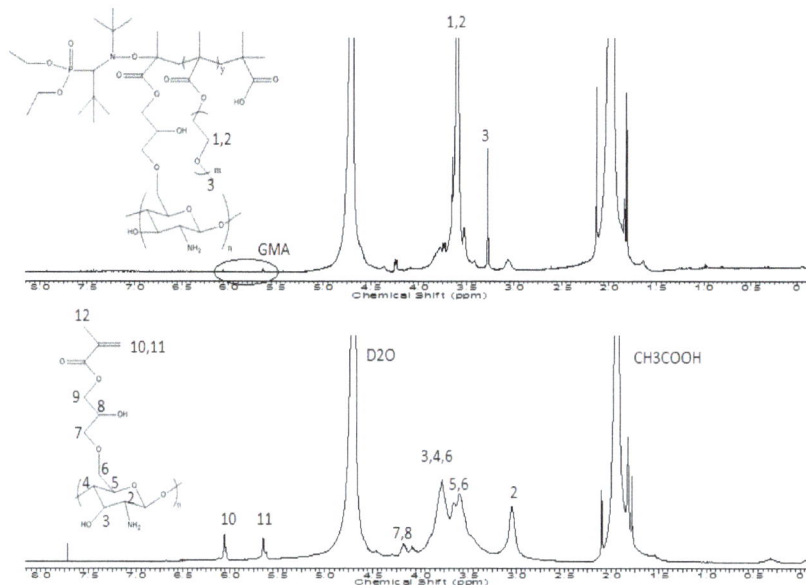

**Figure 8.** $^1$H NMR of CTS-*g*-GMA-PPEGMA (**Top**) and CTS-*g*-GMA (**Bottom**).

*3.4. Adsorption Isotherms*

Although a number of different isotherm models exist that can be used to characterize an adsorption system, the Langmuir and Freundlich models are the most commonly used for the characterization of adsorbents in water and wastewater treatment applications [7]. Hence, these models were employed to estimate the adsorption capacities of chitosan, PPEGMA, CST-*g*-GMA-PPEGMA, and mixture of 50% chitosan and 50% PPEGMA. It is important to establish reliable correlations for the equilibrium curves of the adsorption system. The adsorption isotherms shown in Figure 9 illustrate the adsorption capacity of each material. The equilibrium curves were linearized using the Langmuir and Freundlich models to calculate the model adsorption parameters (Table 1).

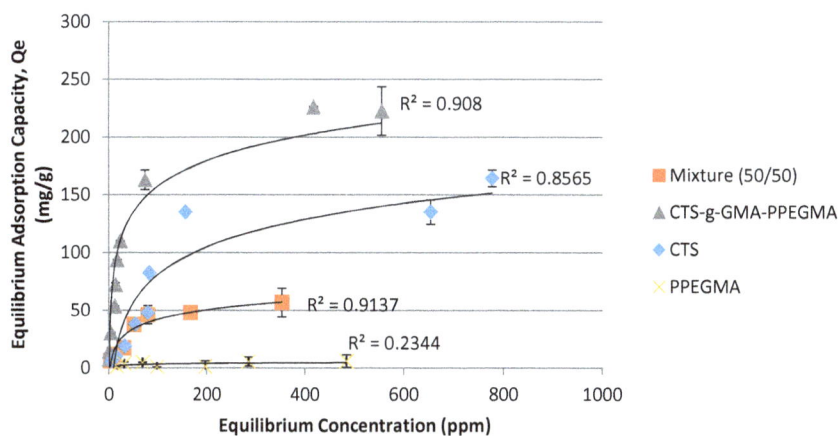

**Figure 9.** Adsorption isotherm.

**Table 1.** Langmuir and Freundlich Isotherm constants.

| Sample | Langmuir | | | Freundlich | | |
|---|---|---|---|---|---|---|
| | $Q_{max}$ (mg/g) | $K$ (L/mg) | $R^2$ | $n$ | $K$ (mg/g) | $R^2$ |
| CTS | 196 | 0.005 | 0.926 | 0.143 | 2.12 | 0.9011 |
| Mixture (50/50) | 64 | 0.021 | 0.9791 | 1.93 | 3.53 | 0.9297 |
| CTS-*g*-GMA-PPEGMA | 232 | 0.039 | 0.9977 | 2.18 | 17.8 | 0.9404 |
| PPEGMA | 7 | 0.003 | 0.1371 | 2.82 | 0.763 | 0.1231 |

*3.5. Langmuir Model*

The Langmuir model describes the formation of a monolayer of adsorbate (dye) on the outer surface of the adsorbent at adsorption equilibrium. The Langmuir model assumes that the surface of the adsorbent is uniform and that the adsorbed molecules do not interact. The model is described by

$$\frac{Q_e}{Q_m} = \frac{C_e K}{1 + K C_e} \tag{2}$$

Equation (2) can be rearranged into its linear form for the determination of the Langmuir adsorption parameters.

$$\frac{C_e}{Q_e} = C_e \frac{1}{Q_m} + \frac{1}{K Q_m} \tag{3}$$

$C_e$ = equilibrium concentration (mg/L)
$Q_e$ = equilibrium adsorption (mg/g)
$Q_m$ = maximum monolayer adsorption capacity (mg/g)
$K$ = Langmuir isotherm constant

According to the Langmuir plot (Figure 10), the maximum adsorption capacities of RO16 were 232, 196, and 64 mg/g for CTS-*g*-GMA-PPEGMA, chitosan, and the 50/50 chitosan-PPEGMA physical mixture, respectively. Pure PPEGMA exhibited negligible adsorption capacities the mixture yielded relatively low adsorption capacities due to its lower chitosan content. Since negligible adsorption was noted with PPEGMA, lower adsorption capacities were expected for the physical mixture. More interestingly, the model showed higher sorption capacity with the grafted chitosan than with chitosan alone. This confirmed that the chitosan modification improved the adsorption efficiency. The high $R^2$ values indicated that the data were a good fit for the Langmuir model.

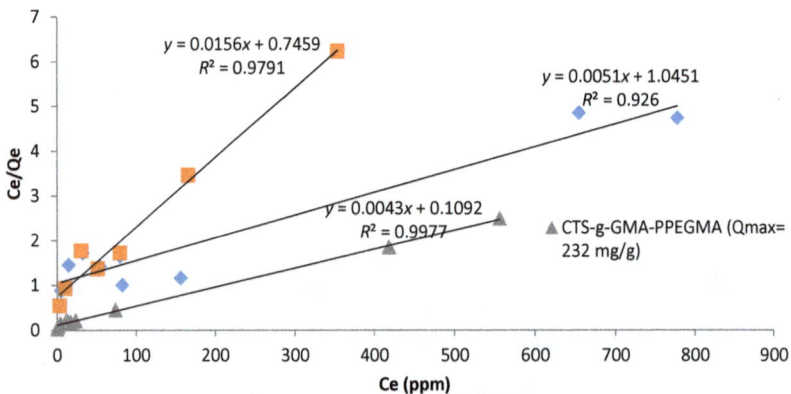

**Figure 10.** Linearized Langmuir isotherm.

From the adsorption isotherm, the maximum adsorption capacities were noted to be 200 mg/g for CTS-*g*-GMA-PPEGMA and 150 mg/g for chitosan, where were different from the Langmuir model predictions. This could suggest that the adsorption of RO16 does not form a monolayer at maximum adsorption, as assumed by Langmuir model, but that adsorption layers are likely heterogeneous. CTS-*g*-GMA-PPEGMA increased the adsorption capacity by 33% according to the adsorption isotherm data, indicating that CTS-*g*-GMA-PPEGMA could be an effective adsorbent.

### 3.6. Freundlich Model

The Freundlich model is based on sorption onto a heterogeneous surface, as described by Equation (3), where *K* and *n* are Freundlich adsorption constants, indicators of adsorption capacity and intensity. $Q_e$ (mg/g) and $C_e$ (mg/L) are the equilibrium sorption capacity and equilibrium concentration values. The Freundlich equation can be linearized in the form of Equation (4). Calculated parameters are summarized in Table 1. The linearized Freundlich plot is shown in Figure 11.

$$Q_e = K\, C_e^{\frac{1}{n}} \tag{4}$$

$$\ln Q_e = \frac{1}{n}\ln C_e + \ln K \tag{5}$$

The computed *n* and *K* values were 2.18 and 17 for CTS-*g*-GMA-PPEGMA and 0.14 and 2.12 chitosan, respectively. Experimental data for pure PPEGMA did not fit either of the isotherm models, which would be consistent with observation that pure PPEGMA exhibited negligible adsorption capacity for RO16. The parameter n is related to the heterogeneity parameter of the sorbent-sorbate system [8], where larger n values typically represent greater expected heterogeneities. For well-fitted data in the Freundlich model, an n value between one and ten would indicate a favorable adsorption process, thus suggesting that, in our study, the grafted copolymer performed more favorably than chitosan in its adsorption of the RO16.

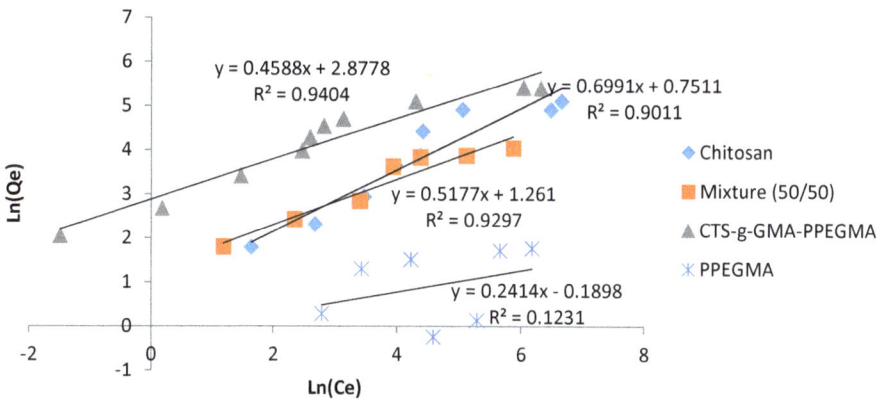

**Figure 11.** Linearized Freundlich isotherm.

### 3.7. Effect of Initial Dye Concentration

Figure 12 displays the equilibrium sorption efficiencies of RO16 for initial dye concentrations ranging from 20 to 1200 ppm (room temperature and neutral pH). The general trend indicated that the highest sorption efficiencies were noted at lower initial concentrations. For CTS-*g*-GMA-PPEGMA, the amount adsorbed started to decrease at 300 ppm, eventually reaching 50% sorption at 1200 ppm. Increasing the initial concentration decreases the total quantity of dye adsorbed because dye ions at lower concentrations would have more access and interactions with the sorption sites on the adsorbent than at higher concentrations. At high concentrations, the adsorption sites would become saturated more readily, thereby limiting access to the sorption sites. In addition, increasing initial concentrations also increased the adsorption capacity of the adsorbent, increasing the driving force facilitated through the concentration gradient.

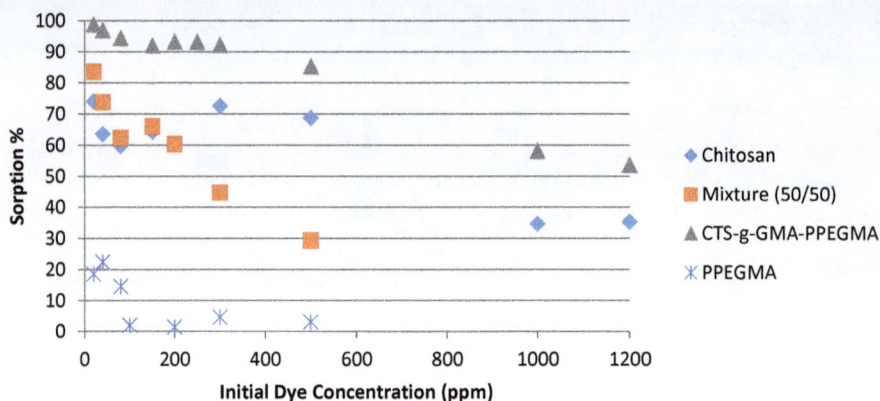

**Figure 12.** Adsorption % at different initial dye concentrations.

## 4. Conclusions

Chitosan was functionalized with the vinyl monomer glycidyl methacrylate and then PPEGMA was grafted to the chitosan backbone. The PPEGMA was prepared by nitroxide-mediated polymerization conducted in an aqueous system, giving a narrowly distributed polymer with controlled molecular weight. The CTS-*g*-GMA-PPEGMA exhibited good adsorption capacity for the removal of the organic dye RO16 from water. CTS-*g*-GMA-PPEGMA proved to be a better adsorbent than chitosan, improving the adsorption efficiency of chitosan by 33%. The estimated Freundlich isotherm parameters derived from the experiments suggested a more favorable adsorption process for the grafted material than for chitosan alone, which was consistent with the experimental observations. The 50/50 chitosan-PPEGMA physical mixture showed a ~50% decrease in adsorption capacity compared to chitosan alone, while negligible adsorption was observed with PPEGMA alone. This work only offers a preliminary outlook on the adsorption capacity of CTS-*g*-GMA-PPEGMA for textile dyes. The results of this study have potential to be useful in real textile wastewater treatment processes. Further research and studies would be required to simulate real textile wastewater properties such as pH, temperature, suspended solids, and recyclability of the adsorbent material. From an adsorbent formulation perspective, exploration of the effects of PPEGMA loading on the CTS and the PPEGMA molecular weight on adsorption behaviour would be beneficial.

**Acknowledgments:** The Natural Sciences and Engineering Research Council of Canada (NSERC), the Ontario Research Chairs Program (MFC), and the Canada Research Chairs Program (PC) provided financial support.

**Author Contributions:** O.G.-V., P.C. and M.F.C. conceived and designed the experiments; B.T. and O.G.-V. performed the experiments and analyzed the data; B.T. and O.G.-V. wrote the paper.

**Conflicts of Interest:** The authors declare no conflict of interest.

## References

1. Umpuch, C.; Sakew, S. Adsorption characteristics of reactive black 5 onto chitosan-intercalated motmorillonite. *Desalin. Water Treat.* **2015**, *53*, 2962–2969. [CrossRef]
2. Chequer, F.M.; Oliveira, G.A.; Ferraz, E.R.A.; Cardoso, J.C.; Zanoni, M.V.B.; Oliveira, D.P. Textile Dyes: Dyeing Process and Environmental Impact. In *Eco-Friendly Textile Dyeing and Finishing*; Günay, M., Ed.; Intech: Sao Paulo, Brazil, 2013; Chapter 6.

3.   Panic, V.V.; Seslija, S.I.; Nesic, A.R.; Velickovic, S.J. Adsorption of azo dyes on polymer materials. *Hem. Ind.* **2013**, *67*, 881–900. [CrossRef]
4.   Liu, L.; Hall, G.; Champagne, P. Effects of environmental factors on the disinfection performance of a wastewater stabilization pond operated in temperate climate. *Water* **2016**, *8*, 5–15. [CrossRef]
5.   Singh, V.; Sharma, A.; Tripathi, D.; Sanghi, R. Poly(methylmehtacrylate) grafted chitosan: An efficient adsorbent for anionic azo dyes. *J. Hazard. Mater.* **2009**, *161*, 955–966. [CrossRef] [PubMed]
6.   García-Valdez, O.; George, S.; Champagne-Hartley, R.; Saldívar-Guerra, E.; Champagne, P.; Cunningham, M.F. Modification of Chitosan with polystyrene and poly(*n*-butyl acrylate) via nitroxide-mediated polymerization and grafting from approach in homogeneous media. *Polym. Chem.* **2015**, *6*, 2827–2836. [CrossRef]
7.   Ozacar, M.; Sengil, I.A. Adsorption of acid dyes from aqueous solutiosn by calcined alunite and granular actuvated carbon. *Adsorption* **2002**, *8*, 301–308. [CrossRef]
8.   Dada, A.; Olalekan, A.; Olatunya, A. Langmuir, freundlich, temkin and dubinin-radushjevich isotherms studies of equilibrium sorption of $Zn^{2+}$ unto phosphoric acid modified rice husk. *J. Appl. Chem.* **2012**, *3*, 38–45.

*processes*

MDPI

*Article*

# Kinetic Control of Aqueous Polymerization Using Radicals Generated in Different Spin States

Ignacio Rintoul

Consejo Nacional de Investigaciones Científicas y Técnicas, Universidad Nacional del Litoral, Santa Fe 3000, Argentina; irintoul@santafe-conicet.gov.ar; Tel.: +54-342-451-1370

Academic Editor: Alexander Penlidis
Received: 3 January 2017; Accepted: 17 March 2017; Published: 24 March 2017

**Abstract:** Background: Magnetic fields can interact with liquid matter in a homogeneous and instantaneous way, without physical contact, independently of its temperature, pressure, and agitation degree, and without modifying recipes nor heat and mass transfer conditions. In addition, magnetic fields may affect the mechanisms of generation and termination of free radicals. This paper is devoted to the elucidation of the appropriate conditions needed to develop magnetic field effects for controlling the kinetics of polymerization of water soluble monomers. Methods: Thermal- and photochemically-initiated polymerizations were investigated at different initiator and monomer concentrations, temperatures, viscosities, and magnetic field intensities. Results: Significant magnetic field impact on the polymerization kinetics was only observed in photochemically-initiated polymerizations carried out in viscous media and performed at relatively low magnetic field intensity. Magnetic field effects were absent in polymerizations in low viscosity media and thermally-initiated polymerizations performed at low and high magnetic field intensities. The effects were explained in terms of the radical pair mechanism for intersystem crossing of spin states. Conclusion: Polymerization kinetics of water soluble monomers can be potentially controlled using magnetic fields only under very specific reaction conditions.

**Keywords:** magnetic field; radical polymerization; quantum chemistry; acrylamide; solution polymerization; photopolymerization; process control

## 1. Introduction

Magnetic field (MF) effects in chemical kinetics have a long tradition. Early in 1929 Bhatagnar observed that the rate of decomposition of hydrogen peroxide is influenced by MF [1]. Afterwards, in 1946, Selwood observed that the efficiency of some catalyst can be increased in the presence of MF [2]. These early works gave birth to the fascinating idea of controlling chemical reactions using MF. The discovery and understanding of nuclear and electronic spin polarization phenomena during chemical reactions in the late 1960s contributed significantly to the development of this idea. Up to now, MF effects in chemical reactions have been observed in a number of situations and have received proper theoretical analysis. However, MF effects in free radical polymerizations has not yet found a practical application [3].

Table 1 summarizes several MF effects observed in polymerization studies. MF effects reveal the possibility to control the kinetics of radical polymerizations and the chain architecture of resulting polymers in a homogeneous, instantaneous, and highly selective way. In addition, MF effects in polymerization reactions can be carried out without physical contact, independently of the temperature, pressure, and agitation degree of the reacting medium, and without modifying recipe formulations, heat and mass transfer conditions, nor any other reaction parameter normally used to control the course of polymerization.

**Table 1.** Summary of MF effects on radical polymerization reported in the literature.

| Monomer | Initiator | System | MF Effect | Ref. |
|---|---|---|---|---|
| AN | AIBN | Bulk | Increase of $R_p$, polymer yield, molar mass, syndiotacticity, crystallinity, and thermal stability of resulting polymers | [4] |
| MMA, ST | AIBN | Bulk | No effect | [4] |
| MMA | MB | H$_2$O-MeOH mixtures | Decrease of the polymer yield and increase of the molar mass polymers | [5] |
| MMA | MB | Aqueous solution | No effect | [5] |
| MMA | BP, AMP, APA, AHC | Aqueous solution | Increase of the initiator efficiency and decrease of the monomer exponent and molar mass | [6] |
| AC, MMA, AM | BP, AMP, APA, AHC | Several solvents | Increase of the initiator efficiency and thermal stability of polymers | [7] |
| MMA, EMA, BMA | AIBN | Bulk | Increase of molar mass and thermal stability of products | [8] |
| ST | AIBN | H$_2$O-EG mixtures | Increase of molar mass and homogeneity of polymers | [9] |
| MMA, ST | BP | Liquid $CO_2$ | Increase of conversion and molar mass | [10] |
| MMA, ST | BP | Cyclohexane | No effect | [10] |
| MMA, EMA, BMA | BP | Bulk | Increase of $R_p$ and molar mass | [11] |
| ST | BK | Emulsion | Increase of $R_p$ and molar mass | [12,13] |
| ST, MMA, AA | BK | Emulsion | Increase of $R_p$ and molar mass | [14] |
| ST, MMA | AIBN | Emulsion | Increase of molar mass and decrease of molar mass distribution | [15] |
| ST | $K_2S_2O_8$ | Emulsion | Decrease of $R_p$ | [16] |
| MMA | BP | 10 different organic solvents | Increase of $R_p$ and conversion and decrease of the induction period for initiation | [17] |
| MMA | TX | Dimethylformamide | Increase of conversion and molar mass | [18] |
| AM | MB | H$_2$O-EG mixtures | Increase of $R_p$ | [19] |
| AM, AA, DADMAC and combinations | $C_{26}H_{27}O_3P$ | H$_2$O-EG mixtures | Increase of $R_p$ of all monomers in homo and copolymerizations. Increase of molar mass of polyAM and copolymer compositions. No effect in the molar mass of polyAA and copolymer compositions | [20,21] |

AA: acrylic acid, AC: vinyl acetate, AHC: 1,1′-azobis(cyclohexane-1-carbonitrile); AIBN: 2,2′-azobisisobutyronitrile; AM: acrylamide, AMP: 2,2′-azobis(2-methylpropionitrile); AN: acrylonitrile; APA: 4,4′-azobis(4-cyanopentanoic acid); BK: benzyl ketone; BMA: butylmethacrylate; BP: benzoyl peroxide, $C_{26}H_{27}O_3P$: phenyl-bis(2,4,6-trimethylbenzoyl)-phosphine oxide; DADMAC: diallyldimethylammonium chloride; EMA: ethylmethacrylate, $K_2S_2O_8$: potassium persulfate; MB: methylene blue; MMA: methylmethacrylate; ST: styrene; TX: thioxanthone.

The aim of this work is to establish some criteria for recipe preparation and reaction conditions needed to study the MF effects on the kinetics of radical polymerization of acrylamide (AM) [22] and to conclude the consequences for the overall rate expression expressed as Equation (1) and the kinetic chain length expressed as Equation (2) [23]:

$$Rp = k_p \cdot [M]^{\alpha} \cdot \left( \frac{f \cdot k_d \cdot [I]}{k_t} \right)^{\beta} \tag{1}$$

$$\nu = \frac{k_p \cdot [M]}{2 \cdot (f \cdot k_d \cdot k_t \cdot [I])^{0.5}} \tag{2}$$

Here $Rp$ is the polymerization rate defined as the negative derivative of the monomer concentration with time. $[M]^{\alpha}$ and $[I]^{\beta}$ are the monomer and initiator concentrations in mol/L powered to their respective reaction orders, $k_p$ and $k_t$ are the propagation and termination rate coefficients in L/mol·s, and $f$ and $k_d$ are the efficiency and decomposition rate of the initiator. In photopolymerization reactions, $f$ is called the quantum yield of the photoinitiator, $\Phi$ and $k_d$ is expressed according to Equation (3):

$$k_d = \varepsilon \cdot I_0 \tag{3}$$

Here $\varepsilon$ is the molar absorptivity of the photoinitiator, in L/mol·cm, and $I_0$ is the light intensity in the polymerization medium in mol/L·s.

Full theoretical description of the interaction between MF and reactants is not the task of this work since it can be consulted in the excellent review paper of Steiner and Ulrich [24]. In any case, a short overview of the fundamentals of three interesting MF phenomena commonly hypothesized as explanations for MF effects in polymerization reactions is presented.

Thermal equilibrium of spin states act at the electron level. This suggests that chemical reactions should be accelerated by magnetically-induced diamagnetic/paramagnetic transitions in the reactive species. If N spins are present in a polymerization medium under a steady MF of intensity $B_0$, $N_{\alpha}$, and $N_{\beta}$ spins will be magnetic moment spin up, $m_s = 0.5$, and down, $m_s = -0.5$, respectively. The conservation law of the total spin value establishes that: $N = N_{\alpha} + N_{\beta}$ and the ratio between $N_{\alpha}$ and $N_{\beta}$ is given by Equation (4) [25]:

$$\frac{N_{\alpha}}{N_{\beta}} = e^{\frac{g \cdot \beta \cdot B_0}{k \cdot T}} \tag{4}$$

Here $g$ is the electron "$g$" factor, $\beta$ is the electronic Bohr magneton, $\beta = 0.92731 \times 10^{-20}$ erg/gauss, $k$ is the Boltzmann constant, $k = 1.38044 \times 10^{-16}$ erg/K, and $T$ is the temperature of the system [25]. Under normal conditions where free radical reactions are carried out immersed in the geomagnetic field (~0.5 Gauss), the ratio $N_{\alpha}/N_{\beta}$ is very close to unity and consequently equal populations of spins up and down can be assumed. Thermodynamically, the magnetic contribution to the free enthalpy of the reaction, $\Delta G_m$, in an externally-applied MF of intensity $B_0$ can be expressed as Equation (5) [24]:

$$DG_m = -\frac{1}{2} \cdot Dc_M \cdot B_0^2 \tag{5}$$

Here $\Delta \chi_M$ is the change of the magnetic susceptibility during the reaction of one molar unit. Therefore, according to Equations (4) and (5), the higher the MF intensities and the lower the temperatures are, the more important the MF effects on $f$ and $k_t$ will be.

MF-induced molecular orientation acts at the molecule level. MF tends to align molecules that present magnetic susceptibility, $\Delta \chi_M \neq 0$. Conversely, temperature tends to randomize the orientation of molecules. Therefore, an average orientation results from the balance between these two opposed effects. Classically, the energy of a molecular dipole oriented with a $\theta$ angle to a MF is given by

Equation (6) [26]. Equation (7) is the Boltzmann form of the average orientation of molecular dipoles, $P(\theta)$, as a function of the MF intensity, the temperature of the system and $\Delta\chi_M$ of the molecules:

$$E = \Delta\chi_M \cdot B \cdot \cos(\theta) \tag{6}$$

$$P(q) = \frac{e^{\left(\frac{Dc_m \cdot B \cdot \cos(q)}{R \cdot T}\right)}}{\int\limits_0^{2p} e^{\frac{Dc_m \cdot B \cdot \cos(q)}{R \cdot T}} dq} \tag{7}$$

Here $P(\theta)$ is the normalized probability to find the molecule oriented with an angle $\theta$ to the direction of an effective MF of magnitude $B$. $B$ is defined according to Equation (8):

$$B = B_0 + B_c \tag{8}$$

Here $B_c$ is the resulting magnetic contribution due to the $B_0$ induced alignment of all molecules. Evidently, it is expected a certain influence of molecular orientation of monomers and growing radicals on $k_p$ and $k_t$.

The radical pair mechanism for spin states acts at the supramolecular level. Initiator molecules are hypothesized to exist in cages formed by solvent and monomer molecules. Eventually, a molecular initiator can decompose, generating a caged radical pair. Caged radical pairs are generated in singlet ($S$) or triplet ($T_+$, $T_0$, $T_-$) spin states from precursors having their respective multiplicity, or when formed by free radical encounters. These spin states describe different electron configurations. Depending on these configurations the caged radical pair may recombine regenerating the initial molecule, undergoing the formation of cage products which generates a new molecule, or the radicals can escape from the cage releasing two free radicals to the reaction medium. Radical pairs in the $S$ state have extremely high probability to undergo recombination reactions and/or formation of cage products. Conversely, radical pairs in any of the three $T$ states cannot recombine. Nevertheless, radicals may pass from one state to another through intersystem crossing mechanisms. The energy associated with the $T_+$ and $T_-$ states increases and decreases proportionally with the MF intensity, while the energy of the $S$ and $T_0$ states are unaffected by the MF. The application of MF splits out the energy levels of the $T$ states diminishing substantially the probability for intersystem crossing to the $S$ state. Therefore, primary caged radical pairs can be quenched in the $T$ state, decreasing the probability of radical recombination. Consequently, more radicals are released to the polymerization medium resulting in an increase of the initiator efficiency leading to an increase of $Rp$. The MF-induced modification of the outcome of caged radical pairs is eventually interpreted as an MF-induced change of $\Phi$ or $f$. Furthermore, when two growing radicals encounter each other in the $T$ state, they cannot recombine. This effect is interpreted as a decrease of $k_t$. Thus, the radicals continue growing increasing $v$.

Finally, $[M]$, $[I]$, $\alpha$, and $\beta$ are not affected by any MF mechanism.

## 2. Materials and Methods

### 2.1. Materials

Ultra-pure AM, four times recrystallized, (AppliChem, Darmstadt, Switzerland) was selected as the monomer. An aqueous dispersion of $C_{26}H_{27}O_3P$, (Ciba Specialty Chemicals, Basel, Switzerland) and potassium persulfate, $K_2S_2O_8$, (Fluka Chemie, Buchs, Switzerland) served as photo- and thermal-initiators. Photochemical decomposition of $C_{26}H_{27}O_3P$ and thermal decomposition of $K_2S_2O_8$ generate radical pairs in triplet and singlet spin states, respectively [27]. The water was of Millipore quality (18.2 $M\Omega \cdot cm$). Ethylene glycol 99% for synthesis (EG) (AppliChem, Darmstadt, Switzerland) was used to vary the viscosity of the polymerization medium. Acetonitrile for high performance liquid chromatography (HPLC) (AppliChem, Darmstadt, Switzerland) served to precipitate the polymer in the withdrawn samples.

## 2.2. Polymer Synthesis

Syntheses were performed in a 100 mL glass reactor (3 cm diameter, 15 cm height) equipped with a UV lamp, stirrer, condenser, gas inlet, and a heating/cooling jacket. The UV lamp had a primary output at 254 nm wavelength with constant and uniform irradiation everywhere in the reaction medium, $I_0 = 5.16 \times 10^{-8}$ mol/L·s. The same reactor, without the UV lamp, was used for thermally-initiated polymerizations. The reactor was entirely placed between the poles of an electromagnet (Bruker-EPRM, Rheinstetten, Germany) for polymerizations carried out in the range $0 < MF < 0.5$ Tesla and in the core of a superconductor magnet (Bruker-UltraShield, Rheinstetten, Germany) for polymerizations carried out in the range $0.5 < MF < 7$ Tesla. A thermostat adjusted the reaction temperature within $\pm 1$ K. Oxygen was removed from the initial monomer solution by purging with $N_2$ ($O_2 < 2$ ppm; Airliquide, Gümligen, Switzerland) during 30 min at 273 K and 0 Tesla of MF intensity. After degassing, the temperature was raised to activate the decomposition of $K_2S_2O_8$ in case of thermally initiated polymerization and the UV lamp was lighted to activate the photodecomposition of $C_{26}H_{27}O_3P$ in case of photopolymerization. Simultaneously, the MF was adjusted to the specified intensity. Complementary experiments were carried out for comparison, without MF, though keeping constant all the other conditions. All reactions were performed isothermally during 60 min continuous purging with $N_2$ and drawing samples of 0.1–0.2 g from the reactor every 5 min for kinetic analysis. Table 2 summarizes the conditions of all polymerizations.

**Table 2.** Summary of polymerization conditions.

| Series | MF Tesla | [AM] mol/L | [Initiator] mol/L | Solvent | Temp. K |
|---|---|---|---|---|---|
| 1 | 0.0 | 0.20 | - | $H_2O$ | 313 |
| 2 | 0.0 | 0.20 | $[C_{26}H_{27}O_3P] = 2 \times 10^{-6}$ | $H_2O$ | 323 |
| 3 | 0.0 | 0.20 | $[K_2S_2O_8] = 2.3 \times 10^{-3}$ | $H_2O$ | 273 |
| 4 | 0.0 7.0 | 0.15 | $[K_2S_2O_8] = 2.3 \times 10^{-3}$ | $H_2O$ | 308 |
| 5 | 0.0 7.0 | 0.15 | $[K_2S_2O_8] = 2.3 \times 10^{-3}$ | 50% EG | 308 |
| 6 | $7.0 < MF < 0.5$ | 0.10 | $[K_2S_2O_8] = 2.3 \times 10^{-3}$ | $H_2O$ | 308 |
| 7 | 0.00 0.11 0.35 0.50 | 0.20 | $[K_2S_2O_8] = 1.2 \times 10^{-2}$ | $H_2O$ | 313 |
| 8A 8B | $0.0 < MF < 0.5$ $0.0 < MF < 0.1$ | 0.20 0.10 | $[K_2S_2O_8] = 1.2 \times 10^{-2}$ $[K_2S_2O_8] = 1.2 \times 10^{-3}$ | $H_2O$ | 313 |
| 9 | 0.0 0.1 | 0.20 | $[C_{26}H_{27}O_3P] = 1 \times 10^{-6}$ | $H_2O$ | 313 |
| 10 | 0.0 0.1 | 0.20 | $[C_{26}H_{27}O_3P] = 1 \times 10^{-6}$ | 50% EG | 313 |

The first three series were performed to demonstrate the absence of side radical generation which could disturb the polymerization path. An initiator-free aqueous AM solution was illuminated with UV light during one hour at 313 K to verify the absence of monomer photolysis (series 1). Another AM solution containing $C_{26}H_{27}O_3P$ was maintained in darkness during one hour at 323 K to demonstrate the absence of thermal decomposition of the photoinitiator (series 2). Finally, AM-$K_2S_2O_8$ was maintained for 1 h at 273 K to prove the absence of $K_2S_2O_8$ decomposition during degassing (series 3). $K_2S_2O_8$ was used within the limiting reaction conditions suitable for radical generation through the monomer-enhanced mechanism [28].

Series 4–9 represent the main experiments. Recipe formulations and reaction conditions without magnetic fields were adjusted to obtain linear conversion curves. Linear conversion paths facilitate the data analysis. Series 4 and 5 were designed to evaluate the effects of 7 Tesla MF intensity in polymerizations initiated with radicals in singlet spin state (i.e., thermally-initiated polymerizations) performed in aqueous monomer solution of relatively low viscosity, $\eta = 1.03 \times 10^{-3}$ Pa·s and in 50 wt % of EG aqueous monomer solution with relatively high viscosity, $\eta = 5.16 \times 10^{-3}$ Pa·s. Series 6–8 were designed to evaluate the effect of MF varying continuously from 7 Tesla to 0.5 Tesla, four MF intensities between 0 and 0.5 Tesla and MF varying continuously from 0 to 0.5 Tesla (series 8A) and from 0 to 0.1 Tesla (series 8B) in polymerizations initiated with radicals in singlet spin state (i.e., thermally-initiated polymerizations) using water as a solvent, respectively. Series 9 and 10 were designed to evaluate the effect of 0.1 Tesla MF intensity in polymerizations initiated with radicals in triplet spin state (i.e., photochemically-initiated polymerizations) using water, $\eta = 1.09 \times 10^{-3}$ Pa·s and 50 wt % EG aqueous solution, $\eta = 5.20 \times 10^{-3}$ Pa·s as solvents, respectively.

### 2.3. Analytics and Instruments Calibration

The dynamic viscosity, $\eta$, of monomer aqueous solutions and monomer solutions with 50 wt % EG was measured at their specified reaction temperatures using a disc viscometer (Brookfield, Middleboro, USA) equipped with a 250 mL thermostatted ($\pm 1$ K) vessel and a disc spindle of 20 mm diameter rotating at 50 rpm. The viscosity of each monomer solution was measured five times. Deviations were within 4%.

The conversion was determined analyzing the residual monomer concentration. It served to calculate $Rp$ and $\nu$ according to a detailed procedure [29]. Briefly, the residual monomer concentration in the samples was monitored using a HPLC system composed of an L-7110 Merck-Hitachi pump (Hitachi, Tokio, Japan) and a SP6 Gynkotek UV detector (Gynkotek, Germering, Germany) operating at $\lambda = 197$ nm. The stationary and mobile phases were LiChrosphere 100 RP-18 (Merck, Darmstadt, Germany) and aqueous solutions containing 5 wt % acetonitrile. The flow rate was 1 mL/min. The HPLC system was calibrated using AM solutions of known concentrations. The concentration as a function of the peak area served as calibration parameter ($r^2 > 0.999$). Figure 1 presents the calibration curve of the HPLC system. The samples were mixed with 4 mL of acetonitrile to precipitate and isolate the polymer from the solution. The non-reacted monomers remained in solution. 20 μL of the supernatant were injected for HPLC analysis.

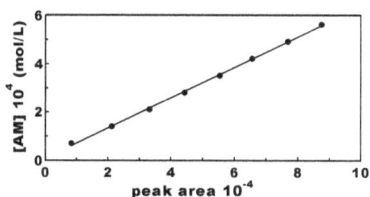

**Figure 1.** HPLC calibration curve. The concentration of standard AM solutions were plotted as a function of the corresponding peak areas. $r^2 > 0.999$.

Due to the limited space between the poles of the electromagnet and in the superconductor bore, it was not possible to simultaneously install both probes to measure the MF and the polymerization reactor there. Consequently, the MF was known indirectly. Probes were installed between the poles of the electromagnet in order to measure the MF for different electrical currents running through the bobbins of the magnet. With such information the calibration curve, MF strength vs. electrical current was determined. The magnetic probes were moved from the gap between the poles and the reactor was installed. The MF was adjusted by setting the electrical current according to the calibration curve presented in Figure 2.

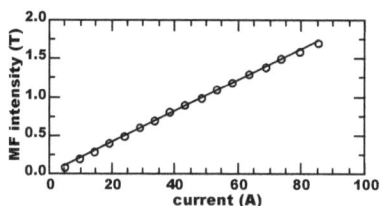

**Figure 2.** Electromagnet calibration. Magnetic field (MF) strength vs. electrical current (Amperes), $r^2 > 0.999$.

In case of polymerizations carried out in the superconductor magnet, the MF intensity was varied moving the reactor along the magnet bore. A magnetic probe was placed at different distances from the top of the magnet in order to determine the calibration curve shown in Figure 3. The MF was adjusted by setting the distance between the reactor and the core of the magnet according to the calibration curve.

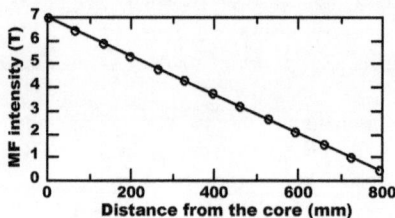

**Figure 3.** Superconductor magnet calibration. Magnetic field (MF) strength vs. distance from the core, $r^2 > 0.999$. The origin was defined at the highest MF intensity (7 Tesla) in the middle of the magnet.

## 3. Results

The absence of polymerization was confirmed for series 1–3 demonstrating that the polymerization path was not disturbed by side radical generation.

Figure 4 shows that thermally-initiated polymerizations carried out at 7 Tesla progressed identically to those performed without MF. However, polymerizations carried out using 50 wt % EG aqueous solution as solvent progressed faster than those performed using pure water as solvent.

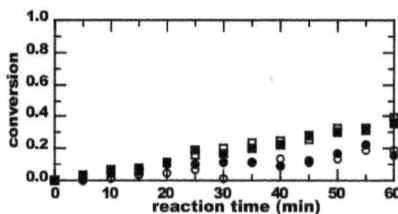

**Figure 4.** Conversion of AM vs. reaction time for polymerizations carried out using water, series 4 (●) and 50 wt % EG aqueous solution, series 5 (■), as solvents. MF intensity: 7 Tesla (full symbols), without MF (empty symbols). [AM] = 0.15 mol/L, [$K_2S_2O_8$] = 2.3 × $10^{-3}$ mol/L, T = 308 K.

Figure 5 shows that thermally-initiated polymerizations using water as solvent still progressed linearly in spite of the fact that the MF intensity varied from 7 to 0.5 Tesla during the 60 min of reaction time.

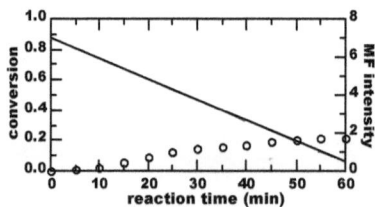

**Figure 5.** Conversion of AM (○) vs. reaction time. The MF intensity (−) varied from 7 Tesla at the beginning of polymerization to 0.5 Tesla after 60 min of reaction. Series 6. [AM] = 0.10 mol/L, [$K_2S_2O_8$] = 2.3 × $10^{-3}$ mol/L, T = 308 K, solvent: water.

Figure 6 shows no substantial differences between thermally-initiated polymerizations using water as solvent when carried out at 0.00, 0.11, 0.35, and 0.50 Tesla.

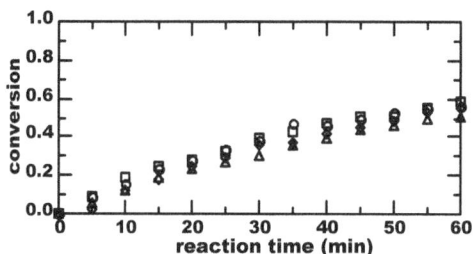

**Figure 6.** Conversion of AM vs. reaction time for polymerizations carried out at different MF intensities. MF = 0 (○), 0.11 (□), 0.35 (◇), 0.50 (△) Tesla. Series 7. [AM] = 0.20 mol/L, [K$_2$S$_2$O$_8$] = 1.2 × 10$^{-2}$ mol/L, T = 313 K, solvent: water.

Figures 7 and 8 show that thermally-initiated polymerizations using water as solvent were not accelerated, nor slowed by any MF intensity in the ranges 0 to 0.5 and 0 to 0.1 Tesla, respectively.

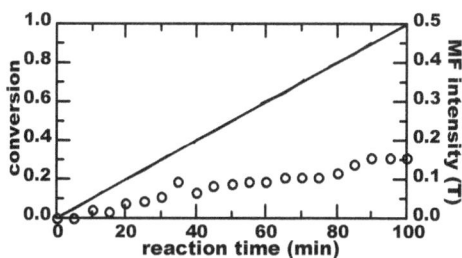

**Figure 7.** Conversion of AM (○) vs. reaction time. The MF intensity (−) varied from 0 Tesla at the beginning of polymerization to 0.5 Tesla after 100 min of reaction. Series 8A. [AM] = 0.20 mol/L, [K$_2$S$_2$O$_8$] = 1.2 × 10$^{-2}$ mol/L, T = 313 K, solvent: water.

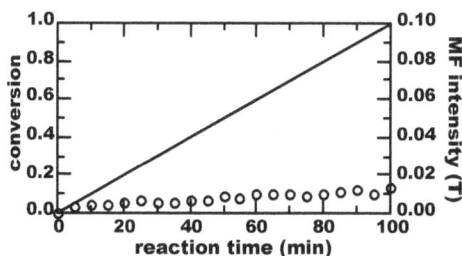

**Figure 8.** Conversion of AM (○) vs. reaction time. The MF intensity (−) varied from 0 Tesla at the beginning of polymerization to 0.1 Tesla after 100 min of reaction. Series 8B. [AM] = 0.10 mol/L, [K$_2$S$_2$O$_8$] = 1.2 × 10$^{-3}$ mol/L, T = 313 K, solvent: water.

Figure 9 shows that aqueous photopolymerizations carried out at 0.1 Tesla of MF intensity progressed slightly faster than aqueous photopolymerizations carried out without MF.

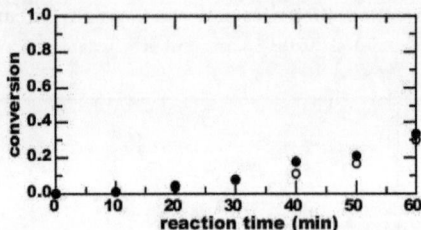

**Figure 9.** Conversion of AM vs. reaction time for polymerizations carried out at 0.1 Tesla of MF intensity (●) and without MF (○). Series 9. [AM] = 0.20 mol/L, [$C_{26}H_{27}O_3P$] = 1 × $10^{-6}$ mol/L, T = 313 K, solvent: water.

Figure 10 shows that the effect of 0.1 Tesla on photopolymerizations was significantly enhanced when the reaction is carried out in a medium with higher viscosity.

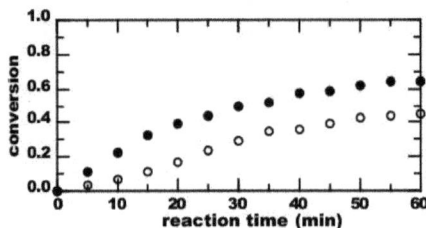

**Figure 10.** Conversion of AM vs. reaction time for polymerizations carried out at 0.1 Tesla of MF intensity (●) and without MF (○). Series 10. [AM] = 0.20 mol/L, [$C_{26}H_{27}O_3P$] = 1 × $10^{-6}$ mol/L, T = 313 K, solvent: 50 wt % EG in water.

## 4. Discussion

Variations of at least 20% in the value of *Rp* in reference to the *Rp* value without MF were considered to evaluate the presence or absence of MF effects.

### 4.1. Thermal Equilibrium of Spin States

Assuming free electrons in the empty space, the ratio $N_\alpha/N_\beta$ can be calculated as 1.03 and 1.0004 at 7 Tesla and 0.1 Tesla, respectively. Such small differences in the population of $N_\alpha$ and $N_\beta$ can hardly be detected since they are within the range of experimental error. However, it is important to note that uncoupled electrons in radical species occurring in polymerization systems are far from being free electrons in the empty space. In any case, the identical polymerization paths of reactions carried out with and without MF observed in Figure 4 (series 4 and 5) and the linear progression of polymerization in Figure 5 (series 6) prove that changes in the thermal equilibrium of spin states due to high MF intensity interactions over uncoupled electrons is insignificant. Any MF effect would be manifested as a change in the slope of the conversion-time plot. As conversion evolved linearly (constant slope) in spite of the fact MF varied, no MF effect could be evidenced when applied in this reaction condition and thus, no MF induced changes can be assigned to *f* and $k_t$.

### 4.2. Magnetically-Induced Molecular Orientation

$\Delta\chi_M$ for AM, $H_2O$ and EG, the main components of the polymerization medium, are reported as $-2.3 \times 10^{-3}$, $-1.6 \times 10^{-3}$, and $-2.6 \times 10^{-3}$ mL/mol, respectively [30]. Introducing the values of $\Delta\chi_M$ into Equation (5), the relative orientation of AM, $H_2O$, and EG molecules under the conditions specified for series 4 were determined and presented in Figure 11. Evidently, the contribution of 7 Tesla of MF intensity to the orientation of either AM, $H_2O$, and EG is minimal. Specifically, the probability to find an AM molecule oriented in the direction of MF is only 1.3% higher than perpendicular, $\pi/2$-radians, to the field. For water and EG it resulted 0.9 and 2%, respectively. Evidently, these small orientations resulted in insignificantly modifying the polymerization rate of AM as it was observed in Figures 4–9 (series 4–9). Therefore, no MF-induced changes can be assigned to $k_p$ and $k_t$.

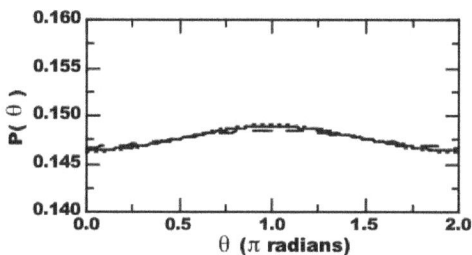

**Figure 11.** Magnetically-induced orientation of AM (−), $H_2O$ (—), and EG (···) at 7 Tesla of MF strength and 308 K.

*4.3. Radical Pair Mechanism*

The polymerization rate increased about 60% in the initial phase when the primary radicals were generated in a triplet spin state through the photochemical dissociation of $C_{26}H_{27}O_3P$ in a relatively high viscosity medium (see Figure 10 (series 10)). However, the effect is very small in the relatively low viscosity medium. The references cited in the introduction explain this effect in terms of the reinforcement of the cage effect due to the reduced mobility of monomers, growing radicals, and solvent molecules in viscous media. Without MF, for the radical pairs initially generated in the T state, spin evolution proceeds in time, passing from the $T_0$ to the S state and, subsequently, undergoes recombination reactions. In addition, $T_+$ and $T_-$ pass to the $T_0$ state to maintain the condition of equal population of spin states. When MF is applied, the energies for $T_+$, $T_0$ and $T_-$ splits out. Consequently, primary radicals are generated preferentially in the $T_+$ state diminishing the possibility for intersystem crossing to $T_0$ and S states. Therefore, the life-time of the primary radicals is significantly increased with higher probability to escape from the cage. This effect can be interpreted as an increase of $\Phi$. Thus, more radicals are released to the medium increasing $Rp$ and decreasing $\nu$. In low viscosity media the effect is less pronounced. Here, the cage effect is very weak and the time needed for the radicals to escape from the cage may be comparable to the time needed for $T_0$–S intersystem crossing. Therefore, approximately the same quantity of radicals is released to the medium independently of their spin states. As a result, very similar $Rp$ and $\nu$ are expected with and without MF.

Interestingly, the augmentation of viscosity due to the formation of polymer molecules in the polymerization medium has no influence on the MF effect. Figure 9 shows no MF increment of $Rp$ with conversion. Thus, it is speculated that the dimension of molecular cages formed by water, EG and polymer are too small, in the order and too large to trap a $C_{26}H_{27}O_3P$ initiator molecule. However, the influences of viscosity induced by molecules of different sizes and its relation with cage dimension and MF effect needs further investigation.

## 5. Conclusions

Polymerization kinetics can be potentially controlled using MF only under very specific reaction conditions. MF effects are significant in systems where primary radical pairs are generated and quenched in a $T_+$ state. Such radicals are produced by photochemical dissociation of the initiator at relatively low MF intensities and in a viscous reaction medium. The viscosity of the reaction medium must be developed by molecules capable to develop a strong molecular cage over the initiator molecules. The combination of these conditions is critical for the observation of MF effects. MF effects can eventually be interpreted as an increase of $\Phi$ and a decrease of $k_t$. Thus, an increase of $Rp$ can be expected since both $\Phi$ and $k_t$ contribute in the same way. However, the effect on $v$ would depend on the resulting competition between the increase of $\Phi$ and the decrease of $k_t$. The modification of the thermal equilibrium of spin states and the molecular orientation induced by MF of 7 Tesla at the temperatures between 308 K and 313 K have negligible effects over the polymerization path.

**Acknowledgments:** The work was supported by the Agencia Nacional de Promoción Científica y Tecnológica, grant PICT 2015 1785 and the Swiss National Science Foundation grants 2000-63395 and 200020-100250. The author thanks Christine Wandrey for her teaching and advice.

**Author Contributions:** I.R. conceived, designed, and performed the experiments, analyzed the data, and wrote the paper.

**Conflicts of Interest:** The author declare no conflict of interest. The founding sponsors had no role in the design of the study; in the collection, analyses, or interpretation of data; in the writing of the manuscript, and in the decision to publish the results.

## References

1. Bhatagnar, S.S.; Mathur, R.N.; Kapur, R.N. The effects of a magnetic field on certain chemical reactions. *Philos. Mag.* **1929**, *8*, 457–473. [CrossRef]
2. Selwood, P.W. Magnetism and catalysis. *Chem. Rev.* **1946**, *38*, 41–82. [CrossRef] [PubMed]
3. Khudyakov, I.V.; Arsu, N.; Jockusch, S.; Turro, N.J. Magnetic and spin effects in the photoinitiation of polymerization. *Des. Mon. Polym.* **2003**, *6*, 91–101. [CrossRef]
4. Bag, D.S.; Maiti, S. Polymerization under magnetic field—II. Radical polymerization of acrylonitrile, styrene and methyl methacrylate. *Polymer* **1998**, *39*, 525–531. [CrossRef]
5. Bag, D.S.; Maiti, S. Polymerization under a magnetic field. VI. Triplet dye-sensitized photopolymerization of acrylamide and methyl methacrylate. *J. Polym. Sci. Part A Polym. Chem.* **1998**, *36*, 1509–1513. [CrossRef]
6. Chiriac, A.P. Polymerization in magnetic field. XVI. Kinetic aspects regarding methyl methacrylate polymerization in high magnetic field. *J. Polym. Sci. Part A Polym. Chem.* **2004**, *42*, 5678–5686. [CrossRef]
7. Chiriac, A.P.; Simionescu, C.I. Some properties of vinyl acetate/methyl methacrylate/acrylamide copolymer synthesized in a magnetic field. *Polym. Test.* **1997**, *16*, 185–192. [CrossRef]
8. Chiriac, A.P.; Simionescu, C.I. Aspects regarding the characteristics of some acrylic and methacrylic polyesters synthesized in a magnetic field. *Polym. Test.* **1996**, *15*, 537–548. [CrossRef]
9. Huang, J.; Song, Q. Effect of polyethylene glycol with sensitizer groups at both ends on the photoinitiated polymerization of styrene in the water phase in the presence of a magnetic field. *Macromolecules* **1993**, *26*, 1359–1362. [CrossRef]
10. Liu, J.; Zhang, R.; Li, H.; Han, B.; Liu, Z.; Jiang, T.; He, J.; Zhang, X.; Yang, G. How does magnetic field affect polymerization in supercritical fluids? Study of radical polymerization in supercritical $CO_2$. *New J. Chem.* **2002**, *26*, 958–961. [CrossRef]
11. Simionescu, C.I.; Chiriac, A.P.; Chiriac, M.V. Polymerization in a magnetic field: 1. Influence of esteric chain length on the synthesis of various poly(methacrylate)s. *Polymer* **1993**, *34*, 3917–3920. [CrossRef]
12. Turro, N.J.; Chow, M.F.; Chung, C.J.; Tung, C.H. An efficient, high conversion photoinduced emulsion polymerization. Magnetic field effects on polymerization efficiency and polymer molecular weight. *J. Am. Chem. Soc.* **1980**, *102*, 7391–7393. [CrossRef]
13. Turro, N.J. Application of weak magnetic fields to influence rates and molecular weight distributions of styrene polymerization. *Ind. Eng. Chem. Prod. Res. Dev.* **1983**, *22*, 272–276. [CrossRef]
14. Turro, N.J.; Chow, M.F.; Chung, C.J.; Tung, C.H. Magnetic field and magnetic isotope effects on photoinduced emulsion polymerization. *J. Am. Chem. Soc.* **1983**, *105*, 1572–1577. [CrossRef]
15. Huang, J.; Hu, Y.; Song, Q. Effect of magnetic field on block copolymerization of styrene and methyl methacrylate by photochemical initiation in micellar solution of poly(ethylene glycol) with sensitizer end group. *Polymer* **1994**, *35*, 1105–1108. [CrossRef]
16. Chiriac, A. Polymerization in a magnetic field. 13 Influence of the reaction conditions in the styrene polymerization. *Rev. Roum. Chim.* **2000**, *45*, 689–695.

17. Chiriac, A.P.; Simionescu, C.I. Polymerization in a magnetic field. X. Solvent effect in poly(methyl methacrylate) synthesis. *J. Polym. Sci. Part A Polym. Chem.* **1996**, *34*, 567–573. [CrossRef]
18. Keskin, S.; Aydin, M.; Khudyakov, I.; Arsu, N. Study of the polymerization of methyl methacrylate initiated by thioxanthone derivatives: a magnetic field effect. *Turk. J. Chem.* **2009**, *33*, 201–207.
19. Vedeneev, A.A.; Khudyakov, I.V.; Golubkova, N.A.; Kuzmin, V.A. External magnetic field effect on the dye-photoinitiated polymerization of acrylamide. *J. Chem. Soc. Faraday Trans.* **1990**, *86*, 3545–3549. [CrossRef]
20. Rintoul, I.; Wandrey, C. Magnetic field effects on the copolymerization of water-soluble and ionic monomers. *J. Polym. Sci. Part A Polym. Chem.* **2009**, *47*, 373–383. [CrossRef]
21. Rintoul, I.; Wandrey, C. Radical homo- and copolymerization of acrylamide and ionic monomers in weak magnetic field. *Macromol. Symp.* **2008**, *261*, 121–129. [CrossRef]
22. Rintoul, I.; Wandrey, C. Magnetic field effects on the free radical solution polymerization of acrylamide. *Polymer* **2007**, *48*, 1903–1914. [CrossRef]
23. Odian, G. Radical Chain Polymerization. In *Principles of Polymerization*, 4th ed.; Wiley-Interscience: Hoboken, NJ, USA, 2004; Chapter 3; pp. 198–346.
24. Steiner, U.; Ulrich, T. Magnetic field effects in chemical kinetics and related phenomena. *Chem. Rev.* **1989**, *89*, 51–147. [CrossRef]
25. Carrington, A.; McLachlan, A.D. *Introduction to Magnetic Resonance, with Applications to Chemistry and Chemical Physics*; Harper International editions: London, UK, 1979.
26. Ayscough, P.B. *Electron Spin Resonance in Chemistry*; Methuen & Co. Ltd.: London, UK, 1967.
27. Sobhi, H.F. Synthesis and Characterization of Acylphospine Oxide Photoinitiators. Ph.D. Thesis, Cleveland State University, Cleveland, OH, USA, May 2008.
28. Rintoul, I.; Wandrey, C. Limit of applicability of the monomer-enhanced mechanism for radical generation in persulfate initiated polymerization of acrylamide. *Lat. Am. Appl. Res.* **2010**, *40*, 365–372.
29. Rintoul, I.; Wandrey, C. Polymerization of ionic monomers in polar solvents: kinetics and mechanism of the free radical copolymerization of acrylamide/acrylic acid. *Polymer* **2005**, *46*, 4525–4532. [CrossRef]
30. Atkins, P.W.; de Paula, J. Statistical thermodynamics. In *Elements of Physical Chemistry*, 6th ed.; Oxford University Press: Oxford, UK, 2013.

![processes logo] *processes*

MDPI

*Article*

# Simultaneous Monitoring of the Effects of Multiple Ionic Strengths on Properties of Copolymeric Polyelectrolytes during Their Synthesis

**Aide Wu [1], Zifu Zhu [1], Michael F. Drenski [2] and Wayne F. Reed [1,\*]**

[1]  Physics Dept., Tulane University, New Orleans, LA 70115, USA; awu2@tulane.edu (A.W.); zzhu@tulane.edu (Z.Z.)

[2]  Advanced Polymer Monitoring Technologies, Inc., 1078 S. Gayoso St., New Orleans, LA 70125, USA; mdrenski@tulane.edu

\*  Correspondence: wreed@tulane.edu; Tel.: +1-504-862-3185

Academic Editor: Alexander Penlidis
Received: 20 February 2017; Accepted: 7 April 2017; Published: 11 April 2017

**Abstract:** A new Automatic Continuous Online Monitoring of Polymerization reactions (ACOMP) system has been developed with multiple light scattering and viscosity detection stages in serial flow, where solution conditions are different at each stage. Solution conditions can include ionic strength (IS), pH, surfactants, concentration, and other factors. This allows behavior of a polymer under simultaneous, varying solution conditions to be monitored at each instant of its synthesis. The system can potentially be used for realtime formulation, where a solution formulation is built up additively in successive stages. It can also monitor the effect of solution conditions on stimuli responsive polymers, as their responsiveness changes during synthesis. In this first work, the new ACOMP system monitored light scattering and reduced viscosity properties of copolymeric polyelectrolytes under various IS during synthesis. Aqueous copolymerization of acrylamide (Am) and styrene sulfonate (SS) was used. Polyelectrolytes in solution expand as IS decreases, leading to increased intrinsic viscosity ($\eta$) and suppression of light scattering intensity due to electrostatically enhanced second and third virial coefficients, $A_2$ and $A_3$. At a fixed IS, the same effects occur if polyelectrolyte linear charge density ($\xi$) increases. This work presents polyelectrolyte response to a series of IS and changing $\xi$ during chemical synthesis.

**Keywords:** ACOMP; online monitoring; copolymeric polyelectrolytes; light scattering; viscosity

---

## 1. Background and Motivation

This work introduces a new version of the Automatic Continuous Online Monitoring of Polymerization reactions (ACOMP) system ("second generation ACOMP") whose aim is to monitor the onset and evolution of stimuli responsive behavior, under multiple simultaneous solution conditions, during the synthesis of stimuli responsive polymers (SRP). SRP is a vast area of modern polymer science and engineering, aimed at producing polymers that can respond to such stimuli as temperature, radiation, and solution conditions such as pH, ionic strength, polymer concentration, presence of such agents as surfactants, nanoparticles, hydrophobic species, etc. The types of responses that can occur in response to these stimuli include coil/globule phase transitions, polymer coil expansion or shrinkage, micellization, aggregation, and other forms of spontaneous self-assembly.

These next-generation materials are expected to have applications in medicine, sensors, self-healing materials, and environmental remediation [1–5]. Hydrogels of poly(N-isopropylacrylamide), for example, have a lower critical solution temperature (LCST), near body temperature, which makes it a candidate for drug delivery applications in which the NIPAM-based polymer releases its medical payload when in contact with targeted tissues [6,7]. While SRP hold much promise, their synthesis is complex and must be tightly controlled. For example, self-assembly of SRP block copolymers into well-defined nanostructures occurs only over a narrow range of compositions, and this ability can be lost with errors in composition as low as two to five percent [8].

A major reason for the development of Second Generation ACOMP (SGA) is to not only monitor synthesis and the onset and evolution of stimuli responsiveness, but also to control the synthetic reactions. Much work in SRP research makes use of controlled radical and other living type reactions [9–14] nucleobase polymers [15,16], and information containing polymers [17,18].

### 1.1. Background on Solution Properties of Polyelectrolytes in Solution

In this first work with the new version of ACOMP a stimuli responsive copolymer produced by simple free radical copolymerization was chosen, rather than the more sophisticated living-type, "click", postpolymer modifications and other routes under current research for producing stimuli responsive polymers.

While the physical properties of polyelectrolytes in solution still present some surprises and puzzles, certain general principles are well established, both experimentally and theoretically. For example, it is well known that the free energy of linear polyelectrolytes in solution is composed of an enthalpic term that expands the polymer coil due to mutual electrostatic repulsion among charge groups, and an entropic term that contracts the coil towards higher probability conformations. The net effect of the free energy balance is a polymer coil that is expanded with respect to a neutral polymer of otherwise identical properties. As the ionic strength of the supporting solution increases, the electrostatic repulsion among charges in the chain decreases and the coil shrinks, which can be measured in many ways, such as by a decrease in polymer reduced viscosity or by an increase in light scattering intensity due to decreased interchain repulsion and excluded volume, and correspondingly reduced second and third virial coefficients $A_2$ and $A_3$, respectively. Similarly, if the linear charge density ($\xi$) is increased at a fixed IS the coil will also expand. $\xi$ can be changed by altering pH for a polybase or polyacid.

Persistence length is a central notion in the theory of polymer conformations and has been examined in detail [19]. The complex problem of excluded volume for polymers has likewise undergone extensive examination [20–23]. The details of the free energy and corresponding calculations of electrostatically enhanced excluded volume and persistence lengths have been treated theoretically and experimentally [24–27]. Similarly, the theory of counterion condensation has been developed and demonstrated experimentally [28–31]. In its simplest statement, counterion condensation predicts that when the repulsive electrostatic energy between charge groups on a polymer chain exceeds thermal energy $k_B T$ (where $k_B$ is Boltzmann's constant), counterions will condense along the chain to lower linear charge density $\xi$ until the repulsive energy is less than or equal to $k_B T$. A recent ACOMP application succeeded in monitoring the cross-over from counterion condensation to non-condensation regimes during copolymer synthesis of anionic and neutral comonomers [32]. In this latter work the authors introduced the term "copolyelectrolyte" as an abbreviation for "copolymer polyelectrolyte".

The current work deals with synthesis of copolyelectrolytes composed of neutral acrylamide (Am) and anionic styrene sulfonate (SS). The reactivity ratios introduced by Mayo and Lewis provide a convenient means of assessing mutual reactivity of comonomers [33]. Given two comonomers, A and B, the reactivity ratio $r_A$ is the ratio of the probability that A will react with another A upon encounter to the probability that A will react with B upon encounter. The reactivity ratio $r_B$ is defined similarly for comonomer B. If $r_A$ and $r_B$ are each zero, then a strictly alternating copolymer will be formed. If $r_A$ and $r_B$ are each infinite, then only homopolymers of A and of B will be produced when the two polymerize in the same reactor. If $r_A = r_B = 1$ then a statistical copolymer is formed whose instantaneous composition only depends on the molar concentrations of A and B, and their respective free radicals, at that instant.

The following approximate values were found using ACOMP for acrylamide and styrene sulfonate copolymerization; $r_{Am} = 0.18$, $r_{SS} = 2.14$ [34]. Since $r_{SS} >> r_{Am}$ it is expected that chains formed early in the copolymerization will be rich in SS, and hence be highly charged. Furthermore, the SS should be consumed rapidly, leaving Am to homopolymerize later in the reaction, after the SS is used up. These trends are clearly seen in ACOMP data.

*1.2. Background on ACOMP*

ACOMP was first introduced in 1998 [35] and was recently reviewed in detail [36]. It has been used in many scenarios including free radical copolymerization [37,38], copolyelectrolytes [39], branching reactions [40], emulsion [41] and inverse emulsion [42] reactions, in batch, semi-batch [43] and continuous reactors [44], for aqueous and organic solvents, for post-polymerization and derivitization reactions [45], and for the controlled radical polymerization routes nitroxide-mediated polymerization (NMP) [46], atom transfer radical polymerization (ATRP) [47], reversible addition fragmentation transfer polymerization (RAFT) [48], and ring opening metathesis polymerization (ROMP) [49]. Simultaneous monitoring of both polymer reaction characteristics and colloid size distributions (latex and monomer droplets) was also achieved in emulsion polymerization [41]. Recently, fully automatic feedback control of free radical polymerization was achieved [50]. The first system controlled was aqueous polymerization of Am. Current work includes simultaneous automatic feedback control of composition and molecular weight during free radical copolymerization, automatic production of multi-modal polymers, and extension of control to living type polymerization. The work presented here is cognate to the latter effort.

ACOMP relies on the continuous extraction, dilution, and conditioning of a small sample stream from the reactor on which measurements by various combinations of detectors are made. By combining simultaneous data from multiple detectors continuous monitoring of salient reaction characteristics can be made, such as kinetics, conversion of comonomers, composition drift, evolution of molecular mass and intrinsic viscosity, and detection of unusual phenomena, such as microgelation and runaway reactions. Typical detectors include light scattering, UV/visible spectrophotometry, viscometry, refractivity, and polarimetry. While ACOMP is not inherently a chromatographic method, its continuous stream can be used in automatic conjunction with gel permeation and other separation techniques, as has been demonstrated [51].

## 2. Second Generation ACOMP (SGA) for Monitoring the Synthesis of Stimuli Responsive Polymers under Varying Solution Conditions

The aims of this work are (1) to introduce the multi-stage serial flow SGA; (2) apply it to the monitoring of copolyelectrolyte synthesis in batch; (3) to qualitatively interpret the results within well understood concepts of polyelectrolyte behavior. It is beyond the scope of this work to develop a complete theoretical formulation for interpreting all the data, such as by electrostatic persistence length, excluded volume theory, and counterion condensation theories.

An earlier version of SGA allowed control of temperature in three detection stages, each stage with a light scattering and viscometer. This was used to study the effects on the lower critical solution temperature (LCST) of copolymerizing n-isopropyl acrylamide (NIPAM) with more hydrophilic comonomers [52,53].

In this work the custom built SGA comprises (i) a system of pumps and a high pressure and a low pressure mixing chamber for continuously withdrawing and diluting reactor content at a low flow rate, typically 0.1 mL/min; (ii) up to seven detector stages in series, each with a custom built single capillary viscometer and a 90° custom built light scattering flow cell; (iii) a high pressure multi-head syringe pump that feeds concentrated NaCl solutions into each stage, thus increasing ionic strength (IS) from stage to stage; (iv) A separate, highly dilute stream at 150 mM NaCl from the low pressure mixing chamber is fed to a diode array UV spectrophotometer, covering the range 190–400 nm (Shimadzu SPD-M20A). From the UV detector, the stream flows through a seven angle light scattering unit

(BI-MwA from Brookhaven Instruments Corp., New York, NY, USA); (v) Data from all the detectors are collected under various protocols and unified in realtime into a master file to allow cross-correlations and calculations using the different signals.

Hence, the SGA combines (i) the features of the original ACOMP to monitor copolymer weight average molecular weight and reduced viscosity, and cumulative and instantaneous copolymer composition with (ii) the ability to interrogate the polymer continuously during synthesis as to its light scattering and viscometric responses to a series of IS.

The SGA system is equipped with two Nexus 6000 high pressure syringe pumps (By Chemyx Inc., Stafford, TX, USA) which infuse NaCl solutions with different concentrations in each syringe into the serial flow, in order to change the ionic strength at each stage. To meet the operation/characterization under high pressure operation condition, stainless steel syringes (Harvard Apparatus, MA, USA) are used to hold the NaCl solutions for injection into the serial flow. The sample extraction and dilution system consists of four Shimadzu AD-VP HPLC pumps. A system diagram is shown in Scheme 1. Pump 1 extracts sample directly from the reactor, Pump 2 and Pump 3 are used for diluting sample with deionized water as a first dilution. A separate HPLC pump delivers this first stage dilution stream through the multi-stage serial dilution stage with the LS/viscometer detection pairs, where the IS is increased at each stage via the syringe pumps. A second dilution of the first stream occurs in a second high pressure mixing chamber producing a highly dilute sample at 150 mM IS, which is sent to a detection stage comprising a UV-vis (Shimadzu SPD-M20A, Kyoto, Japan) and BI-MwA (Brookhaven Instruments Corporation, New York, NY, USA). This highly dilute stream at 150 mM IS allows determination of $M_w$ without interference from the large virial coefficient effects deliberately produced in the more highly concentrated, multi-stage IS detector train. All detectors were at room temperature.

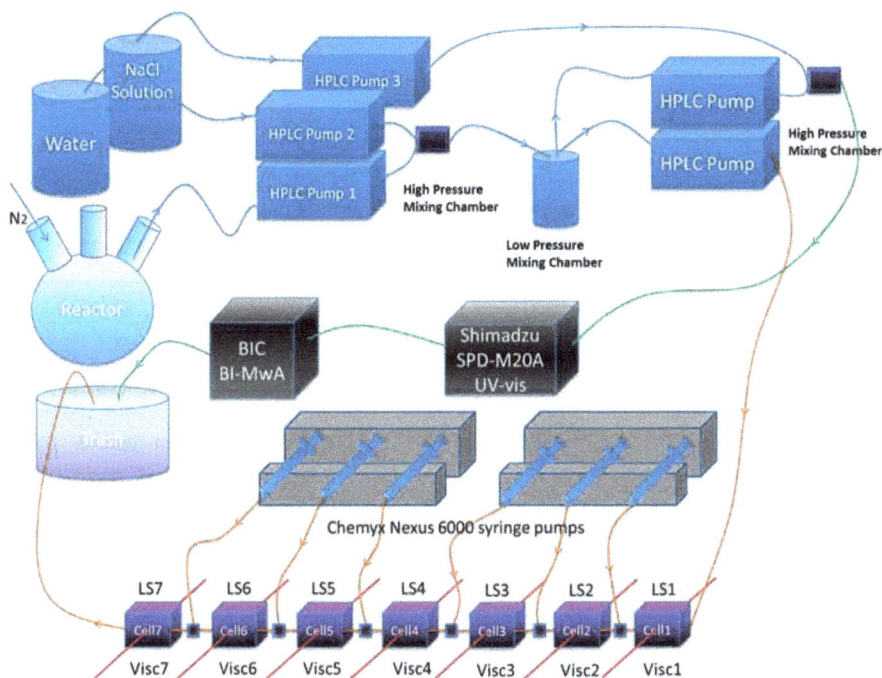

**Scheme 1.** The Second Generation Automatic Continuous Online Monitoring of Polymerization reactions (SGA) instrumentation.

Table 1 provides information on each reaction, including the amount of dilution in the first and second dilution stages. Table 2 shows the syringe pump reservoir strength of each stage, the IS of the serial flow at each stage, and the additional, small dilution that occurs in each stage.

**Table 1.** The information for each of the three reactions.

| Reaction | Type | $Am_0$ (g/cm$^3$) | $SS_0$ (g/cm$^3$) | Initiator (g/cm$^3$) | Second Dilution | First Dilution | $T$ (°C) | Figures |
|---|---|---|---|---|---|---|---|---|
| A | 50/50 batch | $12.9 \times 10^{-3}$ | $37.5 \times 10^{-3}$ | $2.73 \times 10^{-3}$ | 50× | 10× | 65 | Figure 1a,b, Figure 2, Figure 3, Figure 7a |
| B | 50/50 batch | $10.3 \times 10^{-3}$ | $32.5 \times 10^{-3}$ | $5.00 \times 10^{-4}$ | 105× | 15× | 60 | Figure 4a,b, Figure 5, Figure 6 |
| C | Semi-batch | $18.1 \times 10^{-3}$ | $181 \times 10^{-3}$ stock fed into reactor at 0.1 mL/min | $2.73 \times 10^{-3}$ | 50× | 10× | 65 | Figure 7b |

**Table 2.** The flow rates, dilution factors, and ionic strengths in the seven stage SGA detector train.

| | Reaction A | | | Reaction B | | | Reaction C (Semi-Batch) | | |
|---|---|---|---|---|---|---|---|---|---|
| Stage | [NaCl] * (mM) | IS (mM) | Dilution factor | [NaCl] * (mM) | IS (mM) | Dilution factor | [NaCl] * (mM) | IS (mM) | Dilution factor |
| 1 | 0 | 0 | 1 | 0 | 0 | 1 | 0 | 0 | 1 |
| 2 | 2 | 0.095 | 1.05 | 4 | 0.12 | 1.03125 | 2 | 0.095 | 1.05 |
| 3 | 20 | 1 | 1.10 | 40 | 1.3 | 1.0625 | 20 | 1 | 1.10 |
| 4 | 200 | 9.7 | 1.15 | 400 | 13 | 1.09375 | 200 | 9.7 | 1.15 |
| 5 | 1000 | 51 | 1.20 | 5000 | 151 | 1.125 | 1000 | 51 | 1.20 |
| 6 | 2000 | 129 | 1.25 | - | - | - | 2000 | 129 | 1.25 |
| 7 | 5000 | 316 | 1.30 | - | - | - | 5000 | 316 | 1.30 |

* Ionic strength (IS) of reservoir.

Viscosity computations are based on Poisseuille flow of a liquid of viscosity $\eta$ in a capillary of length $L$ and radius $R$ at a flow rate $Q$ (cm$^3$/s), for which the pressure drop $\Delta P$ across the capillary is

$$\Delta P = \frac{8LQ\eta}{\pi R^4} \tag{1}$$

The dilute solution polynomial expansion for viscosity is used to interpret the viscometer data:

$$\eta = \eta_s \left[ 1 + [\eta]c + k_p[\eta]^2 c^2 \right] \tag{2}$$

where $\eta_s$ is the pure solvent viscosity, $[\eta]$ is the intrinsic viscosity of the polymer, $C_p$ is the polymer concentration (in g/cm$^3$) and $k_H$ is a constant related to the hydrodynamic interactions between polymer chains, usually around 0.4 for neutral, coil polymers [54]. The intrinsic viscosity is the extrapolation to zero concentration and zero shear rate of the reduced viscosity $\eta_r$, which is defined according to Equation (2) by

$$\eta_r(t) = \frac{\eta(t) - \eta_s}{\eta_s C_p} = \frac{V(t) - V_s}{(V_s - V_0)C_p} \tag{3}$$

In the second equality $V(t)$ is the time dependent viscometer signal, $V_s$ is the voltage of the pure solvent baseline, and $V_0$ accounts for any voltage offset in the viscometer when the flow rate through it is $Q = 0$. This latter can be written as shown because the viscometer voltage output is directly proportional to $\Delta P$. It is important to realize that because the denominator divides out by solvent baseline voltage, no calibration factor for the capillary viscometer is needed in order to determine $\eta_r$. The specific viscosity $\eta_{sp}$ is another useful quantity computed by

$$\eta_{sp}(t) = \frac{\eta(t) - \eta_s}{\eta_s} = \frac{V(t) - V_s}{V_s - V_0} \tag{4}$$

The absolute excess Rayleigh ratio is determined from the raw light scattering voltages by

$$I_R(t) = \frac{V(t) - V_s}{V_s - V_d} I_{R,toluene} \tag{5}$$

where $V(t)$ is the scattering from the polymer solutions, $V_s$ is the scattering from the pure solvent and $V_d$ is the dark voltage of the detector when there is no incident light on the scattering sample. $I_{R,toluene}$ is the Rayleigh ratio for pure toluene at $T = 25$ °C, which is given for any wavelength by

$$I_R(\text{cm}^{-1}) = 1.069 \times 10^{-5} \left(\frac{677}{\lambda(\text{nm})}\right)^4 \tag{6}$$

$I_R$ measures the fraction of incident light scattered per cm of path through the scattering medium. $I_R$ is used for light scattering analysis using the usual Zimm approach [55]:

$$\frac{Kc}{I_R(q, C_p)} = \frac{1}{MP(q)} + 2A_2C_p + \left[3A_3Q(q) - 4A_2^2MP(q)(1 - P(q))\right]C_p + ...^2 \tag{7}$$

where $K$ is an optical constant, given for vertically polarized incident light by

$$K = \frac{4\pi^2 n^2 (dn/dc)^2}{N_A \lambda^4} \tag{8}$$

where $n$ is the solvent index of refraction, $\lambda$ is the vacuum wavelength of the incident light, $dn/dc$ is the differential refractive index for the polymer in the chosen solvent, and $q$ is the usual scattering wave-vector amplitude, $q = (4\pi n/\lambda)\sin(\theta/2)$, where $\theta$ is the scattering angle.

In the limit as $q \to 0$, and for polydisperse polymers this reduces to

$$\frac{KC_p}{I_R(q = 0, C_p)} = \frac{1}{M_w} + 2A_2C_p + 3A_3C_p^2 + ... \tag{9}$$

where $M_w$ is the weigh average molecular weight and $A_2$ and $A_3$ are complex averages of the virial coefficients.

The individually resolved concentrations of Am and SS can be used to compute the instantaneous mole fractions of Am and SS in a copolymer at any instant according to

$$[F_{inst,Am}] = \frac{d[Am]}{d([Am] + [SS])} \tag{10a}$$

$$[F_{inst,SS}] = \frac{d[SS]}{d([Am] + [SS])} \tag{10b}$$

## 3. Copolymerization

Sodium 4-vinylbenzenesulfonate and acrylamide were used as received from Sigma-Aldrich. The initiator, 2,2′-azobis(2-amidinopropane dihydrochloride (V-50), was from Wako Chemicals USA, Inc. For batch reactions A and B comonomers with molar ratio of 1:1 were fully dissolved in DI water, and the reactions were carried out in a thermostated three neck reactor under continuous $N_2$ purge. The starting comonomers' concentrations and reaction temperatures are listed in Table 1 for Reactions A, B, and C. Linear charge density $\xi$ varies during synthesis due to the high composition drift of the copolymeric system; anionic styrene sulfonate (SS) and electrically neutral acrylamide (Am), which have widely separated reactivity ratios ($r_{Am} = 0.18$, $r_{SS} = 2.14$ in 0.1 M NaCl [39], and with similar high drift at other ionic strengths). The salt, the sodium form of styrene sulfonate remains ionized at the near-neutral pH of the reactor solution. There was no added NaCl in the reactor.

Although a salt is used here (SS) the use of the acid form, sulfonic acid, could lead to different behavior in the polymer, beyond the counterion condensation effect expected even for salts, since $pK_a$ for acids and bases in polymers is different than in their small molecule form. ACOMP has previously used both pH and conductivity probes in the reactor to follow changes in these quantities and would normally be used in such cases.

## 4. Results and Discussion

### 4.1. Typical Raw Data

Figure 1a shows typical raw data for a 50/50 VB/Am copolymerization, where $\eta_{sp}$ and $I_R$ are computed by Equations (4) and (5), respectively. The conditions are given in Table 1. The reaction began at 8000 s. The time scale of the reaction corresponds to earlier findings [39]. Only three IS of the seven measured are shown, in order to keep the figure legible. The trend for viscosity is as expected, namely as IS increases there is a dramatic drop in $\eta_{sp}$ at any given instant, reflecting coil contraction as charges are electrostatically shielded. The light scattering (LS) at low IS, 0.1 mM, is highly suppressed. Also a maximum is reached as the reaction proceeds follow by a steady drop and subsequent small rise late in the reaction, starting at about 16,000 s. The maximum is due to the effect of $A_3$, seen in Equation (9), as the concentration builds. A similar shape and effect is seen at 10 mM, except the overall magnitude of the LS is well above that for 0.1 mM, due to significantly reduced $A_2$ and $A_3$. At 130 mM the $A_3$ effect is completely suppressed and LS increases monotonically as the reaction proceeds. At 130 mM the solution is about 10% more dilute than at 10 mM. The slight rise in LS for 0.1 mM starting at 16,000 s, is also seen for 10 mM and 130 mM LS. This is due to the fact that the high reactivity ratio of SS leads to its complete consumption at about 16,000 s, leaving only Am, which produces more massive homopolymer chains than the copolymer chains. The trend can also be seen in the viscosity data, although less pronounced.

Figure 1b shows the total fractional polymer conversion $f_{total}$, and the instantaneous molar fractions of Am and SS in copolymers produced at a given instant, $[F_{inst,Am}]$ and $[F_{inst,SS}]$, according to Equations (10a,b). This shows that SS is consumed more rapidly, due to its higher reactivity ratio, leaving chains with decreasing fraction of SS as the reaction progresses.

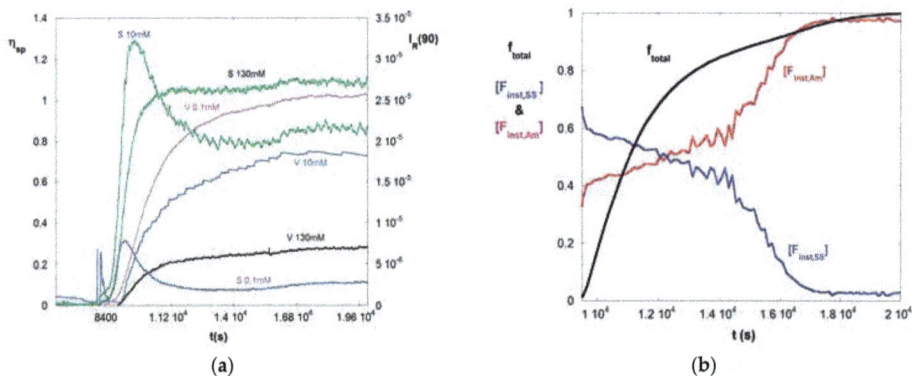

**Figure 1.** (a) A selection of LS (S) and $\eta_{sp}$ (V) data for 50/50 $M/M$ copolymerization of Am and SS shown at three of the seven different IS available; 0.1 mM, 10 mM, and 130 mM for reaction A; (b) Total monomer conversion $f_{total}$, and instantaneous mole fractions $F_{inst,Am}$, and $F_{inst,SS}$ for reaction A.

*4.2. SGA Data Analysis*

Figure 2 shows $Kc/I_R$ vs. $C_p$ for Reaction A up to 15,000 s. $C_p$ is the polymer concentration in the detector train. Also shown are quadratic fits, which yield $A_2$, and $A_3$ for each, according to Equation (9). In these fits $M_w$ was determined by linear extrapolation at low $C_p$ to be 67,000 g/mol $\pm5\%$ and showed low drift through the early and mid-stages of the reaction. It was held constant in the quadratic fit, leaving $A_2$ and $A_3$ as the fit variables. Figure 3 shows these latter values obtained from the quadratic fits.

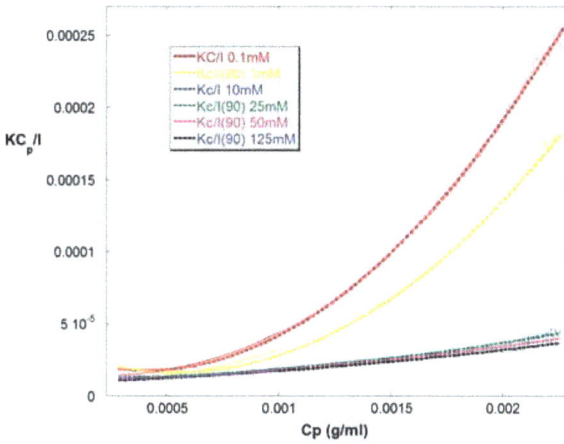

**Figure 2.** $KC_p/I_R(90)$ vs. $C_p$ for reaction A, where $C_p$ is the total polymer concentration in the detector train. Also shown are quadratic fits for $A_2$ and $A_3$ with fixed $M_w = 67{,}000$ g/mol in Equation (9).

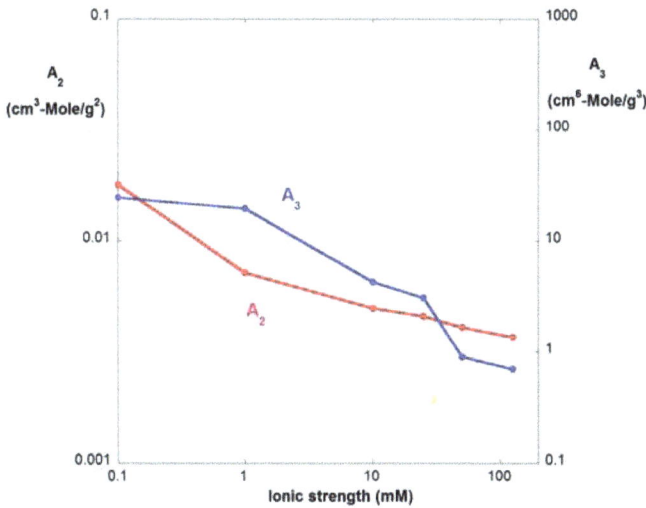

**Figure 3.** $A_2$ and $A_3$ vs. IS, as determined from Figure 2. The $A_3$ axis has two decades for each decade of the $A_2$ axis, suggesting an approximate correspondence to the power law between $A_2$ and $A_3$ of Equation (11) for hard spheres.

Figure 3 shows $A_2$ and $A_3$ vs. IS from these determinations. While both $A_2$ and $A_3$ follow the same expected decreasing trend vs. IS, they do not overlap using the double-$y$ scale of Figure 3. Nonetheless, the representation which brings them into rough parity is two orders of magnitude in $A_3$ for each order of magnitude in $A_2$, which is reminiscent of the relationship between $A_2$ and $A_3$ that Boltzmann found for hard spheres [56].

$$A_3 = \frac{5MA_2^2}{8} \tag{11}$$

While charged, random coil polymers, which the Am/SS resemble, are not expected at all to follow hard sphere behavior the approximate scaling in Equation (11) for the data in Figure 3 is nonetheless suggestive of an underlying relationship between the two and three body interactions of the different morphologies.

Figure 4 shows $[F_{inst,Am}]$ and $[F_{inst,SS}]$ vs. $C_p$ in the highly dilute, last stage detector train for reaction B. $[F_{inst,SS}]$ decreases as the reaction proceeds, similar to Figure 1b, which is shown in the time domain. Also shown is $I_R(90)/KC_p$ vs. $C_p$. Because these data are gathered in the highly diluted last stage, and under an ionic strength of 150 mM, this is a good approximation to the weight average molar mass $M_w$. Although there was a MALS detector in the highly dilute last stage, there was no measurable angular dependence over the $M_w$ range produced so $I_R(90)$ is used for the $M_w$ determination. $M_w$ starts at around $5 \times 10^4$ g/mol and increases linearly until 50% conversion by mass, during the period that $[F_{inst,SS}]$ falls gradually, at which point it levels off, and then increases again as the SS is fully consumed in the last 10% of conversion.

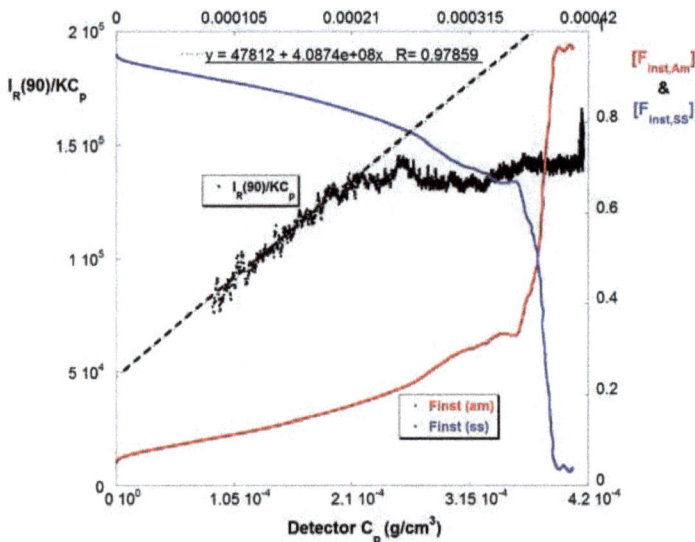

**Figure 4.** The instantaneous comonomer compositions $[F_{inst,SS}]$ and $[F_{inst,Am}]$ (right hand axis) and effective $M_w$ (left hand axis) for Reaction B.

Figure 5 shows $I_R(90)$ vs. $C_p$ in the more concentrated SGA train, for reaction B in Table 1, where $C_p$ is the concentration of copolymer in the SGA detector train. Again, the effect of $A_2$ and $A_3$ is seen as $I_R(90)$ acquires first a negative second derivative and then a maximum.

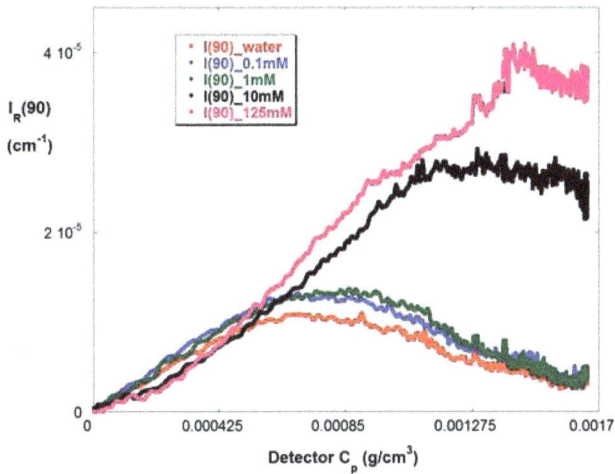

**Figure 5.** $I_R(90)$ at various ionic strengths for reaction B.

Figure 6 shows $\eta_r$ vs. $C_p$ in the dilute train for reaction B at five different ionic strengths. The expected trend is found that at any point in the polymerization reaction $\eta_r$ decreases with increasing IS. The inset shows the final values of $\eta_r$, showing nearly a factor of four drop in $\eta_r$ from low to high IS.

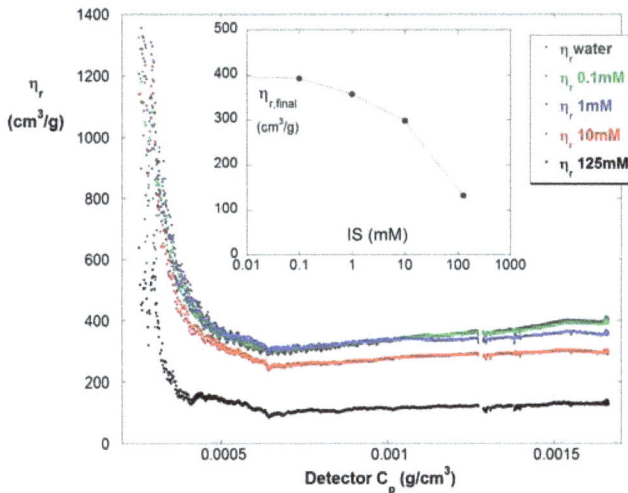

**Figure 6.** $\eta_r$ vs. $C_p$ at the various IS for reaction B.

*4.3. Contrasting Viscosity Behavior in a Low to High $\xi$ Reaction with a High to Low $\xi$ Reaction*

As seen in the batch reactions above, the high reactivity ratio of SS compared to Am results in rapid consumption of SS, which means the copolymer proceeds from high negative $\xi$ which decreases to $\xi = 0$ once the pure polyacrylamide stage is reached during the latter portion of the reaction. A means of starting with $\xi = 0$ and increasing $\xi = 0$ during the reaction is to start with pure Am and feed in a stock of SS (a semi-batch reaction). The semi-batch reaction (reaction C) started with pure Am, had a constant flow of SS into the reactor, and ended with almost 100% SS. It was arranged so that the final polymer concentration was comparable to the final polymer concentration in the batch reaction. The molar masses are also comparable.

The specific viscosity for the batch reaction at three different IS is shown in Figure 7a. $\eta_{sp}$ has a negative second derivative during the entire reaction at the three IS shown. This shows how the highly charged chains at the beginning of the reaction rapidly increase $\eta_{sp}$, but then the increase of $\eta_{sp}$ slows down as the chains become less and less charged as $[F_{inst,Am}]$ increases.

In contrast Figure 7b shows a positive second derivative for $\eta_{sp}$ as SS flows into the reactor. This shows how increasing $\xi$ rapidly increases $\eta_{sp}$ as $[F_{inst,SS}]$ increases. The increase in $[F_{inst,SS}]$ due to semi-batch operation of reaction C is shown in Figure 8, which can be contrasted with Figures 1b and 4 to see the opposite trends in the batch reactions A and B.

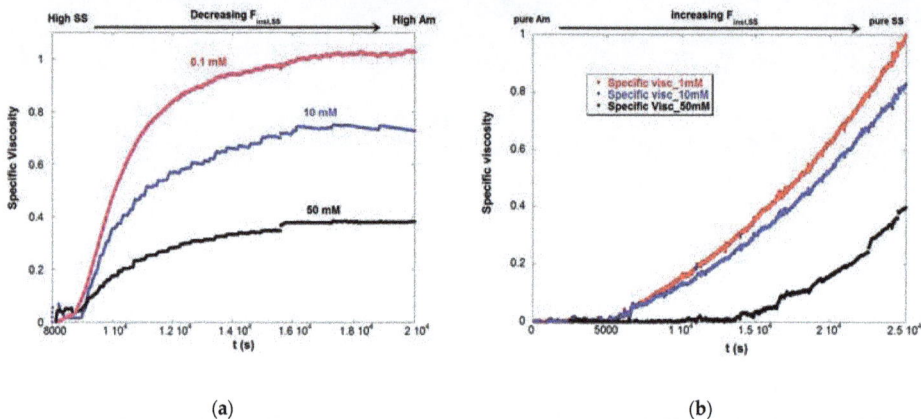

(a)

(b)

**Figure 7.** (a) $\nu_{sp}$ vs. *t* for Reaction A, a batch reaction. The negative second derivative in time reflects the decreasing linear charge density ($\xi$) in the chains as the reaction proceeds. (b) The semi-batch Reaction C, has a positive second derivative in time, showing the effect of increasing $\xi$ as the reaction proceeds. Both reactions ended with the same total polymer concentration.

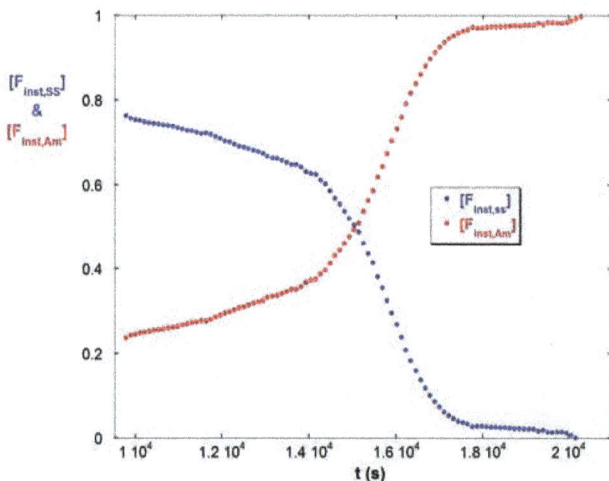

**Figure 8.** $[F_{inst,SS}]$ and $[F_{ionst,Am}]$ for the semi-batch reaction C. The trends are opposite those of batch reactions A and B.

## 5. Summary

A second generation ACOMP system (SGA) has been prototyped and introduced. It is capable of changing solution conditions in serial fashion on a flowing sample stream in up to seven different conditions. A detector stage, consisting of a viscometer and light scattering detector, is present in each successive sample condition, allowing the influence of several solution conditions to be monitored at each instant during the synthesis of polymers.

The first reaction used to demonstrate the capacity of the new system was the copolymerization of a neutral monomer (Am) with an anionic monomer (SS) in both batch and semi-batch processes. The difference in reactivity ratios led to high drift in composition towards chains richer in Am as reactions proceeded, corresponding to lower linear charge density $\xi$. The effects of up to seven different ionic strengths on reduced viscosity and the virial coefficients $A_2$ and $A_3$ were monitored for batch and semi-batch reactions. The $A_2$ and $A_3$ effects, monitored by light scattering, were pronounced at low IS, leading to maxima in scattering during the reaction which then subsided as the reaction progressed further. These strong interparticle effects were largely screened and minimized at high IS. The behavior of reduced viscosity followed the expected trend of decreasing with increasing IS at all points during the reaction. When a semi-batch reaction was run such that SS was increased, and hence also $\xi$, $\eta_r$ increased with a positive second derivative in time, whereas a batch reaction with similar concentrations led to a negative second derivative in time due to $\xi$ decreasing in time.

The SGA should be useful for reactions where stimuli responsive polymers are produced. These include living type copolymerization of block copolymers which exhibit an LCST, and those which can micellize and entrap substances, such as drugs. It may also be possible to develop formulations, since different agents can be added in the seven solution condition system.

It is noted that APMT, Inc. has implemented first generation ACOMP on industrial scale reactors. Hence, scale-up of SG-ACOMP is expected to follow a similar path. It is hoped that, as stimuli responsive polymers become a larger part of the polymer manufacturing sector, SG-ACOMP, or similar, can be integrated into plant design. Similarly, the active, automatic control of conventional polymer synthesis [50] is currently being adapted for living type reactions, on which the synthesis of many stimuli responsive polymers relies.

**Acknowledgments:** This work was supported by NSF EPS-1430280 and Louisiana Board of Regents.

**Author Contributions:** Aide Wu improved the instrumentation and carried out most of the experiments. Michael F. Drenski contributed to the design and build of the original SGA system and guided many of the improvements. Zifu Zhu helped build the original SGA system, together with Michael F. Drenski, and carried out the semi-batch reaction and reaction A in the work. Wayne F. Reed conceived the idea of SG-ACOMP and the multiple syringe pump approach to changing conditions in serial flow, and wrote most of this manuscript.

**Conflicts of Interest:** The authors declare no conflict of interest.

## References

1.  Zhuang, J.; Gordon, M.R.; Ventura, J.; Li, L.; Thayumanavan, S. Multi-stimuli responsive macromolecules and their assemblies. *Chem. Soc. Rev.* **2013**, *42*, 7421–7435. [CrossRef] [PubMed]
2.  Jeong, B.; Gutowska, A. Lessons from nature: Stimuli-responsive polymers and their biomedical applications. *Trends Biotechnol.* **2002**, *20*, 305–311. [CrossRef]
3.  Nath, N.; Chilkoti, A. Creating "smart" surfaces using stimuli responsive polymers. *Adv. Mater.* **2002**, *14*, 1243–1247. [CrossRef]
4.  Liu, F.; Urban, M.W. Recent advances and challenges in designing stimuli-responsive polymers. *Prog. Polym. Sci.* **2010**, *35*, 3–23. [CrossRef]
5.  Galaev, I.; Mattiasson, B. *Smart Polymers: Applications in Biotechnology and Biomedicine*; CRC Press: Boca Raton, FL, USA, 2007.
6.  Chung, J.E.; Yokoyama, M.; Yamato, M.; Aoyagi, T.; Sakurai, Y.; Okano, T. Thermo-responsive drug delivery from polymeric micelles constructed using block copolymers of poly(*n*-isopropylacrylamide) and poly(butylmethacrylate). *J. Control. Release* **1999**, *62*, 115–127. [CrossRef]

7. Zhang, J.; Peppas, N.A. Synthesis and characterization of pH-and temperature-sensitive poly(methacrylic acid)/poly(*n*-isopropylacrylamide) interpenetrating polymeric networks. *Macromolecules* **2000**, *33*, 102–107. [CrossRef]

8. Jain, S.; Bates, F.S. On the origins of morphological complexity in block copolymer surfactants. *Science* **2003**, *300*, 460–464. [CrossRef] [PubMed]

9. Matyjaszewski, K.; Sumerlin, B.S.; Tsarevsky, N.V.; Chiefari, J. *Controlled Radical Polymerization*; American Chemical Society: Washington, DC, USA, 2016.

10. Ballard, N.; Mecerreyes, D.; Asua, J.M. Redox active compounds in controlled radical polymerization and dye-sensitized solar cells: Mutual solutions to disparate problems. *Chem. A Eur. J.* **2015**, *21*, 18516–18527. [CrossRef] [PubMed]

11. Mastan, E.; Li, X.; Zhu, S. Modeling and theoretical development in controlled radical polymerization. *Prog. Polym. Sci.* **2015**, *45*, 71–101. [CrossRef]

12. Chan, N.; Cunningham, M.F.; Hutchinson, R.A. Copper-mediated controlled radical polymerization in continuous flow processes: Synergy between polymer reaction engineering and innovative chemistry. *J. Polym. Sci. Part A Polym. Chem.* **2013**, *51*, 3081–3096. [CrossRef]

13. Barner-Kowollik, C.; Delaittre, G.; Gruendling, T.; Paulöhrl, T. Elucidation of reaction mechanisms and polymer structure: Living/controlled radical polymerization. In *Mass Spectrometry in Polymer Chemistry*; Wiley-VCH Verlag GmbH & Co. KGaA: Weinheim, Germany, 2011; pp. 373–403.

14. Flores, J.D.; Abel, B.A.; Smith, D.; McCormick, C.L. Stimuli-responsive polymers via controlled radical polymerization. In *Monitoring Polymerization Reactions*; John Wiley & Sons: New York, NY, USA, 2013; pp. 45–58.

15. Hemp, S.T.; Long, T.E. DNA-inspired hierarchical polymer design: Electrostatics and hydrogen bonding in concert. *Macromol. Biosci.* **2012**, *12*, 29–39. [CrossRef] [PubMed]

16. Cheng, S.; Zhang, M.; Dixit, N.; Moore, R.B.; Long, T.E. Nucleobase self-assembly in supramolecular adhesives. *Macromolecules* **2012**, *45*, 805–812. [CrossRef]

17. Roy, R.K.; Meszynska, A.; Laure, C.; Charles, L.; Verchin, C.; Lutz, J.-F. Design and synthesis of digitally encoded polymers that can be decoded and erased. *Nat. Commun.* **2015**, *6*, 7237. [CrossRef] [PubMed]

18. Lutz, J.-F.; Ouchi, M.; Liu, D.R.; Sawamoto, M. Sequence-controlled polymers. *Science* **2013**, *341*, 1238149. [CrossRef] [PubMed]

19. Landau, L.D.; Lifshitz, E.M. *Statistical Physics*, 3rd ed.; Pergammon Press: Oxford, UK; New York, NY, USA, 1980.

20. Flory, P.J. *Principles of Polymer Chemistry*; Cornell University Press: New York, NY, USA, 1953.

21. Yamakawa, H. *Modern Theory of Polymer Solutions*; Harper & Row: New York, NY, USA, 1971.

22. McMillan, W.G., Jr.; Mayer, J.E. The statistical thermodynamics of multicomponent systems. *J. Chem. Phys.* **1945**, *13*, 276–305. [CrossRef]

23. Hansen, J.-P.; McDonald, I.R. *Theory of Simple Liquids*; Elsevier: Amsterdam, The Netherlands, 1990.

24. Odijk, T. Polyelectrolytes near the rod limit. *J. Polymer Sci. Polym. Phys. Ed.* **1977**, *15*, 477–483. [CrossRef]

25. Skolnick, J.; Fixman, M. Electrostatic persistence length of a wormlike polyelectrolyte. *Macromolecules* **1977**, *10*, 944–948. [CrossRef]

26. Sorci, G.A.; Reed, W.F. Electrostatically enhanced second and third virial coefficients, viscosity, and interparticle correlations for linear polyelectrolytes. *Macromolecules* **2002**, *35*, 5218–5227. [CrossRef]

27. Reed, W.; Ghosh, S.; Medjahdi, G.; Francois, J. Dependence of polyelectrolyte apparent persistence lengths, viscosity, and diffusion on ionic strength and linear charge density. *Macromolecules* **1991**, *24*, 6189–6198. [CrossRef]

28. Manning, G.S. Limiting laws and counterion condensation in polyelectrolyte solutions II. Self-diffusion of the small ions. *J. Chem. Phys.* **1969**, *51*, 934–938. [CrossRef]

29. Oosawa, F. *Polyelectrolytes*; Marcel Dekker Inc.: New York, NY, USA, 1971.

30. Wilson, R.W.; Bloomfield, V.A. Counterion-induced condesation of deoxyribonucleic acid. A light-scattering study. *Biochemistry* **1979**, *18*, 2192–2196. [CrossRef] [PubMed]

31. Hinderberger, D.; Spiess, H.W.; Jeschke, G. Dynamics, site binding, and distribution of counterions in polyelectrolyte solutions studied by electron paramagnetic resonance spectroscopy. *J. Phys. Chem. B* **2004**, *108*, 3698–3704. [CrossRef]

32. Kreft, T.; Reed, W.F. Experimental observation of crossover from noncondensed to counterion condensed regimes during free radical polyelectrolyte copolymerization under high-composition drift conditions. *J. Phys. Chem. B* **2009**, *113*, 8303–8309. [CrossRef] [PubMed]

33. Mayo, F.R.; Lewis, F.M. Copolymerization. I. A basis for comparing the behavior of monomers in copolymerization; the copolymerization of styrene and methyl methacrylate. *J. Am. Chem. Soc.* **1944**, *66*, 1594–1601. [CrossRef]

34. Kreft, T.; Reed, W.F. Predictive control of average composition and molecular weight distributions in semibatch free radical copolymerization reactions. *Macromolecules* **2009**, *42*, 5558–5565. [CrossRef]

35. Florenzano, F.H.; Strelitzki, R.; Reed, W.F. Absolute, on-line monitoring of molar mass during polymerization reactions. *Macromolecules* **1998**, *31*, 7226–7238. [CrossRef]

36. Reed, W.F. Automated continuous online monitoring of polymerization reactions (acomp) and related techniques. In *Encyclopedia of Analytical Chemistry*; John Wiley & Sons, Ltd.: New York, NY, USA, 2006.

37. Çatalgil-Giz, H.; Giz, A.; Alb, A.M.; Öncül Koç, A.; Reed, W.F. Online monitoring of composition, sequence length, and molecular weight distributions during free radical copolymerization, and subsequent determination of reactivity ratios. *Macromolecules* **2002**, *35*, 6557–6571. [CrossRef]

38. Alb, A.M.; Enohnyaket, P.; Drenski, M.F.; Head, A.; Reed, A.W.; Reed, W.F. Online monitoring of copolymerization involving comonomers of similar spectral characteristics. *Macromolecules* **2006**, *39*, 5705–5713. [CrossRef]

39. Alb, A.M.; Paril, A.; Catalfil-Giz, H.; Giz, A.; Reed, W.F. Evolution of composition, molar mass, and conductivity during the free radical copolymerization or polyelectrolytes. *J. Phys. Chem. B* **2007**, *111*, 8560–8566. [CrossRef] [PubMed]

40. Farinato, R.S.; Calbick, J.; Sorci, G.A.; Florenzano, F.H.; Reed, W.F. Online monitoring of the final, divergent growth phase in the step-growth polymerization of polyamines. *Macromolecules* **2005**, *38*, 1148–1158. [CrossRef]

41. Alb, A.M.; Reed, W.F. Simultaneous monitoring of polymer and particle characteristics during emulsion polymerization. *Macromolecules* **2008**, *41*, 2406–2414. [CrossRef]

42. Alb, A.M.; Farinato, R.; Calbick, J.; Reed, W.F. Online monitoring of polymerization reactions in inverse emulsions. *Langmuir* **2006**, *22*, 831–840. [CrossRef] [PubMed]

43. Kreft, T.; Reed, W.F. Predictive control and verification of conversion kinetics and polymer molecular weight in semi-batch free radical homopolymer reactions. *Eur. Polym. J.* **2009**, *45*, 2288–2303. [CrossRef]

44. Grassl, B.; Reed, W.F. Online polymerization monitoring in a continuous reactor. *Macromol. Chem. Phys.* **2002**, *203*, 586–597. [CrossRef]

45. Paril, A.; Alb, A.M.; Reed, W.F. Online monitoring of the evolution of polyelectrolyte characteristics during postpolymerization modification processes. *Macromolecules* **2007**, *40*, 4409–4413. [CrossRef]

46. Mignard, E.; Leblanc, T.; Bertin, D.; Guerret, O.; Reed, W.F. Online monitoring of controlled radical polymerization: Nitroxide-mediated gradient copolymerization. *Macromolecules* **2004**, *37*, 966–975. [CrossRef]

47. Mignard, E.; Lutz, J.-F.; Leblanc, T.; Matyjaszewski, K.; Guerret, O.; Reed, W.F. Kinetics and molar mass evolution during atom transfer radical polymerization of *n*-butyl acrylate using automatic continuous online monitoring. *Macromolecules* **2005**, *38*, 9556–9563. [CrossRef]

48. Alb, A.M.; Serelis, A.K.; Reed, W.F. Kinetic trends in raft homopolymerization from online monitoring. *Macromolecules* **2008**, *41*, 332–338. [CrossRef]

49. Alb, A.M.; Enohnyaket, P.; Craymer, J.F.; Eren, T.; Coughlin, E.B.; Reed, W.F. Online monitoring of ring-opening metathesis polymerization of cyclooctadiene and a functionalized norbornene. *Macromolecules* **2007**, *40*, 444–451. [CrossRef]

50. McAfee, T.; Leonardi, N.; Montgomery, R.; Siqueira, J.; Zekoski, T.; Drenski, M.F.; Reed, W.F. Automatic control of polymer molecular weight during synthesis. *Macromolecules* **2016**, *49*, 7170–7183. [CrossRef]

51. Alb, A.M.; Drenski, M.F.; Reed, W.F. Simultaneous continuous, nonchromatographic monitoring and discrete chromatographic monitoring of polymerization reactions. *J. Appl. Polym. Sci.* **2009**, *113*, 190–198. [CrossRef]

52. McFaul, C.A.; Alb, A.M.; Drenski, M.F.; Reed, W.F. Simultaneous multiple sample light scattering detection of LCST during copolymer synthesis. *Polymer* **2011**, *52*, 4825–4833. [CrossRef]

53. McFaul, C.A.; Drenski, M.F.; Reed, W.F. Online, continuous monitoring of the sensitivity of the LCST of nipam-Am copolymers to discrete and broad composition distributions. *Polymer* **2014**, *55*, 4899–4907. [CrossRef]

54. Huggins, M.L. The viscosity of dilute solutions of long-chain molecules. *J. Am. Chem. Soc.* **1942**, *64*, 2716–2718. [CrossRef]
55. Zimm, B.; Stein, R.; Doty, P. Polymer bull. 1 (1945) 90; bh zimm. *J. Chem. Phys.* **1948**, *16*, 1093. [CrossRef]
56. Hansen, J.P.; McDonald, I.R. *Theory of Simple Liquids*; Acdamic Press: New York, NY, USA, 1986.

*processes*

MDPI

*Article*

# Biodegradable and Biocompatible $P_{DL}LA$-$PEG_{1k}$-$P_{DL}LA$ Diacrylate Macromers: Synthesis, Characterisation and Preparation of Soluble Hyperbranched Polymers and Crosslinked Hydrogels

**Alan Hughes [1], Hongyun Tai [1,\*], Anna Tochwin [1] and Wenxin Wang [2]**

[1]   School of Chemistry, Bangor University, Deiniol Road, Bangor, Gwynedd LL57 2UW, UK; alan.hughes@bangor.ac.uk (A.H.); annatoffi.at@gmail.com (A.T.)
[2]   Charles Institute of Dermatology, University College Dublin, Dublin 4, Ireland; wenxin.wang@ucd.ie
\*   Correspondence: h.tai@bangor.ac.uk; Tel.: +44-(0)1248-382383; Fax: +44-(0)1248-370528

Academic Editor: Alexander Penlidis
Received: 5 March 2017; Accepted: 17 April 2017; Published: 20 April 2017

**Abstract:** A series of $P_{DL}LA$-$PEG_{1k}$-$P_{DL}LA$ tri-block co-polymers with various compositions, i.e., containing 2–10 lactoyl units, were prepared via ring opening polymerisation of D,L-lactide in the presence of poly (ethylene glycol) (PEG) ($M_n$ = 1000 g·mol$^{-1}$) as the initiator and stannous 2-ethylhexanoate as the catalyst at different feed ratios. $P_{DL}LA$-$PEG_{1k}$-$P_{DL}LA$ co-polymers were then functionalised with acrylate groups using acryloyl chloride under various reaction conditions. The diacrylated $P_{DL}LA$-$PEG_{1k}$-$P_{DL}LA$ (diacryl-$P_{DL}LA$-$PEG_{1k}$-$P_{DL}LA$) were further polymerised to synthesize soluble hyperbranched polymers by either homo-polymerisation or co-polymerisation with poly(ethylene glycol) methyl ether methylacrylate (PEGMEMA) via free radical polymerisation. The polymer samples obtained were characterised by $^1$H NMR (proton Nuclear Magnetic Resonance), FTIR (Fourier Transform Infra-red spectroscopy), and GPC (Gel Permeation Chromatography). Moreover, the diacryl-$P_{DL}LA$-$PEG_{1k}$-$P_{DL}LA$ macromers were used for the preparation of biodegradable crosslinked hydrogels through the Michael addition reaction and radical photo-polymerisation with or without poly(ethylene glycol) methyl ether methylacrylate (PEGMEMA, $M_n$ = 475 g·mol$^{-1}$) as the co-monomer. It was found that fine tuning of the diacryl-$P_{DL}LA$-$PEG_{1k}$-$P_{DL}LA$ constituents and its combination with co-monomers resulted in hydrogels with tailored swelling properties. It is envisioned that soluble hyperbranched polymers and crosslinked hydrogels prepared from diacryl-$P_{DL}LA$-$PEG_{1k}$-$P_{DL}LA$ macromers can have promising applications in the fields of nano-medicines and regenerative medicines.

**Keywords:** poly (ethylene glycol); D,L-lactide; macromers; triblock co-polymers; hyperbranched polymers; biodegradable hydrogels

## 1. Introduction

Hydrogels are cross-linked polymer networks that have a high affinity for water but do not dissolve in it, thus retaining their three dimensional structures [1]. Hydrogels have very wide applicability in the biomedical and pharmaceutical fields including tissue engineering [2–4], diagnostics, and drug delivery [5,6]. Hydrogels are biocompatible because they cause minimal tissue irritation with little cell adherence when in contact with the extracellular matrix due to their hydrophilicity and soft nature [7]. In tissue engineering and regenerative medicine, there are many advantages if hydrogels can form in situ; for example, an injectable macromer system can occupy an irregular defect and then crosslink to attain a permanent feature. This would be desirable, particularly, if cells or molecular signals could be incorporated in the injection mixture [8]. Solidification of materials in situ can then be achieved by either physical or chemical means [9,10].

Hydrogels based on poly (ethylene glycol) (PEG) and polyester block co-polymers are especially useful as many are known to be temperature-sensitive in aqueous solutions as they undergo a phase transition from the sol to the gel state at higher temperatures [11–13]. PEG is a synthetic polymer and is soluble in water, non-toxic, and can be eliminated from the body depending on its size. Incorporation of lactate segments into the PEG polymer backbone can introduce biodegradability thus facilitating degradation of the materials in vivo [14,15]. Various polyesters have been used to form the hydrophobic blocks to PEG based polymers, including poly(lactide) (PLA) [16], poly($\varepsilon$-caprolactone) (PCL) [17], and poly(glycolic-co-lactic acid) (PLGA) [18], via ring opening polymerisation of lactide, $\varepsilon$-caprolactone, or glycolide with PEG, respectively.

Co-polymers based on PEG and PLA are of interest because of their biocompatibility and biodegradability [19]. PLA–PEG block co-polymers have been synthesized by various means including the ring opening polymerisation of lactide in the presence of PEG. Deng et al. first reported the use of ring opening catalysts in the co-polymerisation of D,L-lactide with PEG [20]. The use of stannous 2-ethylhexanoate as a catalyst at 180 °C through bulk polymerisation was first reported by Zhu [21]. These co-polymers have a tri-block nature. Such co-polymers using PEG with different chain lengths have been reported and used to prepare biodegradable hydrogels. It appears that PEG-PLA tri-block co-polymers with a short domain of PEG at molecular weights below 6000 g·mol$^{-1}$ are rarely studied. Co-polymers with varying segment lengths and distributions have been synthesized and it is found that their properties were influenced by these variations [22]. Lee et al. prepared P$_L$LA–PEG multi-block co-polymers [12] with long blocks of PEG ($M_n$ = 2000–10,000 g·mol$^{-1}$) and P$_L$LA ($M_n$ = 2000–4500 g·mol$^{-1}$). Luo made P$_L$LA–PEG multi-block co-polymers with shorter PEG ($M_n$ = 600, 2000 g·mol$^{-1}$) and PLA ($M_n$ = 1000, 2000, 3000 g·mol$^{-1}$) [23]. To enable P$_{DL}$LA-PEG-P$_{DL}$LA co-polymers to undergo free radical polymerisation, they were further functionalized with vinyl groups at the terminal ends. Sawhney and Hubbell first described the synthesis of synthetic co-polymers which are hydrolytically degradable and cross-linkable consisting of a hydrophilic PEG central domain which was then co-polymerised with poly(lactic acid) (PLA) and end-capped with acrylate groups [14]. Since then a variety of such macromers have been created in which PEG is modified with various hydrolytically degradable ester moieties and terminal acrylate/methacrylate groups [24]. Varying the PEG molecular weight was found to influence the degree of gel swelling and other mechanical properties including degradation. Hubbell further prepared hydrogels by crosslinking acrylated multi-arm PEGs with thiol compounds via a Michael-type addition [25].

In this work, we prepared tri-block P$_{DL}$LA-PEG-P$_{DL}$LA diacrylate macromers with a relatively short PEG length ($M_n$ = 1000 g·mol$^{-1}$), which were used in further polymerisation with or without other PEG based co-monomers via free radical polymerisation to prepare both soluble hyperbranched polymers and crosslinked biodegradable hydrogels. Generally, other reported syntheses have been performed using relatively large PLA fractions to primarily change the hydrophobic/hydrophilic balance [26]. This study also gives details of the swelling behavior of the prepared hydrogels. The swelling curves show a reduction in the swelling ratio due to degradation after maximum swelling is achieved and is seen to commence first in the material with the highest lactoyl content. The degradation is hypothesized to be a result of hydrolysis of the lactoyl domains in the material. The low critical solution temperatures (LCSTs) of these co-polymers and its derived macromer materials can be tailored to be on the order of physiological temperatures.

The polymer samples (linear block and hyperbranched) obtained were characterised by $^1$H NMR, Gel Permeation Chromatography (GPC), and FTIR. Moreoever, the diacryl-P$_{DL}$LA-PEG-P$_{DL}$LA marcromers were utilized to prepare crosslinked biodegradable hydrogels. The experiments demonstrated the challenges of working with these materials due to the unstable nature of ester functional groups within the P$_{DL}$LA-PEG-P$_{DL}$LA co-polymers and the macromers. Photo-polymerisation and the Michael addition reaction were used for the preparation of crosslinked hydrogels. To the best of our knowledge, this is the first report on the study of such macromers with PEG 1000 (g·mol$^{-1}$) as the core and short poly-lactoyl terminal domains conferring on the co-polymer an amphiphilic character of varying degree. Soluble hyperbranched polymers and crosslinked hydrogels prepared from the diacryl-P$_{DL}$LA-PEG-P$_{DL}$LA macromers can be used as nanocarriers or depot systems for drug delivery [27] and tissue engineering applications [28].

## 2. Results and Discussion

In this study, the molecular weight of the hydrophilic PEG was chosen as 1000 g·mol$^{-1}$, to which hydrophobic lactoyl terminal domains were added to both ends of the PEG chain. Diacryl-P$_{DL}$LA-PEG$_{1k}$-P$_{DL}$LA macromers were prepared in two steps (Scheme 1). The first step was the co-polymerisation of PEG and D,L-Lactide. The co-polymer was subsequently acrylated using acryloyl chloride to produce the macromer (Figure S1 in the Supplementary Materials). An alternative method was also attempted by using a one-pot two-stage procedure, in which the P$_{DL}$LA-PEG$_{1k}$-P$_{DL}$LA co-polymer was formed as an intermediate and the second acrylation step, without separation and purification of the P$_{DL}$LA-PEG$_{1k}$-P$_{DL}$LA co-polymer from the reaction mixture, was then performed. Both methods demonstrated a similar yield and controllability, thus the one-pot method was adopted (Table 1).

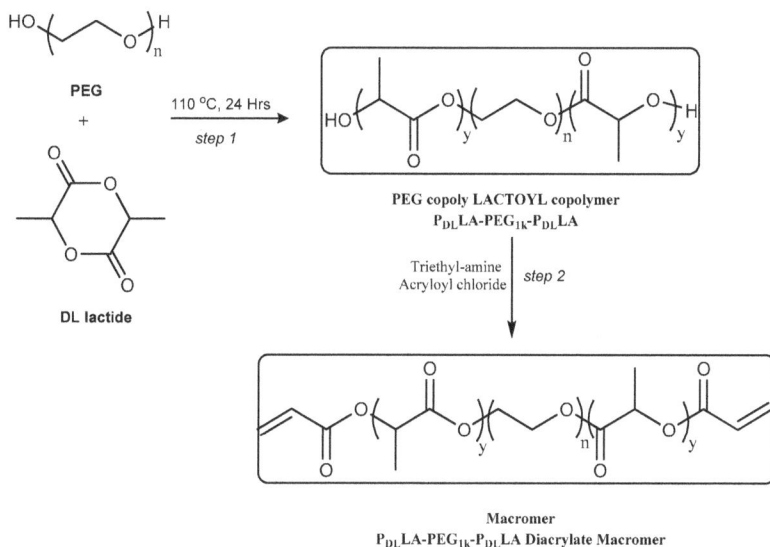

**Scheme 1.** Two-step synthesis of diacryl-P$_{DL}$LA-PEG$_{1k}$-P$_{DL}$LA macromers.

**Step 1: Synthesis of P$_{DL}$LA-PEG$_{1k}$-P$_{DL}$LA co-polymers.** Telechelic P$_{DL}$LA-PEG$_{1k}$-P$_{DL}$LA co-polymers were synthesised via a ring opening polymerisation (ROP) of the cyclic-diester monomer D,L-lactide. The ROP was initiated by the $\alpha$ and $\omega$ hydroxyl terminal groups of the PEG catalysed by stannous 2-ethylhexanoate and consisted of the step-wise addition of the lactide. It was important to eliminate any water from the reaction mixture to avoid hydrolysis of the products. PEG and D,L-lactide were dried as described in the experimental section. A range of P$_{DL}$LA-PEG$_{1k}$-P$_{DL}$LA co-polymers were prepared by varying the ratio of D,L-lactide to PEG (Table 1).

The degree of polymerisation was determined by $^1$H NMR (see Figures S2 and S3 in the Supplementary Materials) and it was found to be dependent on the feed molar ratio of the reactants, i.e., the PEG and the D,L-lactide (see Table 1). Typically, the PEG signal –CH$_2$–CH$_2$–O– was found at δ = 3.65, and the methyl signal for the lactoyl group was found at δ = 1.6, thus for Table 1 entry 6 the integrations were found to be 1.48 and 1.0, respectively. Each lactoyl methyl residue contains three protons and if we assume equi-molar addition at both ends of the PEG block, the total number of protons for the lactoyl signal is equal to 6 $m$ where $m$ is the number of lactoyl residues on each end of the co-polymer. Note that PEG $M_n$ = 1000 g·mol$^{-1}$ consists of, on average, $n$ = 22.32 repeating residues (–O–CH$_2$–CH$_2$) with an additional H and an –OH at each terminal of the linear molecule, i.e., an additional 18 g·mol$^{-1}$. Therefore, the number of lactoyl units in the co-polymer, $m$, can be calculated by Equation (1):

$$m = \frac{4 \times 22.32 \times 1.0}{6 \times 1.48} = 10.05 \tag{1}$$

The above co-polymer was named 1KL10, i.e., 10 lactoyl residues on each end of the PEG domain. The PEG/Lactoyl ratios in the co-polymers were generally found to be lower than the ratio in the feed. This was in part due to the loss of any residual water from the pre-weighed PEG. Also the conversion of lactide was not complete as unreacted lactide vaporised on the cooler parts of the reaction vessel and thus was effectively removed from the reaction mixture and subsequently removed during purification. If PEG of higher molecular weight is used, the relative hydroxyl content will decline as there are two moles of hydroxyl groups per mole of linear PEG. The present study used PEG of molecular weight of 1000 (g·mol$^{-1}$) which is relatively low compared to most other studies involving this reaction. Thus, with increasing molar mass of PEG for a given weight of PEG there are fewer hydroxyl end groups, which function as initiation sites with the consequent lowering of the conversion ratio of the lactide under the same reaction conditions. The experimental entries show that longer reaction time reduced the lactoyl content in the co-polymer. For example, comparing entries 6 and 7 (in Table 1), the longer reaction time resulted in lower lactoyl content despite otherwise similar conditions. This may be result of cleavage of the lactoyl chain under these conditions. It should also be noticed from Table 1 that the co-polymer yield was variable and often low, as is illustrated in entries 5 and 6 (Table 1), with a great disparity in yield between the two similar products. This was possibly due to the failure to obtain optimal precipitation during the extraction step when a fine suspension resulted which could not be isolated by either filtration or centrifugation. Moreover, P$_{DL}$LA-PEG$_{1k}$-P$_{DL}$LA co-polymers demonstrated poor water solubility (Table 1) due to the incorporation of hydrophobic PLA with hydrophilic PEG. As the mass fraction of the PLA in the co-polymer increases, its water solubility decreases. We can see (Table 1) that the co-polymer 1KL3 was soluble whereas those with higher lactide content were insoluble. These results are in agreement with the findings of Sawhney [14]. In general, a high PEG/PLA ratio and a low molar mass confer water solubility. After synthesis via ROP, the P$_{DL}$LA-PEG$_{1k}$-P$_{DL}$LA products were recovered, either by dissolution in dichloromethane and precipitation in anhydrous ether which was then dried under reduced pressure after filtration, or by dissolving the co-polymer mixture in ice-cold water and then gradually increasing its temperature above to its cloud point at about 52 °C at which the co-polymer precipitated out of the solution.

**Step 2: Diacryl-P$_{DL}$LA-PEG$_{1k}$-P$_{DL}$LA Macromers**. The P$_{DL}$LA-PEG$_{1k}$-P$_{DL}$LA co-polymer, with both α and ω hydroxyl end-groups, was end-capped with acrylate groups which could then be further used for cross linking. The degree of acrylation was determined by $^1$H NMR (see Figure S3 in Supplementary Materials). Vinyl proton NMR signals typically appeared at δ = 5.9, 6.2, and 6.4 ppm. Thus, for entry 3 (Table 2), where the integration of the acrylate signal is 0.04 and for the methyl is 1, the total number of vinyl protons per mole of the macromer was determined as $x$ where

$$x = \frac{0.04 \times 89.28}{1} = 3.57 \tag{2}$$

As the vinyl groups each comprise three protons at each end of the macromer chain (i.e., a total of six per molecule), the given the macromer is 59.5% acrylated on average. This would translate to a value for the maximum molecular weight for the di-acrylated macromer of $M_n$ = 2549 g·mol$^{-1}$ (i.e., for each di-acrylated molecule).

The quantity of DCM solvent, i.e., the concentration of the reactants is critical to the rate of acrylation. Too much DCM will re-dissolve the triethylamine hydrochloride salt. It was noted that the precipitate of the salt correlated with high acrylation. It is presumed that if the solvent for the acrylation is in excess, this reduces the rate of the acrylation reaction by reducing the concentration of the reactants. Thus a minimal amount of solvent is used, sufficient for dissolving the co-polymer.

For the macromer 1KL3 (Table 2, entries 11 and 13), it can be seen that the greater degree of acrylation occurs for entry 13 (80.3%) with a lower acryloyl chloride ratio of 1:1.2:2.5 compared to entry 11 (14.9%) synthesised with a ratio 1:4:4. This situation is reversed with macromer 1KL8 with its relatively higher lactoyl content. Thus, for entries 5 and 10, the entry 5 (acrylation 33.1%) was synthesised with a lower acryloyl chloride ratio of 1:1.2:2.5 (cf. 75.7% for entry 10). However, excessive ratios should be avoided because of the possibility of causing cleavage of the polymer chain especially as the reaction is exothermic and because of the need to maintain the reaction at 0 °C at least at the early stage. To ensure that the highest possible conversion to acrylate was achieved, the reaction was allowed to run for 24 h in accordance with the procedure outlined by Zhang et al. [29]. The ice temperature affected the product quality and if not kept low, a deep yellow colour resulted from the exothermic reaction between the acryloyl chloride and the diol groups of the co-polymer.

The $P_{DL}LA$-$PEG_{1k}$-$P_{DL}LA$ Co-Polymer and Diacryl-$P_{DL}LA$-$PEG_{1k}$-$P_{DL}LA$ Macromer showed thermoresponsive behaviors. Varying the ratios of the hydrophilic/hydrophobic domains within the co-polymer results in changing the phase transition temperature in the aqueous solution; the more hydrophilic the co-polymer, the higher the LCST. The LCST for macromer 1KL10 (Mac-1KL10) and co-polymer 1KL10 (Co-1KL10), respectively, were determined. The Mac-1KL10 and the Co-1KL10 were dissolved in deionized water, and the change in absorbency with increasing temperature was measured at a wavelength of 550 nm using UV-Vis spectrophotometry (Figure 1). The LCST was found to be around 34 °C and 40 °C, respectively. These findings are consistent with LCSTs determined by visual observation when the temperatures for this transition were found to be 27 °C and 32 °C, respectively.

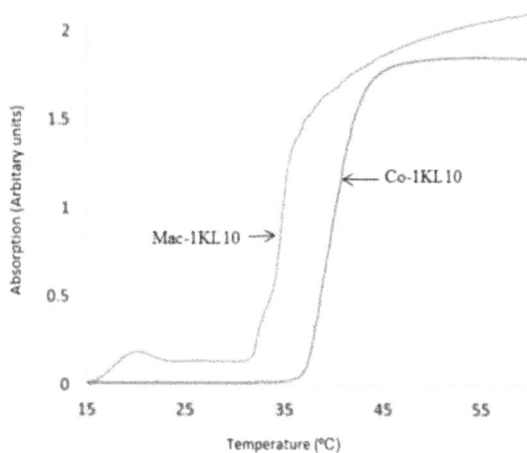

**Figure 1.** Low critical solution temperatures are determined by UV absorption curves for $P_{DL}LA$-$PEG_{1k}$-$P_{DL}LA$ 1KL10 (Co-1KL10) and diacryl-$P_{DL}LA$-$PEG_{1k}$-$P_{DL}LA$ (Mac-1KL10).

Moreover, the FTIR spectra for the co-polymer and its macromer (see Figure S4 in the Supplementary Materials) show strong absorption at 3510 cm$^{-1}$ for the PEG precursor due to the terminal hydroxyl group and this signal is reduced due to acrylation although not eliminated. A strong absorption at 1756 cm$^{-1}$ for the 1KL3 confirms the presence of the ester due to the lactoyl moieties. A weak signal for the –C=C– in the region 1680–1640 cm$^{-1}$ can also be seen.

Table 1. Experimental results and reaction conditions for the synthesis of $P_{DL}LA$-$PEG_{1K}$-$P_{DL}LA$ co-polymers.

| Entry | PEG:Lactide Feed Molar Ratio | Co-Polymer Composition [a] | $M_n$ [b] (g·mol$^{-1}$) | Yield (%) | Temp. [c] (°C) | RT [d] (h) | Solubility [e] (in Water) | Appearance |
|---|---|---|---|---|---|---|---|---|
| 1 | 1:8 | 1KL6.8 | 1980 | 49.9 | 110 | 24 | insoluble | slight yellow, viscous |
| 2 | 1:8 | 1KL7.6 | 2100 | 30.4 | 110 | 24 | insoluble | yellow, viscous |
| 3 | 1:8 | 1KL6.0 | 1870 | 49.3 | 110 | 14 | insoluble | yellow, viscous |
| 4 | 1:9 | 1KL9.0 | 2300 | - | 110 | 24 | insoluble | slight yellow, semi-solid |
| 5 | 1:8 | 1KL9.9 | 2430 | 30.1 | 130 | 17 | insoluble | slight yellow, semi-solid |
| 6 | 1:8 | 1KL10.2 | 2470 | 91.7 | 130 | 17 | insoluble | slight yellow, viscous |
| 7 | 1:8 | 1KL7.3 | 2050 | 53.5 | 130 | 20 | insoluble | brown, viscous |
| 8 | 1:9 | 1KL6.7 | 1960 | 55.7 | 130 | 11 | insoluble | slight yellow, viscous |
| 9 | 1:8 | 1KL7.5 | 2080 | 12.0 | 130 | 20.5 | insoluble | slight yellow, semi-solid |
| 10 | 1:8 | 1KL9.0 | 2300 | 20.4 | 130 | 24 | insoluble | yellow, semi- solid |
| 11 | 1:3 | 1KL2.7 | 1390 | - | 130 | 24 | soluble | yellow, viscous |

[a] $P_{DL}LA$-$PEG_{1K}$-$P_{DL}LA$ Co-polymer composition, determined by $^1H$ NMR; [b] Molecular Weight determined by $^1H$ NMR; [c] Reaction Temperature; [d] Reaction Time; [e] Determined by the addition of 100 mg to 500 µL of distilled water at room temperature.

Table 2. Experimental results and reaction conditions for acrylations of the co-polymers.

| Entry | Co-Polymer | $M_n$ [a] (g·mol$^{-1}$) | Mole Ratio of Reactants [b] | Acryl. [c] (%) | Macromer Conc. [d] (g/L) | RT [e] (h) | Precipitate Formed [f] | Appearance of Acrylated Co-Polymers |
|---|---|---|---|---|---|---|---|---|
| 1 | 1KL6 | 1920 | 1:1.2:1.7 | 36.3 | 0.055 | 21 | No | white, viscous |
| 2 | 1KL7 | 2050 | 1:2.1:4.4 | 28.6 | 0.1313 | 24 | No | yellow semi-viscous |
| 3 | 1KL10 | 2550 | 1:2.5:4.3 | 59.5 | 0.22 | 24 | Yes | yellow solid |
| 4 | 1KL10 | 2510 | 1:2.5:4.3 | 78.5 | 0.299 | 24 | Yes | yellow solid |
| 5 | 1KL8 | 2200 | 1:1.2:2.5 | 33.1 | 0.06 | 17 | No | white semi-viscous |
| 6 | 1KL8 | 2160 | 1:1.2:2.5 | 7.5 | 0.058 | 24 | No | white semi-viscous |
| 7 | 1KL8 | 2150 | 1:1.2:2.5 | 0.1 | 0.05 | 24 | No | white semi-viscous |
| 8 | 1KL9 | 2310 | 1:1.2:2.5 | 0.9 | 0.05 | 24 | Yes | pale yellow semi-viscous |
| 9 | 1KL8 | 2250 | 1:4:4 | 70.0 | 0.03 | 22 | No | slight yellow solid |
| 10 | 1KL8 | 2260 | 1:4:4 | 75.7 | 0.1 | 22 | Yes | slight yellow solid |
| 11 | 1KL3 | 1450 | 1:4:4 | 14.9 | 0.069 | 24 | Yes | white waxy solid |
| 12 | 1KL6 | 1930 | 1:1.2:2.5 | 53.8 | 0.057 | 24 | Yes | white waxy solid |
| 13 | 1KL3 | 1440 | 1:1.2:2.5 | 80.3 | 0.08 | 24 | Yes | white waxy solid |

[a] $M_n$ determined by $^1H$ NMR; [b] Mole ratio is the ratio of –OH:Triethylamine:Acryloyl chloride; [c] Acrylation content, calculated by $^1H$ NMR; [d] Macromer concentration; [e] Reaction Time (hours); [f] Observation on the formation of precipitates during acrylation.

## 3. Preparation of Soluble Hyperbranched Polymers from Free Radical Polymerisation (FRP) of Diacryl-$P_{DL}LA$-$PEG_{1k}$-$P_{DL}LA$ Macromer in the Presence and Absence of Co-Monomer PEGMEMA

To prepare branched polymers, free radical polymerisations were performed using the diacryl-$P_{DL}LA$-$PEG_{1k}$-$P_{DL}LA$ macromer alone (homo-polymerisation, entries 1 and 2 in Table 3) and also with another co-monomer, i.e., PEGMEMA (co-polymerisation, entries 3–5 in Table 3).

Polymerisation was found to occur in both DMF and chloroform (entries 1 and 2 in Table 3). However, the conversion was higher in DMF and the PDI was given as 1.03 which indicates a controlled polymerisation mechanism under the reaction conditions used. GPC revealed that higher molecular weights were obtained in chloroform but the conversion was lower given that the temperature and reaction time were comparable. This difference may reflect the difference in solubility and polarity between the two solvents as it is observed that the chloroform was found to be a better solvent but the DMF yielded a slightly translucent solution. The co-polymerisations with PEGMEMA (entries 3–5 in Table 3) were observed, evidenced by the GPC data, however, the molecular weights obtained were not as high as for the homo-polymerisations of the macromer (see entries 2 and 3) when the reactions were undertaken in the same solvent and under similar experimental conditions. If the concentration was increased, this caused a lower molecular weight product to be synthesised despite doubling the reaction time (see entries 4 and 5 in Table 3). It was also noted that all polymerisations were to a greater or lesser degree conducted under heterogeneous conditions. Entries 4 and 5 in Table 3 were conducted in THF and the monomer to solvent ratios was generally lower, reflecting better solubility than those entries using DMF and chloroform.

**Table 3.** Free Radical Polymerisation of diacryl-$P_{DL}LA$-$PEG_{1k}$-$P_{DL}LA$ Macromer and poly(ethylene glycol) methyl ether methylacrylate (PEGMEMA).

| Entry | Macromer | $M/S$ [a] | $M/P$ [b] | Solvent | $M_w$ [c] (KDa) | PDI [d] | Conv. [e] (%) | Gel [e] | Temp. (°C) [f] | RT [g] (h) |
|-------|----------|-----------|-----------|---------|------------------|---------|---------------|---------|----------------|------------|
| 1 | 1KL3 | 1:3 | 1:0 | DMF | 224 | 1.03 | 29.3 | Yes | 65 | 23 |
| 2 | 1KL3 | 1:3 | 1:0 | Chloroform | 551 | 2.0 | 19.4 | Yes | 55 | 20 |
| 3 | 1KL3 | 1:3 | 1:1 | Chloroform | 269 | 1.04 | 13.4 | Yes | 55 | 20 |
| 4 | 1KL10 | 1:1.5 | 1:9 | THF | 47 | 1.49 | 1.04 | Yes | 50 | 21 |
| 5 | 1KL11 | 1:0.5 | 1:9 | THF | 37 | 1.44 | 5.00 | Turbid | 50 | 44 |

[a] Weight/Volume Ratio of Total Monomers to solvent; [b] Molar Ratio of Macromer to PEGMEMA; [c] Weight average molecular weight, determined by Gel Permeation Chromatography (GPC); [d] Polydispersity Index PDI, determined by GPC; [e] Gelation observed by inspection upon termination of the reaction; [f] Reaction temperature; [g] Reaction time (hours).

The polydispersity (PDI) of the branched polymers obtained from free radical polymerisation of the diacryl-$P_{DL}LA$-$PEG_{1k}$-$P_{DL}LA$ macromer with or without PEGMEMA were generally fairly narrow (less than 2, see Table 3). This may be due to the fact that the GPC samples were obtained upon gelation and, therefore, represent a soluble extract of the reaction mixture. Figure 2 gives a GPC overlay of the FRP homo-polymerisation of Mac-1KL3 (entry 2 in Table 3) as a comparison with its precursor, the co-polymer 1KL3 (Co-1KL3).

These experimental results show that soluble (most likely hyperbranched) polymers can be synthesized by free radical polymerisation of the diacryl-$P_{DL}LA$-$PEG_{1k}$-$P_{DL}LA$ macromer at a relatively low yield (less than 30%), before gelation occurs. However, it was found to be challenging using the diacryl-$P_{DL}LA$-$PEG_{1k}$-$P_{DL}LA$ macromer as a multifunctional monomer to prepare soluble hyperbranched polymers by radical polymerisations, including controlled radical polymerisation approaches such as atom transfer radical polymerisation and reversible addition-fragmentation chain transfer polymerisation. This is because the diacryl-$P_{DL}LA$-$PEG_{1k}$-$P_{DL}LA$ macromer is unstable, thus the polymerisation conditions should be carefully considered and chosen, including reaction temperature, initiators, and solvents. Nevertheless, the soluble hyperbranched polymers prepared from the diacryl-$P_{DL}LA$-$PEG_{1k}$-$P_{DL}LA$ macromers could have potential as nanocarriers to deliver drugs for nanomedicine applications or could be used to fabricate hydrogels for regenerative medicine applications [28].

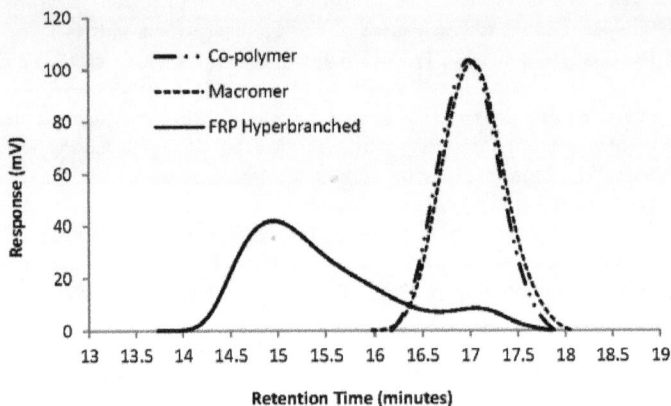

**Figure 2.** Homo-polymerisation of Mac-1KL3 (Entry 2 in Table 3) to produce the hyperbranched polymer. Overlay of GPC traces from the Refractive Index (RI) detector for the resulting hyperbranched polymer, Macromer, and precursor co-polymer, for comparison.

*3.1. Synthesis of Chemical Crosslinked Hydrogels from Diacryl-$P_{DL}LA$-$PEG_{1k}$-$P_{DL}LA$ Macromers via Michael-Addition Reaction*

The diacrylate macromer with its vinyl functionality is able to undergo a Michael addition type reaction when it is used with a suitable reagent containing thiol functional groups. The macromer should be capable of a homo-polymerisation when used on its own or possibly in the presence of another vinyl functional monomer such as the PEGMEMA monomer, where co-polymerisation should occur. Therefore, a number of reactions were performed (Table 4). It can be seen that the homo-polymerisation of the macromer occurred in all the solvents used, i.e., water, DMSO, and PBS (entries 1–3 and 6). However, the reactions with PEGMEMA were unsuccessful (entries 4, 7, and 8). The Michael addition requires the presence of a base to act as a catalyst. The basic conditions were supplied by the use of PBS buffer (pH 7.4), triethylamine, or sodium hydroxide. The reactions with the macromer only resulted in a white gel indicating that the reaction occurred. Co-polymerisation with PEGMEMA was also attempted but did not yield any evidence of reaction other than in the solvent THF. This could be explained by the lower reactivity of the methacrylate groups in PEGMEMA with thiol groups for undertaking the Michael addition reaction, compared to the acrylate groups in diacryl-$P_{DL}LA$-$PEG_{1k}$-$P_{DL}LA$ macromers.

**Table 4.** Synthesis of hydrogels from the Michael-addition of Macromer 1KL11.

| Entry | Mac-1KL11/PEGMEMA Weight Ratio | Solvent | Base | QT/M [a] | Gel [b] |
|-------|-------------------------------|---------|------|----------|---------|
| 1 | 1:0 | water | TEA | 1.2:1 | Yes |
| 2 | 1:0 | water | NaOH | 1.2:1 | yes |
| 3 | 1:0 | DMSO | TEA | 1.8:1 | Yes |
| 4 | 10:90 | DMSO | TEA | 1.3:1 | No |
| 5 | 10:90 | THF | TEA | 1.3:1 | Yes |
| 6 | 1:0 | PBS | PBS | 1.9:1 | Yes |
| 7 | 50:50 | PBS | TEA | 1.5:1 | No |
| 8 | 50:50 | DMSO | TEA | 1.5:1 | No |

[a] Mole ratio of the pentaerythritol tetrakis (3-mercaptopropionate) (QT) and the macromer (M); [b] Gel is defined as a gelation as observed by visual inspection and is unable to flow when the tube is inverted.

## 3.2. Photocrosslinked Hydrogels from Macromers: Synthesis, Swelling, and Degradation

The diacryl-P$_{DL}$LA-PEG$_{1k}$-P$_{DL}$LA macromer can be used to prepare hydrogels as either a homo-polymer or as a co-polymer, with for example PEGMEMA, which is also water soluble. With this in mind, a swelling study was undertaken on the hydrogels prepared from the photo-polymerisation of macromer 1KL3 and on hydrogels prepared from the co-polymerisation of macromer 1KL11 and PEGMEMA (Table 5).

Photo-polymerisation was conducted at 25 °C by irradiation with UV light. A 1% aqueous solution of the initiator Irgacure 2959 was prepared, and the macromer was then dissolved at different ratios of Macromer/PEGMEMA to the concentrations shown. The reaction mixture was then subjected to UV exposure. After a given exposure time the resulting solids were gently washed with de-ionised water and dried in a vacuum oven.

**Table 5.** Synthesis of hydrogels from the photo-polymerisation of diacryl-P$_{DL}$LA-PEG$_{1k}$-P$_{DL}$LA Macromers.

| Entry | Macromer | [Macromer]:[PEGMEMA] (Weight Percentage Ratio) | Concentration [a] (%) |
|-------|----------|-----------------------------------------------|-----------------------|
| 1 | 1KL3 | 100:0 | 30 |
| 2 | 1KL3 | 100:0 | 54 |
| 3 | 1KL11 | 100:0 | 50 |
| 4 | 1KL11 | 10:90 | 50 |
| 5 | 1KL11 | 5:95 | 50 |

[a] Concentration of the Macromer/PEGMEMA in distilled water ($w/w$ %)

In Figure 3, plots are shown of the Swelling Ratio vs. Time for the dried cross-linked macromer gel 1KL3 in water, and it is seen that the 30% hydrogel becomes more swollen than the 54% hydrogel. This may be a result of the hydrogel structure being more "open" as a consequence of a less crosslinked structure being formed when less macromer was used. An open structure with lower crosslinking density is likely to be more water absorbent and have a higher swelling ratio. Both curves have maximum swelling in approximately 3 days, and then they both start to degrade. This indicates that the 54% hydrogel is not capable of absorbing more water which would be the case if the material were merely slower in uptake due to its density of crosslinking. It is also observed that after being fully swollen, degradation occurs but this is more rapid for the 30% hydrogel which is again a consequence of its higher water content. The 30% hydrogel mass, with its fewer crosslinking points and consequent lower mechanical strength, is likely to disintegrate more rapidly if these links are broken as a result of hydrolysis.

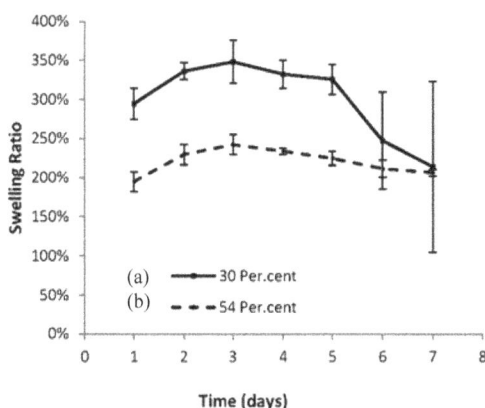

**Figure 3.** Swelling curves for Mac-1KL3 in water at room temperature (20 °C) (a) 30% (b) 54%. (The experiments were performed in triplicate and the swelling ratio at each time point was the average of three experimental data) (Entries 1 and 2 in Table 5).

Figure 4c shows the swelling curve for a 50% concentration for the UV cross-linked gel macromer 1KL11 (entry 3 in Table 5) which has a greater lactoyl component than that for macromer 1KL3 (entry 2 in Table 5). It is observed that 1KL11 has a higher swelling ratio which could be due to its low crosslinking density compared to 1KL3, although the hydrogel from 1KL11 has a higher lactoyl content which generally confers greater hydrophobicity to the macromer. This explanation is given more credence by the fact that 1KL3 undergoes a faster degradation than 1KL11 (entry 3 in Table 5) which is presumably due to the greater exposure of its lactoyl domains to water and therefore a greater likelihood of hydrolytic cleavage.

It may be seen from Figure 4 that both the macromer and the co-polymers of PEGMEMA absorbed water to achieve swelling. Of the co-polymers, Figure 4 shows that hydrogels with the higher PEGMEMA content (Figure 4a) had the highest swelling ratio, followed by the other co-polymer (Figure 4b), and finally the macromer only (Figure 4c) exhibited the least swelling. It is thought that the lower concentration of the macromer crosslinking agent in Figure 4a leads to a more open structure with a lower crosslinking density. This would allow for ready access of water to the hydrophilic regions of the cross linked structure, i.e., the PEG domains contained within the macromer and the PEGMEMA. This would give rise to the highest swelling ratio of the three materials. The homo-polymerised macromer (Figure 4c) formed hydrogels with the lowest swelling ratios. This, in line with the above argument, is due to its dense crosslinked structure which hinders water access to the interior structure. Conversely, the hydrophobic regions, due to the presence of the lactoyl domains within the macromer, are found in the highest ratio in Figure 4c and contribute to the lowest degree of swelling. The lactoyl content is reduced further in Figure 4b and is at its lowest value in Figure 4a, which has the highest swelling ratio.

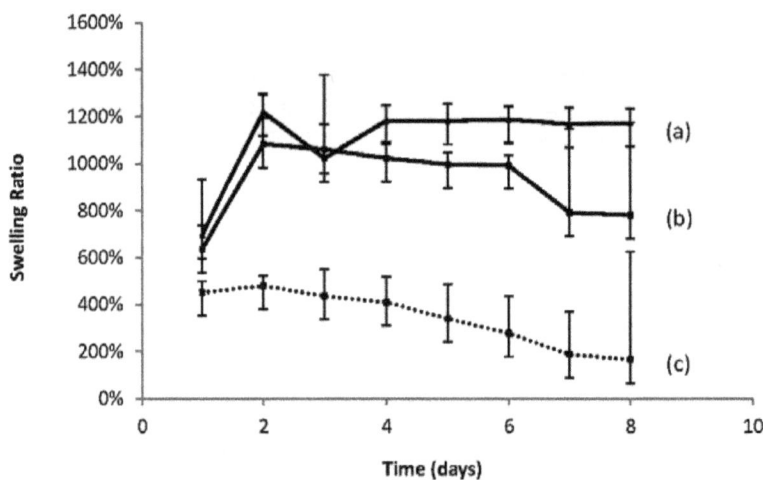

**Figure 4.** Swelling curves for the three hydrogels used in the swelling study, i.e., (**a**) Macromer: PEGMEMA = 5:95 (entry 5 in Table 5) ($w/w$); (**b**) Macromer: PEGMEMA = 10: 90 ($w/w$) (entry 4 in Table 5); (**c**) Macromer only (entry 3 in Table 5).

The swelling curves also show a reduction in the swelling ratio after approximately 2 days for all three entries, indicating the hydrogels begin to degrade after the swelling reaches a maximum. The degradation is due to the hydrolysis of the lactoyl domains in the material. The macromer with the highest lactoyl content (Figure 4c) is seen to first commence degradation after two days. The material with the lowest lactoyl content (Figure 4a) is seen to be the last to commence degradation.

## 4. Experimental (Materials and Methods)

### 4.1. Materials

Poly(ethylene glycol) (PEG) ($M_n$ = 1000 g·mol$^{-1}$) was obtained from Polysciences Inc. (Europe GmbH, Germany). The following chemicals were purchased from Sigma Aldrich and used as received: D,L-lactide (3,6-dimethyl-1,4-dioxane-2,5-dione), stannous 2-ethylhexanoate, triethylamine, acryloyl chloride, poly(ethylene glycol) methyl ether methylacrylate (PEGMEMA, $M_n$ = 475 g·mol$^{-1}$) (containing inhibitors 100 ppm MEHQ and 200 ppm BHT), 1,1′-azobis (cyclohexanecarbonitrile) (ACHN), Irgacure 2959 (2-hydroxy-4′-(2-hydroxyethoxy)-2-methylpropiophenone), lithium bromide (LiBr), dry IR-grade potassium bromide (KBr), pentaerythritol tetrakis (3-mercaptopropionate) (QT) cross-linker, deuterated chloroform (CDCl$_3$), dimethyl sulfoxide-d6 (DMSO-d6), tetrahydrofuran (THF), dichloromethane (DCM), toluene, and dimethylformamide (DMF).

### 4.2. Synthesis of $P_{DL}LA$-$PEG_{1k}$-$P_{DL}LA$ Co-Polymers

The PEG was dried either by heating under a stream of dry nitrogen gas with stirring at 150 °C for 3 h or was dried by azeotropic distillation with toluene. D,L-lactide was dried either by adding it to the dry melt of the PEG and heating for 30 min under a stream of nitrogen gas at 150 °C or was recrystallized in ethyl acetate before use. A typical procedure for the synthesis of the co-polymer e.g., 1KL2.7 (entry 11 in Table 1) is described below: a total of 30 g (30 mmol) of dried PEG was placed in a two neck flask and heated under a stream of nitrogen at 150 °C for 3 h with constant stirring to which 12.97 g (90 mmol) of D,L-lactide was then added and heating was continued for another 30 min under nitrogen to remove any water impurity. The stannous 2-ethylhexanoate (194 μL (0.6 mM)) was transferred to the reaction flask and the temperature reduced to 130 °C. The reaction was left to proceed for 24 h and was then terminated by turning off the heat and allowing to cool. A stock solution of stannous 2-ethylhexanoate was prepared to enable accurate delivery of the correct amount into the reaction vessel. After the reaction, the resulting co-polymer was then dissolved in dichloromethane, precipitated in anhydrous ether, then dried in a vacuum oven at room temperature. The PEG with its α,ω-dihydroxy end groups acted as a ring opening reagent to initiate the polymerisation of the D,L-lactide [14]. Other co-polymers were prepared by varying the ratio of D,L-lactide to PEG (see Table 1).

### 4.3. Synthesis of Diacryl-$P_{DL}LA$-$PEG_{1k}$-$P_{DL}LA$ Macromers

The above $P_{DL}LA$-$PEG_{1k}$-$P_{DL}LA$ co-polymers were end capped with acrylate moieties to introduce vinyl functionalities and thus form polymerisable diacryl-$P_{DL}LA$-$PEG_{1k}$-$P_{DL}LA$ macromers. In a typical procedure, a total of 12 g (8.38 mM) of $P_{DL}LA$-$PEG_{1k}$-$P_{DL}LA$ co-polymer was dissolved in 100 mL of dichloromethane in a 250 mL two neck round bottom reaction flask and cooled in an ice bath to 0 °C. Triethylamine, 2.8 mL (20.11 mM) was added to the mixture under constant stirring, and then 3.39 mL (41.90 mM) of acryloyl chloride was slowly added. The reaction mixture was maintained at 0 °C for several hours and allowed to continue at room temperature for another 12 h. The mixture was then filtered to remove the white precipitate of triethylamine hydrochloride and the macromer was separated by dropwise addition into a large excess of anhydrous diethyl ether. It was then redissolved in dichloromethane and re-precipitated out of a large excess of dry hexane and dried under vacuum at room temperature overnight. Alternatively, to separate out the macromer product, the mixture was dissolved in ice cold water (5–8 °C), and the resulting solution was then heated to 80 °C and caused the co-polymer to precipitate, thus leaving the unreacted monomers in the solution. The macromer was isolated by removing the supernatant liquid and was dried under vacuum at room temperature.

*4.4. Preparation of Soluble Hyperbranched Polymers Using Diacryl-P$_{DL}$LA-PEG$_{1k}$-P$_{DL}$LA Macromers via Free Radical Polymerisation*

Hyperbranched polymers were prepared by homo-polymerisation of the diacryl-P$_{DL}$LA-PEG$_{1k}$-P$_{DL}$LA macromer and also by the co-polymerisation of this macromer with PEGMEMA via free radical polymerisation (FRP). Typically, for the homo-polymerisation of 1KL10, 1.64 g (0.714 mmol) of macromer (prepared according to entry 3 in Table 2) was dissolved in 2 mL of DMF to which 0.004 g (0.24 wt %) of the initiator ACHN was then added. The mixture was purged with nitrogen for 15 min to remove any dissolved oxygen. The mixture was then heated to 65 °C and GPC samples were taken at suitable intervals. For the co-polymerisation of the macromer and PEGMEMA, the molar ratio of macromer to PEGMEMA was varied according to Table 3. Samples were withdrawn at required intervals for GPC analysis.

*4.5. Characterisations of P$_{DL}$LA-PEG$_{1k}$-P$_{DL}$LA Co-Polymers, Diacryl-P$_{DL}$LA-PEG$_{1k}$-P$_{DL}$LA Macromers, and Hyperbranched Polymers*

The co-polymer and macromer structures were determined by $^1$H NMR spectroscopy. The spectra were obtained on a Brucker 500 MHz NMR and analysed using MestReNova-Lite software (Version 11.0). Deuterated chloroform (CDCl$_3$) with tetramethylsilane (TMS) as an internal standard was used as the solvent. Gel permeation chromatography (GPC) was used to determine the size of the macromers and polymers as this method separates analytes on the basis of size/hydrodynamic volume. Chromatograms were recorded on a PL-GPC 50 Plus Integrated GPC/SEC System from Agilent Technologies. The number average molecular weight ($M_n$), the weight average molecular weight ($M_w$), and the polydispersity index ($M_w/M_n$) were determined using an RI (refractive index) detector. The columns (PLgel Mixed-C column 300 mm in length, two in series) were eluted using THF and calibrated with poly (methyl methacrylate) standard. All calibrations and analyses were performed at a flow rate of 1 mL/min at 40 °C. FTIR spectra were recorded on a PerkinElmer Spectra 100 FTIR Spectrometer. Samples were cast as thin films on sodium chloride disks from a chloroform solution or were prepared as KBr disks. Thermal responsive properties of the polymers were studied by measuring the low critical solution temperatures (LCSTs) of the polymers in aqueous solutions (0.1 $w/v$ % concentration in de-ionised water) using a Cary 100 UV-Vis spectrophotometer. The parameters for the measurements used were the heating rate as 1 °C/min, data collection rate as 0.06 °C/point, absorbance wavelength at 550 nm, and the temperature range between 15–60 °C.

*4.6. Preparation of Biodegradable Hydrogels from Linear Diacryl-P$_{DL}$LA-PEG-P$_{DL}$LA Macromers*

Biodegradable hydrogels were prepared using the diacryl-P$_{DL}$LA-PEG$_{1k}$-P$_{DL}$LA macromer by homo-polymerisation and co-polymerisation with PEGMEMA via the Michael addition reaction and free radical photopolymerisation.

4.6.1. Michael Addition Method

Chemically cross-linked hydrogels were prepared using diacryl-P$_{DL}$LA-PEG$_{1k}$-P$_{DL}$LA macromers and PEGMEMA by means of the Michael-addition reaction between their acrylate groups and the thiol functional groups in QT cross-linker (Table 4). A typical homo-polymerisation involves the addition of 0.5 g (0.019 mmol) of macromer ($M_n$ = 2693.43 g·mol$^{-1}$) to a vial containing 500 μL of phosphate buffered solution (pH 7.4). Then 171.6 μL of QT was added in a stoichiometric molar ratio of vinyl group to thiol of 1:1. Triethylamine (TEA, 42.13 μL) was then added, and the sample was incubated at 37 °C for 2 h. Upon completion of the reaction, a transparent gel was formed. Co-polymerisation with the monomer PEGMEMA was undertaken using various ratios of macromer to PEGMEMA according to Table 4.

### 4.6.2. Photo-Crosslinking Method

A 1% (*w*/*v*) stock solution of the photo-initiator Irgacure 2959 in de-ionised water was prepared. The homo-polymerisation of Macromer 1KL3 was performed by dissolving it in Irgacure stock solution to a final concentration of 54% and 30%. Respectively, 1 mL of each solution was placed into a small glass vial to form a layer with thickness of about 1 mm which was exposed to UV light (2.3 mW/cm$^2$) overnight at room temperature. Photo-polymerisation of Macromer 1KL11 and PEGMEMA was performed using the same procedures, and the ratio of macromer to PEGMEMA was varied according to Table 5.

### 4.7. Swelling Studies

For the macromer 1KL3 hydrogel, two samples were prepared at 54% and 30% concentration (Table 5), and then were tested for their swelling characteristics (Figure 3). The dry weights of the materials were determined ($w_0$), and then hydrogels were formed from the co-polymers by the addition of de-ionised water and were allowed to soak at room temperature (25 °C). The excess surface water was gently removed by means of a fine pipette and gently touching the surface with tissue paper. The weight of the swollen gel was taken ($w_s$). The swelling ratio was defined as:

$$Swelling\ Ratio\ \% = \frac{(w_s - w_0)}{w_0} \times 100 \tag{3}$$

where $w_s$ is the weight of the swollen hydrogel and $w_0$ is the weight of the dried hydrogel.

Readings were taken every 24 h and the results were tabulated and presented as a plot of swelling ratio vs. time in days. The tests were performed in triplicate.

### 5. Conclusions

PEG lactoyl triblock co-polymers P$_{DL}$LA-PEG$_{1k}$-P$_{DL}$LA were synthesised using PEG with a relatively small molecular weight of 1000 g·mol$^{-1}$. Vinyl functionality was then incorporated to obtain diacryl-P$_{DL}$LA-PEG$_{1k}$-P$_{DL}$LA. The LCSTs, being on the order of physiological temperatures, of some of these co-polymers and their derivative macromer materials may be worthy of further study for their clinical potential. The macromers were further used to synthesize soluble hyperbranched polymers and crosslinked hydrogels via UV irradiation and the Michael Addition reaction. The results demonstrated that the degradation and swelling properties of the prepared hydrogels can be tailored by varying the composition and topology of the PEG and PLA co-polymers and showed that degradation is seen to commence first in the material with the highest lactoyl content. These degradable hydrogels could be used in regenerative medicine and drug delivery applications.

**Supplementary Materials:** The following are available online at www.mdpi.com/2227-9717/5/2/18/s1, Figure S1: General molecular structure of diacryl-PDLLA-PEG1k-PDLLA. Figure S2: 1H NMR of PDLLA-PEG1k-PDLLA Co-polymer 1KL10. Figure S3: 1H NMR and chemical structure of diacryl-PDLLA-PEG1k-PDLLA Macromer 1KL10 (Mac-1KL10). Figure S4: FT-IR Spectra for Co-1KL10 (red) and Mac-1KL10 (blue).

**Acknowledgments:** EPSRC and Bangor University are gratefully acknowledged for financial support. We also thank the Waterford Institute of Technology for conducting the LCST measurements using a UV-vis spectrophotometer through the WINSS collaboration project, which is partially funded by the ERDF Interreg programme.

**Author Contributions:** A.H. and H.T. conceived and designed the experiments; A.H. performed the experiments. All authors contributed on data analysis, discussions and writing the paper.

**Conflicts of Interest:** The authors declare no conflict of interest.

### References

1.  Hoffman, A.S. Hydrogels for biomedical applications. *Adv. Drug Deliv. Rev.* **2012**, *64*, 18–23. [CrossRef]

2.  Balakrishnan, B.; Banerjee, R. Biopolymer-based hydrogels for cartilage tissue engineering. *Chem. Rev.* **2011**, *111*, 4453–4474. [CrossRef] [PubMed]
3.  Vermonden, T.; Censi, R.; Hennink, W.E. Hydrogels for protein delivery. *Chem. Rev.* **2012**, *112*, 2853–2888. [CrossRef] [PubMed]
4.  Nguyen, K.T.; West, J.L. Photopolymerizable hydrogels for tissue engineering applications. *Biomaterials* **2002**, *23*, 4307–4314. [CrossRef]
5.  Nam, K.; Watanabe, J.; Ishihara, K. Modeling of swelling and drug release behavior of spontaneously forming hydrogels composed of phospholipid polymers. *Int. J. Pharm.* **2004**, *275*, 259–269. [CrossRef] [PubMed]
6.  Bos, G.W.; Jacobs, J.J.L.; Koten, J.W.; Van Tomme, S.; Veldhuis, T.; van Nostrum, C.F.; Den Otter, W.; Hennink, W.E. In situ crosslinked biodegradable hydrogels loaded with il-2 are effective tools for local il-2 therapy. *Eur. J. Pharm. Sci.* **2004**, *21*, 561–567. [CrossRef] [PubMed]
7.  Park, H.; Park, K. Biocompatibility issues of implantable drug delivery systems. *Pharm. Res.* **1996**, *13*, 1770–1776. [CrossRef] [PubMed]
8.  Hou, Q.P.; De Bank, P.A.; Shakesheff, K.M. Injectable scaffolds for tissue regeneration. *J. Mater. Chem.* **2004**, *14*, 1915–1923. [CrossRef]
9.  Van Tomme, S.R.; Storm, G.; Hennink, W.E. In situ gelling hydrogels for pharmaceutical and biomedical applications. *Int. J. Pharm.* **2008**, *355*, 1–18. [CrossRef] [PubMed]
10. Hennink, W.E.; van Nostrum, C.F. Novel crosslinking methods to design hydrogels. *Adv. Drug Deliv. Rev.* **2012**, *64*, 223–236. [CrossRef]
11. Mao, H.L.; Shan, G.R.; Bao, Y.Z.; Wu, Z.L.L.; Pan, P.J. Thermoresponsive physical hydrogels of poly(lactic acid)/poly(ethylene glycol) stereoblock copolymers tuned by stereostructure and hydrophobic block sequence. *Soft Matter* **2016**, *12*, 4628–4637. [CrossRef] [PubMed]
12. Jeong, B.M.; Lee, D.S.; Shon, J.I.; Bae, Y.H.; Kim, S.W. Thermoreversible gelation of poly(ethylene oxide)biodegradable polyester block copolymers. *J. Polym. Sci. Part A—Polym. Chem.* **1999**, *37*, 751–760. [CrossRef]
13. Wu, Y.L.; Chen, X.H.; Wang, W.Z.; Loh, X.J. Engineering bioresponsive hydrogels toward healthcare applications. *Macromol. Chem. Phys.* **2016**, *217*, 175–188. [CrossRef]
14. Sawhney, A.S.; Pathak, C.P.; Hubbell, J.A. Bioerodible hydrogels based on photopolymerized poly(ethylene glycol)-co-poly(alpha-hydroxy acid) diacrylate macromers. *Macromolecules* **1993**, *26*, 581–587. [CrossRef]
15. Martens, P.J.; Bryant, S.J.; Anseth, K.S. Tailoring the degradation of hydrogels formed from multivinyl poly(ethylene glycol) and poly(vinyl alcohol) macromers for cartilage tissue engineering. *Biomacromolecules* **2003**, *4*, 283–292. [CrossRef] [PubMed]
16. Wang, Q.; Wang, C.D.; Du, X.; Liu, Y.; Ma, L.X. Synthesis, thermosensitive gelation and degradation study of a biodegradable triblock copolymer. *J. Macromol. Sci. Part A—Pure Appl. Chem.* **2013**, *50*, 200–207. [CrossRef]
17. Bae, S.J.; Suh, J.M.; Sohn, Y.S.; Bae, Y.H.; Kim, S.W.; Jeong, B. Thermogelling poly(caprolactone-*b*-ethylene glycol-*b*-caprolactone) aqueous solutions. *Macromolecules* **2005**, *38*, 5260–5265. [CrossRef]
18. Cho, H.; Kwon, G.S. Thermosensitive poly-(D,L-lactide-co-glycolide)-block-poly(ethylene glycol)-block-poly-(D,L-lactide-co-glycolide) hydrogels for multi-drug delivery. *J. Drug Target.* **2014**, *22*, 669–677. [CrossRef] [PubMed]
19. Li, C.H.; Ma, C.Y.; Zhang, Y.; Liu, Z.H.; Xue, W. Blood compatibility evaluations of poly(ethylene glycol)-poly(lactic acid) copolymers. *J. Biomater. Appl.* **2016**, *30*, 1485–1493. [CrossRef] [PubMed]
20. Deng, X.M.; Xiong, C.D.; Cheng, L.M.; Xu, R.P. Synthesis and characterization of block copolymers from D,L-lactide and poly(ethylene glycol) with stannous chloride. *J. Polym. Sci. Part C—Polym. Lett.* **1990**, *28*, 411–416. [CrossRef]
21. Zhu, K.J.; Lin, X.Z.; Yang, S.L. Preparation, characterization, and properties of polylactide (pla) poly(ethylene glycol) (peg) copolymers—A potential-drug carrier. *J. Appl. Polym. Sci.* **1990**, *39*, 1–9. [CrossRef]
22. Chen, X.H.; McCarthy, S.P.; Gross, R.A. Synthesis and characterization of L-lactide-ethylene oxide multiblock copolymers. *Macromolecules* **1997**, *30*, 4295–4301. [CrossRef]
23. Luo, W.J.; Li, S.M.; Bei, J.Z.; Wang, S.G. Synthesis and characterization of poly(L-lactide)-poly(ethylene glycol) multiblock copolymers. *J. Appl. Polym. Sci.* **2002**, *84*, 1729–1736. [CrossRef]
24. Cho, E.; Kutty, J.K.; Datar, K.; Lee, J.S.; Vyavahare, N.R.; Webb, K. A novel synthetic route for the preparation of hydrolytically degradable synthetic hydrogels. *J. Biomed. Mater. Res. Part A* **2009**, *90*, 1073–1082. [CrossRef] [PubMed]

25. Metters, A.; Hubbell, J. Network formation and degradation behavior of hydrogels formed by michael-type addition reactions. *Biomacromolecules* **2005**, *6*, 290–301. [CrossRef] [PubMed]
26. Velthoen, I.W.; van Beek, J.; Dijkstra, P.J.; Feijen, J. Thermo-responsive hydrogels based on highly branched poly(ethylene glycol)-poly(L-lactide) copolymers. *React. Funct. Polym.* **2011**, *71*, 245–253. [CrossRef]
27. Wicki, A.; Witzigmann, D.; Balasubramanian, V.; Huwyler, J. Nanomedicine in cancer therapy: Challenges, opportunities, and clinical applications. *J. Control. Release* **2015**, *200*, 138–157. [CrossRef] [PubMed]
28. Grinstaff, M.W. Dendritic polymers and hydrogels for biomedical applications. *Nanomed.-Nanotechnol. Biol. Med.* **2006**, *2*, 308. [CrossRef]
29. Zhang, Y.L.; Won, C.Y.; Chu, C.C. Synthesis and characterization of biodegradable network hydrogels having both hydrophobic and hydrophilic components with controlled swelling behavior. *J. Polym. Sci. Part A—Polym. Chem.* **1999**, *37*, 4554–4569. [CrossRef]

*processes*

MDPI

*Article*

# Kinetics of the Aqueous-Phase Copolymerization of MAA and PEGMA Macromonomer: Influence of Monomer Concentration and Side Chain Length of PEGMA

Iñaki Emaldi [1], Shaghayegh Hamzehlou [1], Jorge Sanchez-Dolado [2] and Jose R. Leiza [1,*]

[1] POLYMAT and Kimika Aplikatua Saila, Kimika Zientzien Fakultatea, University of the Basque Country UPV/EHU, Joxe Mari Korta Zentroa, Tolosa Hiribidea 72, 20018 Donostia-San Sebastián, Spain; inaki.emaldi@ehu.eus (I.E.); shaghayegh.hamzehlou@ehu.eus (S.H.)

[2] Sustainable Construction Division, Tecnalia Research & Innovation, Parque Tecnológico de Bizkaia, c/Geldo, Edificio 700, 48160 Derio, Spain; jorge.dolado@tecnalia.com

[*] Correspondence: jrleiza@ehu.eus; Tel.: +34-943-01-5329

Academic Editor: Alexander Penlidis
Received: 15 March 2017; Accepted: 14 April 2017; Published: 20 April 2017

**Abstract:** An in situ nuclear magnetic resonance spectroscopy (NMR) technique is used to monitor the aqueous-phase copolymerization kinetics of methacrylic acid (MAA) and poly(ethylene glycol) methyl ether methacrylate (PEGMA) macromonomers. In particular, the study analyses the effect of the number of ethylene glycol (EG) groups along the lateral chains of PEGMA and is carried out under fully ionized conditions of MAA at different initial monomer ratios and initial overall monomer concentrations (5–20 wt % in aqueous solution). The composition drift with conversion indicates that PEGMA macromonomer is more reactive than MAA. Individual monomer consumption rates show that the rates of consumption of both monomers are not first order with respect to overall concentration of the monomer. The reactivity ratios estimated from the copolymerization kinetics reveal, that for the short PEGMA, the reactivity ratios $r_{MAA}$ and $r_{PEGMA}$ increase with the solids content ($SC$). A totally different trend is obtained for the longer PEGMA, whose reactivity ratio ($r_{PEGMA23}$) decreases with solids content, whereas the reactivity ratio of MAA remains roughly constant.

**Keywords:** aqueous-phase copolymerization; polyethylene glycol methacrylate monomers; reactivity ratios; ionization degree; solids content

## 1. Introduction

During the last few years, the use of monomers with a polyethylene glycol (PEG) side chain in the synthesis of polymeric materials (as comonomers in aqueous-phase solution copolymerization [1–3] or as reactive stabilizers in several heterogeneous polymerization systems) [4–7] has gained increasing attention. Water-soluble polymers are used in a variety of applications including coatings, cosmetics, antiflocculants, textiles, superabsorbers and water treatment [8–10]. These materials are generally produced via free-radical (co)polymerization in aqueous solution [9,11,12]. Although of great industrial importance, understanding of the aqueous polymerization kinetics is far from reaching the understanding achieved in organic solvents. One major difference found between aqueous and organic solvents is that the propagation rate constant of water-soluble monomers, in addition to temperature, depends on other variables such as monomer concentration, pH and ionic strength of the aqueous medium, which makes the kinetics substantially more complex than in organic systems, especially when dealing with copolymerization processes.

Thus, in the last twenty years, several research groups have systematically analysed the effect of monomer concentration on the propagation rate coefficient, $k_p$, of the most common water-soluble monomers including acrylic acid (AA) [13–16], acrylamide (AM) [17,18], methacrylic acid (MAA) [19–21], *N*-vinyl pyrrolidone (NVP) [22] and *N*-vinyl formamide (NVF) [23,24]. Very recently, Smolne et al. [1] have also determined the propagation rate coefficient of poly(ethylene glycol) methyl ether methacrylate (PEGMA) in aqueous phase at different concentrations.

All these studies shared a common finding: the $k_p$ of these monomers decrease with increasing the concentration of monomer. This decrease is attributed to a reduction in the Arrhenius pre-exponential factor, which is related to entropic factors (independent of temperature) due to the influence of competitive hydrogen bonding between the transition-state structure and side groups of the monomer and water. The reduction can be up to one order of magnitude going from bulk to very dilute conditions [1,19,21]. In addition to the effect of the monomer concentration on the $k_p$, the ionization degree also affects the kinetics of acidic water-soluble monomers like AA and MAA. For instance, for MAA, one of the monomers used in this work, Lacík et al. [19–21,25] have determined the effect of the ionization degree (from non-ionized to fully ionized conditions) and found that the $k_p$ significantly reduces on going from non-ionized form to the fully ionized form, and that this decrease diminishes as MAA concentration increases.

For PEGMA monomers, there are no reports on the effect of the pH of aqueous media on the propagation rate coefficient. However, it has been found that for AM [18] and NVP [22], the pH does not affect the propagation rate coefficient.

Copolymerization is generally implemented in order to obtain specific properties which are not attainable by homopolymers. Water-soluble poly(MAA-*co*-PEGMA) copolymers present comb-like structure, where the size of the lateral chain can be tuned by the use of PEGMA with different numbers of ethylene oxide (EO) groups. This class of comb copolymers under alkali conditions presents an anionic backbone and an uncharged side chain and are effective dispersants and/or lubricants of inorganic particles. The backbone is considered to drive adsorption mainly through electrostatic interactions with surfaces, while the side chains are chosen to be non-absorbing and to induce steric hindrance among adsorbed layers [26,27]. These water-soluble copolymers have found a tremendous success in cementitious formulations where the comb copolymer dispersants are known as polycarboxylate ether (PCE) superplasticizers (SPs). They are mainly used to produce concrete of greater strength and durability by making it possible to reduce the water content without sacrificing rheological properties. Two synthetic routes can be used to produce these comb copolymers: one route is the partial esterification of polymethacrylic acid (PMAA) with methoxy polyethylene glycol (MPEG) of different lengths employing acid or base catalyst and vacuum to remove water [28,29]. This method provides highly uniform MPEG chains with a statistical distribution of the PEG side chains along a PMAA backbone. The second route is via free-radical copolymerization of MAA and PEGMA, which is a much common route of producing this type of PCEs [30–32].

The composition and sequence distribution of the comonomers in the chain is expected to have a tremendous impact on the performance of the comb copolymer chains in the cementitious formulations. Therefore, it is of paramount importance to control the composition of the chains during the polymerization reaction in order to produce homogeneous copolymers and hence be able to understand their adsorption behaviour when implemented in cementitious formulations as SPs.

The reactivity ratios of the comonomers (MAA and PEGMA) are the key parameters to understand the type of copolymer chains produced during the copolymerization and thus the necessary parameters to develop control strategies aimed at controlling the instantaneous composition along the reaction. Unfortunately, the information about the reactivity ratios of the MAA/PEGMA comonomer pair in aqueous solution available in the literature is scarce [33–35] and reports for conditions where the MAA is fully ionized (basic conditions) are not available. Smith and Klier [33] found that the reactivity ratios measured in deuterated water (D$_2$O) for the non-ionized MAA were very close to 1 for both monomers ($r_{MAA}$ = 1.03 and $r_{PEGMA}$ = 1.02); however, by changing 50% of D$_2$O by ethanol (D$_2$O/Ethanol = 50/50, a continuous phase used in dispersion polymerization), the reactivity ratios measured for the monomers increased ($r_{MAA}$ = 2.0 and $r_{PEGMA}$ = 3.6). Obviously this has an impact on the distribution of the monomer sequence in the chain that is random in water and more "blocky" in the case of water/ethanol.

Krivorotova et al. [35] have recently analyzed the copolymerization of MAA and PEGMA macromonomers carried out by conventional and controlled (RAFT, reversible addition fragmentation chain transfer) free-radical polymerization (FRP) in the mixture $D_2O$/dioxane initiated by azobisisobutyronitrile (AIBN). MAA was also non-ionized in the experimental conditions investigated in this work. Interestingly, the authors considered two PEGMA macromonomers differing in the number of ethylene glycol (EG) in side chain (PEGMA5 and PEGMA45). The authors found that for PEGMA5, the reactivity ratios for the conventional free-radical copolymerization did not vary with overall monomer conversion (up to 60% conversion) and the values were close for both monomers, but below 1 ($r_{PEGMA}$ = 0.81; $r_{MAA}$ = 0.60). Although slightly different (the ellipsoids at 95% confidence interval did not overlap), the reactivity ratios in RAFT copolymerization conditions were similar to FRP ($r_{PEGMA}$ = 0.59; $r_{MAA}$ = 0.68). In contrast, for PEGMA45 the results were substantially different and in FRP the reactivity ratios substantially change with conversion; $r_{MAA}$ decreased (from 1.83 at 10% to 1.25 at 60% conversion) and $r_{PEGMA45}$ increased (from 0.31 at 10% to 1.55 at 60% conversion). For RAFT copolymerization, this effect of the conversion was not found and the values of the reactivities were smaller for both monomers and closer between them.

In this work, the copolymerization of PEGMA macromonomers (with different EG side chain lengths) and MAA in aqueous solution under conditions where MAA is fully ionized is investigated. The principal aim of this study is to understand the copolymerization kinetics of these two monomers that, as discussed above, have shown dependence of the homopropagation rate constant on monomer concentration (MAA and PEGMA), and also of the pH (MAA) of the aqueous phase. We aim at determining the reactivity ratios of the comonomer system under fully ionized conditions of the MAA and at varying initial monomer concentrations (solids content) for two PEGMAs with increasing number of EG groups in the side chain (PEGMA5 and PEGMA23). The knowledge of the reactivity ratios will help in developing an accurate and predictive mathematical model of the copolymerization process that is needed to develop advanced control strategies for the control of the microstructure of these copolymers.

The article is organized as follows: first, the in situ proton nuclear magnetic resonance spectroscopy ($^1$H-NMR) technique used to monitor the copolymerizations of MAA and PEGMA in basic conditions is described. Second, the approach used to estimate the reactivity ratios is briefly described and, finally, the results for the two systems studied (MAA/PEGMA5 and MAA/PEGMA23) are presented and compared in the context of recent literature works for other copolymerization systems presenting similar features.

## 2. Materials and Methods

### 2.1. Materials

Methacrylic Acid (MAA) 99% with 250 ppm hydroquinone monomethyl ether (MEHQ) as inhibitor and Polyethylene glycol methyl ether methacrylate (PEGMA) (average Mn 300 g/mol "PEGMA5") with 100 ppm MEHQ and 300 ppm butylated hydroxyl toluene (BHT) as inhibitors were purchased from Sigma-Aldrich (St. Louis, MO, USA) and used as received. Visiomer MPEG 1000 MA W (Methoxy polyethylene glycol 1000 methacrylate 50% water solution "PEGMA23") with 200 ppm MEHQ as inhibitor was kindly supplied by Evonik Industries (Essen, Germany) Industries and used as received. Potassium persulfate (KPS) >99%, sodium bicarbonate (NaHCO$_3$) and sodium hydroxide (NaOH) >98% in pellets were also purchased from Aldrich and used as received. Deuterium oxide (D$_2$O) >99.9% was purchased from Euriso-Top (Saint-Aubin Cedex, France). Mili-Q quality water was employed to prepare the solutions.

### 2.2. Copolymerization Reactions

Aqueous-phase (mixture of H$_2$O/D$_2$O) free-radical copolymerization reactions of MAA and PEGMA were carried out in NMR tubes and the copolymerization reactions were in situ monitored.

Comonomer solutions were prepared at different solids contents (5–20 wt %) with different monomer molar ratios (MAA/PEGMA5: 3/1, 2/1, 1/1 and MAA/PEGMA23: 9/1, 6/1, 3/1). Comonomer mixture solutions were neutralized with a 30% NaOH solution and kept at a pH between 7 and 9, ensuring total ionization of the carboxylic monomer in all cases. $NaHCO_3$ was used as buffer at 1/1 mol ratio with respect to the initiator. All reactions were initiated with a water-soluble thermal initiator (potassium persulfate, KPS) at a concentration of 1 wt % based on monomer. Higher MAA/PEGMA monomer mol ratios were employed for the longer PEGMA (PEGMA23) because the difference in molecular weight between comonomers will lead to very low initial weight fraction of methacrylic acid ($W_{MAA,0}$) and hence very low rate of polymerization [21]. Note that the monomer molar ratios expressed here are the nominal values; the actual monomer molar ratios were calculated from the vinyl peak areas of each monomer before starting the reaction ($t_0$ in the NMR tube).

The liquid $^1$H-NMR spectra were recorded on a Bruker 500 AVANCE (500 MHz) equipped with a $Z$ gradient Broadband observe (BBO) probe. The kinetic study of the reaction was conducted at 343 K using $^1$H spectra with suppress of the solvent using WATERGATE sequence. The spectrum was recorded every 2 min during the first 16 min and then every 10 or 15 min for 3 h. Monomer conversion of MAA and PEGMA were calculated based on the evolution of the peaks corresponding to the vinyl protons of MAA ($\delta$, 5.60, 5.25, ppm) and PEGMA ($\delta$, 6.10, 5.70, ppm). The peak related to the methoxy group protons ($\delta$, 3.30, ppm), which is present in the PEGMA macromonomer side chain, was selected as internal reference [36]. Figure 1 illustrates the change on the intensity of the MAA and PEGMA peaks over copolymerization time. The spectra were processed by the software MestRe Nova 9.0 (MestreLab Research Chemistry Software solutions, Santiago de Compostela, Spain).

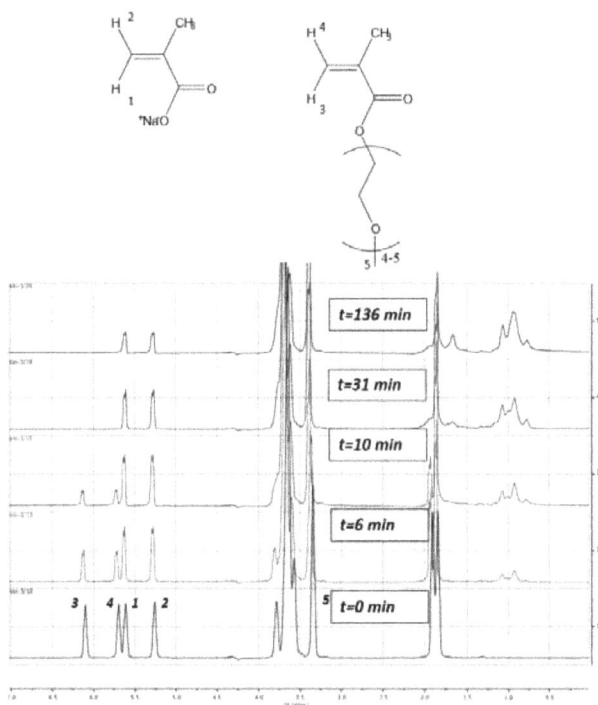

**Figure 1.** Time evolution of the proton nuclear magnetic resonance ($^1$H-NMR) spectra of the aqueous-solution copolymerization of MAA and PEGMA5 carried out at 70 °C ($SC$ = 10 wt % MAA/PEGMA5 = 1/1). MAA = methacrylic acid; PEGMA = poly(ethylene glycol) methyl ether methacrylate; $SC$ = solids content.

### 2.3. Estimation of Reactivity Ratios

Free-radical copolymerization leads to the formation of copolymers in which the distribution of the monomers in the chain is governed by kinetics. There is considerable experimental evidence [37–39] showing that, in many copolymerization systems, propagation depends on the nature of the monomer and on the last two units of the growing chain. This is referred to as penultimate model. Nevertheless, copolymer composition can be well described by considering a model in which the reactivity of the propagation reaction is governed by the nature of the monomer and the terminal unit of the polymer radical (terminal model). In this work, the later assumption was considered to describe the copolymerization of MAA and PEGMA macromonomers. The reactivity ratios were estimated using the evolution of individual comonomer conversions over overall conversion following the method developed by De la Cal et al. [40,41]. This method is briefly described below:

Methacrylic acid (MAA) and polyethylene glycol methacrylate (PEGMA) will be represented with letters A and B, respectively. The material balances for each of the monomers in a batch reactor, considering that terminal model kinetics is applied, can be written as:

$$\frac{d[A]}{dt} = -R_{pA} = -\left(k_{pAA}P_A + k_{pBA}P_B\right)[A][R^*] \tag{1}$$

$$\frac{d[B]}{dt} = -R_{pB} = -\left(k_{pAB}P_A + k_{pBB}P_B\right)[B][R^*] \tag{2}$$

In Equations (1) and (2) [$i$] is concentration of monomer $i$ (mol/L), $R_{pi}$, the polymerization rate of monomer $i$ (mol/L·s), $k_{pij}$, the propagation rate constant of radicals of terminal unit $i$ with monomer $j$ (L/mol·s), $P_i$ the probability of finding active chain with ultimate unit of type $i$, and $R^*$ is the total concentration of radicals (mol/L).

Considering the Quasi-Steady-State assumption (QSSA) is fulfilled, the probabilities are defined as follows:

$$P_A = \frac{k_{pBA}[A]}{k_{pBA}[A] + k_{pAB}[B]} \tag{3}$$

$$P_B = 1 - P_A \tag{4}$$

MAA conversion and overall conversion are defined as:

$$X_A = \frac{[A]_0 - [A]}{[A]_0} \tag{5}$$

$$X_T = \frac{([A]_0 - [A]) + ([B]_0 - [B])}{[A]_0 + [B]_0} \tag{6}$$

where $[A]_0$ and $[B]_0$ are initial concentration of MAA and PEGMA, respectively. Thus:

$$dX_A = -\frac{d[A]}{[A]_0} \tag{7}$$

$$dX_T = \frac{-d[A] - d[B]}{[A]_0 + [B]_0} \tag{8}$$

$$\frac{dX_A}{dX_T} = \frac{[A]_0 + [B]_0}{[A]_0} \frac{R_{pA}}{R_{pA} + R_{pB}} = \frac{[A]_0 + [B]_0}{[A]_0} \left( \frac{1 + r_A \frac{[A]}{[B]}}{2 + r_A \frac{[A]}{[B]} + r_B \frac{[B]}{[A]}} \right) \tag{9}$$

where $r_A$ and $r_B$ are the reactivities of MAA and PEGMA defined as:

$$r_A = \frac{k_{pAA}}{k_{pAB}} \tag{10}$$

$$r_B = \frac{k_{pBB}}{k_{pBA}} \tag{11}$$

To integrate Equation (9), concentrations can be expressed as a function of $X_A$ and $X_T$ employing Equations (5) and (6):

$$\frac{[A]}{[B]} = \frac{[A]_0(1 - X_A)}{[B]_0 - X_T([A]_0 + [B]_0) + [A]_0 X_A} \tag{12}$$

The cumulative composition can be determined as a function of the individual conversion of MAA, $X_A$, and the overall conversion as follows:

$$Y_A = \frac{[A]_0 X_A}{([A]_0 + [B]_0) X_T} \tag{13}$$

The reactivity ratios $r_A$ and $r_B$ can be estimated using a parameter estimation algorithm that minimizes the objective function of Equation (14), where $Y_{Aexp}$ is the experimentally measured cumulative composition referred to MAA determined by in situ $^1$H-NMR, and $Y_{Acal}$ is the theoretically determined cumulative composition calculated using the set of Equations (9)–(13) and the initial concentrations of the comonomers. The subscript $i$ makes reference to the experiment, and subscript $j$ to the sample number of each of the experiments used in the estimation procedure. The only parameters of the model are the reactivity ratios.

$$J = \left[ \sum_{i=1}^{N} \sum_{j=1}^{Pi} \left( Y_{Aexp} - Y_{Acal} \right)^2 \right] \tag{14}$$

Parameter estimation was carried out using a direct-search estimation algorithm employing subroutine DBCPOL and the subroutine for solving ordinary differential equations DIVPRK from IMSL library.

## 3. Results and Discussion

The aqueous-phase solution copolymerization of two comonomer systems was studied: MAA-co-PEGMA5 and MAA-co-PEGMA23; namely, polyethylene glycol methyl methacrylate monomers with 5 and 23 ethylene glycol (EG) units.

### 3.1. MAA-co-PEGMA5 (5 Ethylene Glycol Units)

Copolymerizations of MAA and PEGMA5 were carried out in NMR tubes at different comonomer ratios (MAA/PEGMA: 3/1, 2/1 and 1/1) and different solids content (SC: 5, 10, 15 and 20 wt %).

Figure 2 shows that conversions of both comonomers increase with solids content (for MAA, the effect was more clear at higher MAA/PEGMA ratios), and that PEGMA5 is more reactive than MAA at fully ionized conditions. The effect of the solids content on the conversion indicates that the dependence of the polymerization rate on the monomer conversion is not first order for any of the two monomers. This has been reported for the aqueous solution homopolymerization of MAA [19–21,25,42]. It has been found that the propagation rate coefficient of MAA was a function of the concentration of the monomer in the aqueous phase and of the ionization degree. Under fully ionized conditions (like in this work) the propagation rate coefficient of MAA increases with monomer weight fraction, but under non-ionized and partially ionized the propagation rate constant decreases with increasing monomer fraction.

Recently, Smolne and co-workers have studied the propagation and termination kinetics of PEG-lated methacrylates in aqueous solution [1]. Similar to other water-soluble monomers, it has been also found that the kinetics of the polymerization is affected by the monomer weight fraction in the aqueous solution. The propagation rate coefficient increased as the solids content decreased, which follows the same trend as the polymerization of non-ionized or partially ionized MAA. In the copolymerizations of Figure 2, the trends of the overall propagation rate constants for each monomer are like for the fully ionized MAA; namely, the overall propagation rate increases with solids content.

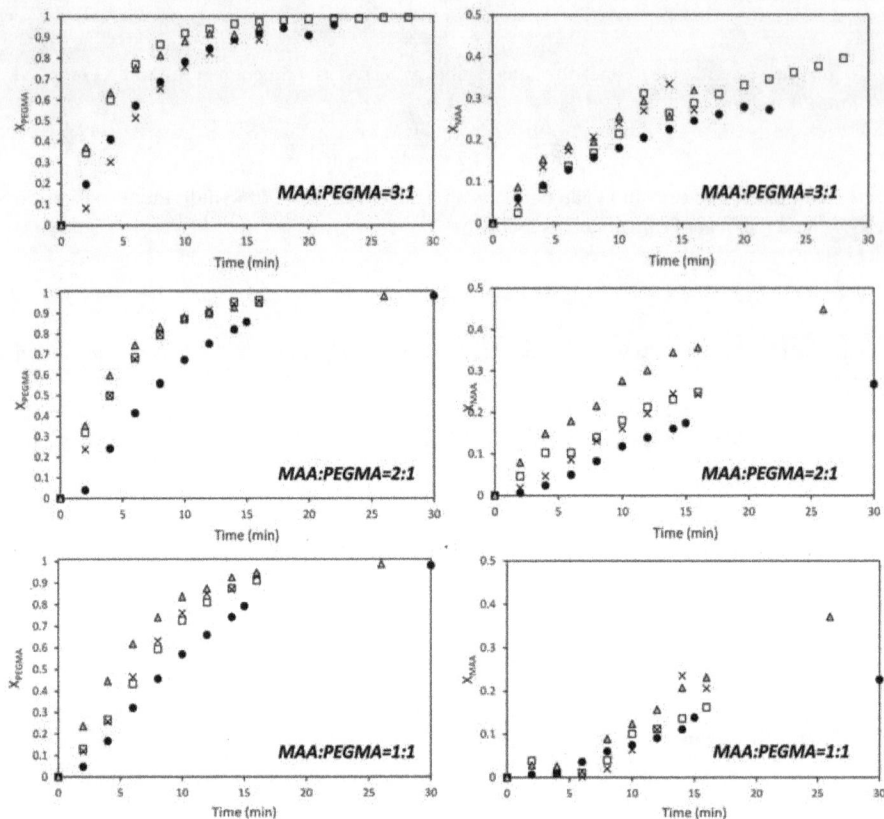

**Figure 2.** Time evolution of the conversion of MAA (right) and PEGMA 5 (left) for different monomer ratios (MAA/PEGMA as indicated in the figures) at different solids content (*SC*): ●—5 wt %; □—10 wt %; ×—15 wt %; Δ—20 wt %.

The faster conversion of the PEGMA5 macromonomer can be attributed to a substantial difference in reactivity ratios. As discussed in the introduction, there are no reports for the reactivity ratios of this comonomer system calculated in aqueous phase at fully ionized conditions of the MAA monomer. However, reactivity ratios of AA and AM in aqueous phase at fully ionized conditions have been investigated by several groups [8,43–46]. Preusser et al. have done an extensive experimental analysis including copolymerization in a broad range of initial monomer concentrations and degrees of ionization and have determined the reactivity ratios. They found that the reactivity ratio of each monomer is a function of initial weight fraction of monomer and of the ionization degree and they provided an empirical equation that captures well the combined effect of both variables [8]. The reactivity ratios for the MAA/PEGMA5 comonomer system were estimated from these data using the algorithm presented above. Figure 3 shows the comparison between the cumulative compositions of MAA determined experimentally from the NMR data (dots) and the estimated one (lines) for the estimated reactivity ratios at each solids content. Note that the reactivity ratios were initially estimated at each solids content because of the effects observed on the individual monomer conversions shown in Figure 2, and also because a similar effect was found for the AA and AM comonomer system [8].

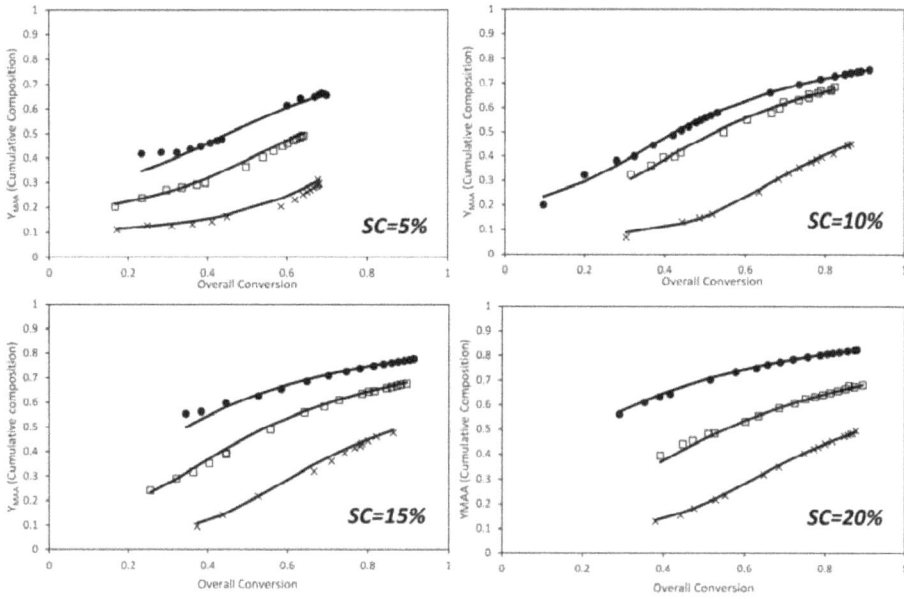

**Figure 3.** Conversion evolution of the cumulative composition of MAA for the in situ nuclear magnetic resonance (NMR) experiments. (Dots) experimental results, (lines) model predictions for the estimated reactivity ratios at each solids content (see Figure 4). Molar ratios: x—1/1; □—2/1; ●—3/1.

Figure 4 shows the estimated reactivity ratios at each solids content. It can be seen that the reactivity ratio of each comonomer increases with solids content (in the range studied in this work; note that the solids content analysed is limited by the viscosities produced during this polymerization carried out at 70 °C). This unexpected dependence of the reactivity with solids content has been recently found for the aqueous-phase copolymerization of AA and AM [8], a comonomer system that shared similar features on the dependence of the propagation rate coefficients with initial monomer concentration and ionization degree as the monomers investigated in this work.

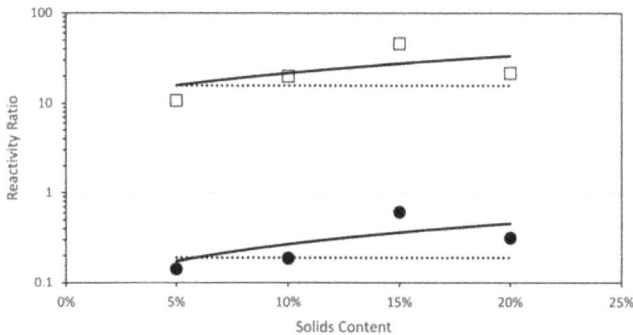

**Figure 4.** Reactivity ratios for the copolymerization of fully ionized MAA and PEGMA5 as determined by a global fit to the combined data (dashed line) and by fitting data at each initial solids content (squares and circles). A linear fit of the reactivity ratios estimated at each solids content is included (solid line) □ $r_{PEGMA}$; ● $r_{MAA}$.

Figure 4 plots the reactivity ratios estimated at each initial solids content and, for comparison purpose, the reactivity ratio estimated to fit globally all the experimental data gathered at different initial concentrations and monomer ratio is included. It was found that fitting the experimental data with just a pair of reactivity ratios (global fitting) was not appropriate, because the value of the objective function was almost one order of magnitude higher as compared with the objective functions of any of the individual fittings. The reactivity ratios estimated at each solids content were fitted with a linear equation that is also drawn in Figure 4 and detailed in the following equations. Note that the regressions of the linear fits are not excellent and hence the equations below should be considered only as a reasonable approximation.

$$r_{MAA} = 0.07 + 1.89 \cdot SC_0 \tag{15}$$

$$r_{PEGMA} = 9.89 + 117.21 \cdot SC_0 \tag{16}$$

where $SC_0$ is the initial weight fraction of comonomers in the copolymerization formulation ($0 < SC_0 < 1$).

### 3.2. MAA-co-PEGMA23 (23 Ethylene Glycol Units)

In this part of the work, we aimed to assess the effect of a longer lateral chain of the PEGMA macromonomer on the reactivity ratios with MAA under fully ionized conditions.

As in the previous set of reactions, the polymerizations were carried out under fully ionized conditions of the MAA; however, for this pair of monomers lower solids content were employed (5, 7.5 and 10 wt %) to avoid excessively high viscosities in the NMR tube and hence inhomogeneity in the reaction medium and unreliable results.

According to Krivorotova et al. [35], in a mixture of $D_2O/$Dioxane and with non-ionized MAA increasing the size of the lateral chain of the PEGMA (from PEGMA5 to PEGMA45), substantial differences were found on the reactivity ratios of this comonomer pair. Whereas the reactivities were found independent of conversion for the short PEGMA, for the long PEGMA the reactivity of MAA (that was higher than that of PEGMA45) decreased and that of PEGMA45 increased making them to be close at higher conversion.

Figure 5 shows the time evolution of the individual conversions of MAA and PEGMA23 monitored by in situ $^1$H-NMR. The trends are very similar to those observed for PEGMA5; namely, PEGMA23 macromonomer is more reactive than the MAA at fully ionized conditions, and the dependence of the polymerization rate on the solids content is not of first order for any of the two monomers. For the sake of comparison, the reactivity ratios of this pair has been estimated using the approach described above at each solids content and also using the whole set of data.

Figure 6 shows the comparison between the cumulative compositions of MAA experimentally determined from the NMR (dots) and the estimated cumulative compositions (lines) calculated for the estimated reactivity ratios at each solids content. Fitting of the cumulative composition for the set of experiments carried out at different solids content and monomer ratios is excellent.

The estimated reactivity ratios at different solids content for the combined data are shown in Figure 7. There are noticeable differences with respect to the short PEGMA. First, none of the reactivity ratios increase with solids content; on the contrary, $r_{PEGMA}$ decreases and $r_{MAA}$ is very similar for all the solids content. Indeed, the single pair of the reactivity ratios obtained by globally fitting all the data (the dashed lines) is very close to the values obtained in the individual fitting except for the values at 10% solids content. The linear fitting of the $r_{PEGMA}$ and $r_{MAA}$ calculated at each solids content provide the following empirical equations for dependence of the reactivity ratios with the solids content.

$$r_{MAA} = 0.14 - 0.14 \cdot SC_0 \tag{17}$$

$$r_{PEGMA} = 26.56 - 153.12 \cdot SC_0 \tag{18}$$

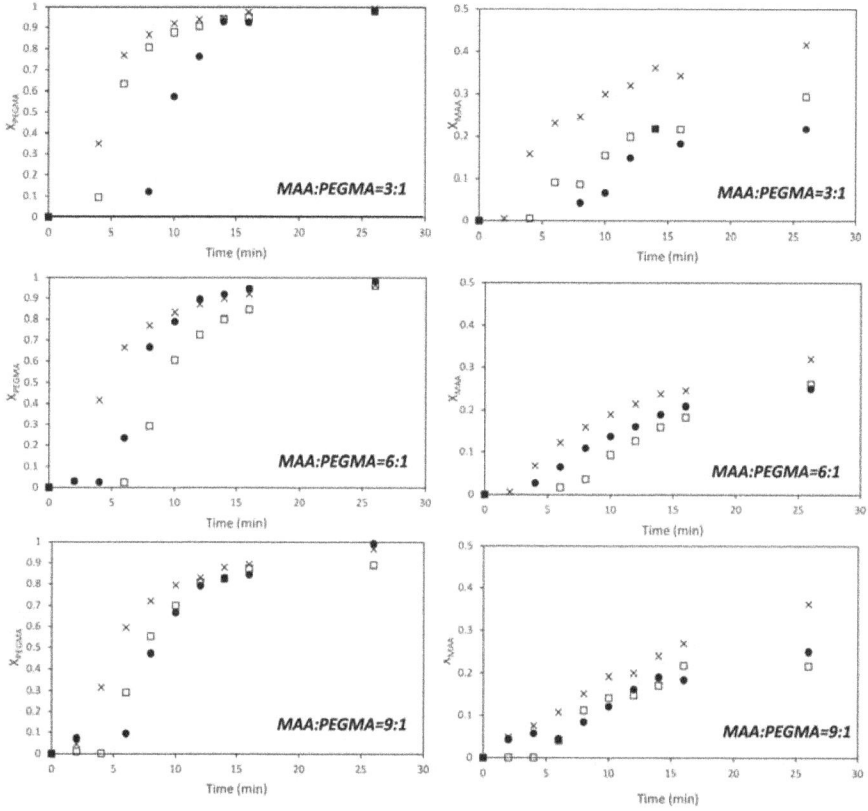

**Figure 5.** Time evolution of the conversion of MAA (right) and PEGMA 23 (left) at monomer ratios (MAA/PEGMA23 as indicated in the figure) at different solids content. ●—5 wt %; □—7.5 wt %; ×—10 wt %.

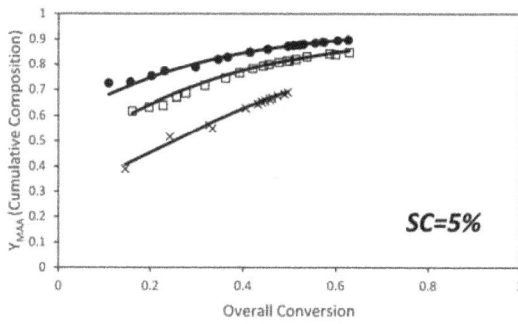

**Figure 6.** *Cont.*

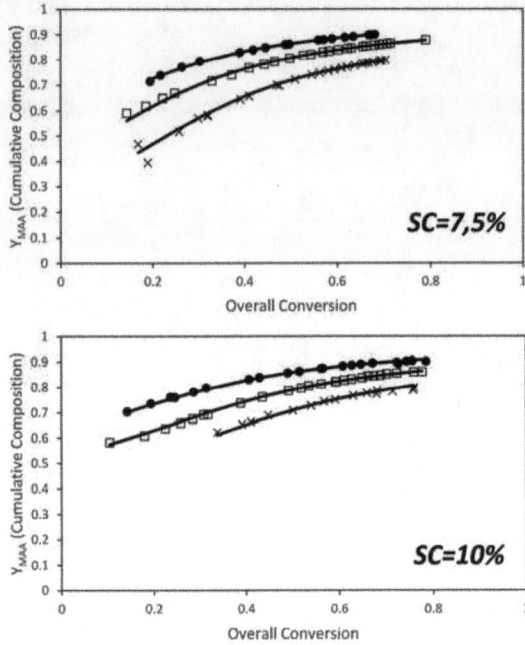

**Figure 6.** Conversion evolution of the cumulative composition of MAA for the in situ NMR experiments. (Dots) experimental results, (lines) model predictions for the estimated reactivity ratios. Molar ratios: ×—3/1; □—6/1; ●—9/1.

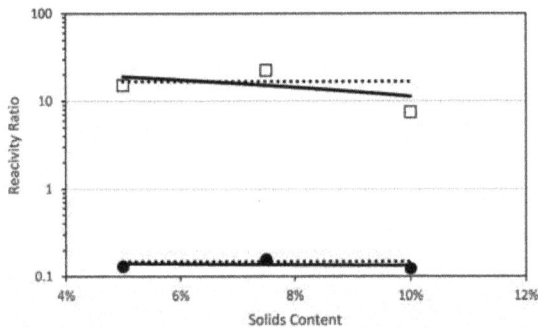

**Figure 7.** Reactivity ratios for the copolymerization of fully ionized MAA and PEGMA23 as determined by a global fit to the combined data (dashed line) and by fitting data at each initial solids content (squares and circles). A linear fit of the reactivity ratios at each solids content is included (solid line) □ $r_{PEGMA}$; ● $r_{MAA}$.

## 4. Conclusions

The aqueous-phase copolymerization kinetics of MAA and PEGMA macromonomers (with two lateral chains of different number of EG groups, PEGMA5 and PEGMA23) under fully ionized conditions of MAA at different initial monomer ratios and initial overall monomer concentrations (solids content) were monitored by in situ NMR. The rate of consumption of both monomers indicates that PEGMA macromonomer is more reactive than MAA, and that the rates of consumption of both monomers are not first order with respect to the concentration of the monomer in agreement with recent results that demonstrated that, like for other water-soluble monomers, the propagation rate coefficient of these monomers depends on the concentration of the monomer.

The reactivity ratios estimated from the copolymerization kinetics revealed that this dependency of the propagation rate coefficient on the monomer concentration was also translated to the reactivity ratios of these monomers. Thus, it was found that for the short PEGMA the reactivity ratios $r_{MAA}$ and $r_{PEGMA}$ increased with the solids content. Interestingly, this trend was not maintained for the longer PEGMA, whose reactivity ($r_{PEGMA23}$) decreased with conversion, whereas the reactivity ratio of MAA remained roughly constant with the longer PEGMA.

For both systems, empirical linear fits of the reactivity ratios with the solids content were calculated that can be easily implemented in modelling and control schemes aimed at controlling the microstructure of the comb-like copolymers.

**Acknowledgments:** This work has been carried out in the framework of the BASKRETE initiative under the umbrella of the EUSKAMPUS project. Iñaki Emaldi acknowledges the funding provided by EUSKAMPUS Fundazioa, POLYMAT and TECANLIA for his scholarship. Shaghayegh Hamzehlou and Jose Ramon Leiza acknowledge the funding provided by MINECO (CTQ 2014-59016P) and Basque Government (IT-999-16). Jorge Sánchez Dolado acknowledges the funding for the GEI Green Concrete Project given by the Basque Government (2015 Emaitek Program). The authors also thank the discussion with José Carlos de la Cal on the estimation of the reactivity ratios and they are grateful to José Ignacio Miranda and the SGIker Gipuzkoa Unit (UPV/EHU) for the NMR facilities.

**Author Contributions:** I.E. and S.H. performed the experiments. I.E., J.R.L. and J.S.-D. conceived the experiments. I.E., S.H. and J.R.L. wrote the paper.

**Conflicts of Interest:** The authors declare no conflict of interest.

## References

1. Smolne, S.; Weber, S.; Buback, M. Propagation and termination kinetics of poly(ethylene glycol) methyl ether methacrylate in aqueous solution. *Macromol. Chem. Phys.* **2016**, *217*, 2391–2401. [CrossRef]
2. Neugebauer, D. Graft copolymers with poly(ethylene oxide) segments. *Polym. Int.* **2007**, *56*, 1469–1498. [CrossRef]
3. Robinson, D.N.; Peppas, N.A. Preparation and characterization of pH-responsive poly (methacrylic acid-g-ethylene glycol) nanospheres. *Macromolecules* **2002**, *35*, 3668–3674. [CrossRef]
4. Tuncel, A. Emulsion copolymerization of styrene and poly(ethylene glycol) ethyl ether methacrylate. *Polymer* **2000**, *41*, 1257–1267. [CrossRef]
5. Halake, K.; Birajdar, M.; Kim, B.S.; Bae, H.; Lee, C.; Kim, Y.J.; Kim, S.; Kim, H.J.; Ahn, S.; An, S.Y.; et al. Recent application developments of water-soluble synthetic polymers. *J. Ind. Eng. Chem.* **2014**, *20*, 3913–3918. [CrossRef]
6. Brown, R.; Stitzel, B.; Sauer, T. Steric stabilization by grafting and copolymerization of water soluble oligomers and polymers. *Macromol. Chem. Phys.* **1995**, *196*, 2047–2064. [CrossRef]
7. Colombo, C.; Gatti, S.; Ferrari, R.; Casalini, T.; Cuccato, D.; Morosi, L.; Zucchetti, M.; Moscatelli, D. Self-assembling amphiphilic PEGylated block copolymers obtained through RAFT polymerization for drug-delivery applications. *J. Appl. Polym. Sci.* **2016**, *133*, 1–8. [CrossRef]
8. Preusser, C.; Ezenwajiaku, I.H.; Hutchinson, R.A. The combined influence of monomer concentration and ionization on acrylamide/acrylic acid composition in aqueous solution radical batch copolymerization. *Macromolecules* **2016**, *49*, 4746–4756. [CrossRef]
9. Amjad, Z. *Water Soluble Polymers. Solution properties and applications*; Kluwer Academic Press: New York, NY, USA, 2002.
10. Dautzenberg, H.; Jaeger, W.; Kötz, J.; Phillip, B.; Seidel, C.; Stscherbina, D. *Polyelectrolytes: Formation, Characterisation and Application*; Hauser Publishers: Munich, Germany, 1994.
11. Yu., K.E. *Water Soluble Poly-N-Vinylamides Synthesis and Physiochemical Properties*; John Wiley and Sons Ltd.: Chichester, UK, 1998.
12. Kricheldorf, H.R. *Handbook of Polymer Synthesis*; Mercel Dekker Inc.: New York, NY, USA, 1991.

13. Lacik, I.; Beuermann, S.; Buback, M. PLP-SEC study into the free-radical propagation rate coefficients of partially and fully ionized acrylic acid in aqueous solution. *Macromol. Chem. Phys.* **2004**, *205*, 1080–1087. [CrossRef]

14. Lacík, I.; Beuermann, S.; Buback, M. Aqueous phase size-exclusion-chromatography used for PLP-SEC studies into free-radical propagation rate of acrylic acid in aqueous solution. *Macromolecules* **2001**, *34*, 6224–6228. [CrossRef]

15. Lacík, I.; Beuermann, S.; Buback, M. PLP-SEC study into free-radical propagation rate of nonionized acrylic acid in aqueous solution. *Macromolecules* **2003**, *36*, 9355–9363. [CrossRef]

16. Cutie, S.S.; Smith, P.B.; Henton, D.E.; Staples, T.L.; Powell, C. Acrylic acid polymerization kinetics. *J. Polym. Sci. Part A Polym. Chem.* **1997**, *35*, 2029–2047. [CrossRef]

17. Seabrook, S.A.; Tonge, M.P.; Gilbert, R.G. Pulsed laser polymerization study of the propagation kinetics of acrylamide in water. *J. Polym. Sci. Part A Polym. Chem.* **2005**, *43*, 1357–1368. [CrossRef]

18. Lacík, I.; Chovancova, A.; Uhelsk, L.; Preusser, C.; Hutchinson, R.A.; Buback, M. PLP-SEC studies into the propagation rate coefficient of acrylamide radical polymerization in aqueous solution. *Macromolecules* **2016**, *49*, 3244–3253. [CrossRef]

19. Beuermann, S.; Buback, M.; Hesse, P.; Lacík, I. Free-radical propagation rate coefficient of nonionized methacrylic acid in aqueous solution from low monomer concentrations to bulk polymerization. *Macromolecules* **2006**, *39*, 184–193. [CrossRef]

20. Beuermann, S.; Buback, M.; Hesse, P.; Kukućková, S.; Lacík, I. Propagation rate coefficient of non-ionized methacrylic acid radical polymerization in aqueous solution. The effect of monomer conversion. *Macromol. Symp.* **2007**, *248*, 41–49. [CrossRef]

21. Lacík, I.; Učňová, L.; Kukučková, S.; Buback, M.; Hesse, P.; Beuermann, S. Propagation rate coefficient of free-radical polymerization of partially and fully ionized methacrylic acid in aqueous solution. *Macromolecules* **2009**, *42*, 7753–7761. [CrossRef]

22. Stach, M.; Lacík, I.; Chorvát, D.; Buback, M.; Hesse, P.; Hutchinson, R.A.; Tang, L. Propagation rate coefficient for radical polymerization of *N*-vinyl pyrrolidone in aqueous solution obtained by PLP-SEC. *Macromolecules* **2008**, *41*, 5174–5185. [CrossRef]

23. Stach, M.; Lacík, I.; Kasák, P.; Chorvát, D.; Saunders, A.J.; Santanakrishnan, S.; Hutchinson, R.A. Free-radical propagation kinetics of *N*-vinyl formamide in aqueous solution studied by PLP-SEC. *Macromol. Chem. Phys.* **2010**, *211*, 580–593. [CrossRef]

24. Santanakrishnan, S.; Hutchinson, R.A.; Učňová, L.; Stach, M.; Lacík, I.; Buback, M. Polymerization kinetics of water-soluble *N*-vinyl monomers in aqueous and organic solution. *Macromol. Symp.* **2011**, *302*, 216–223. [CrossRef]

25. Buback, M.; Hesse, P.; Hutchinson, R.A.; Lacík, I.; Kasák, P.; Stach, M.; Utz, I. Kinetics and modeling of free-radical batch poymerization of nonionized methacrylic acid in aqueous solution. *Ind. Eng. Chem.* **2008**, *47*, 8197–8204. [CrossRef]

26. Flatt, R.; Schober, I. *Superplasticizers and the Rheology of Concrete 7. Understanding the Rheology of Concrete*; Roussel, N., Ed.; Woodhead Publishing Limited: Oxford, UK, 2011.

27. Marchon, D.; Sulser, U.; Eberhardt, A.; Flatt, R.J. Molecular design of comb-shaped polycarboxylate dispersants for environmentally friendly concrete. *Soft Matter* **2013**, *9*, 10719–10728. [CrossRef]

28. Guicquero, J.P.; Maitrasse, P.; Mosquet, M.A.; Alphonse, S. A Water Soluble or Water Dispersible Dispersing Agent. FR2776285, 19 March 1998.

29. Lei, L.; Plank, J. Synthesis and properties of a vinyl ether-based polycarboxylate superplasticizer for concrete possessing clay tolerance. *Ind. Eng. Chem. Res.* **2014**, *53*, 1048–1055. [CrossRef]

30. Plank, J.; Pöllmann, K.; Zouaoui, N.; Andres, P.R.; Schaefer, C. Synthesis and performance of methacrylic ester based polycarboxylate superplasticizers possessing hydroxy terminated poly(ethylene glycol) side chains. *Cem. Concr. Res.* **2008**, *38*, 1210–1216. [CrossRef]

31. Plank, J.; Sakai, E.; Miao, C.W.; Yu, C.; Hong, J.X. Chemical admixtures—Chemistry, applications and their impact on concrete microstructure and durability. *Cem. Concr. Res.* **2015**, *78*, 81–99. [CrossRef]

32. Pourchet, S.; Liautaud, S.; Rinaldi, D.; Pochard, I. Effect of the repartition of the PEG side chains on the adsorption and dispersion behaviors of PCP in presence of sulfate. *Cem. Concr. Res.* **2012**, *42*, 431–439. [CrossRef]

33. Smith, B.L.; Klier, J. Determination of monomer reactivity ratios for copolymerizations of methacrylic acid with poly(ethylene glycol) monomethacrylate. *J. Appl. Polym. Sci.* **1998**, *68*, 1019–1025. [CrossRef]
34. Georges, S.; Hamaide, T. Determination of the reactivity ratios of the methacrylic acid-polyethylene glycol methacrylate system by aqueous SEC. *Bul. Stiint. al Univ. "Politehnica" din Timisoara* **2001**, *46*, 187–191.
35. Krivorotova, T.; Vareikis, A.; Gromadzki, D.; Netopilík, M.; Makuška, R. Conventional free-radical and RAFT copolymerization of poly(ethylene oxide) containing macromonomers. *Eur. Polym. J.* **2010**, *46*, 546–556. [CrossRef]
36. Najafi, V.; Ziaee, F.; Kabiri, K.; Mehr, M.J.Z.; Abdollahi, H.; Nezhad, P.M.; Jalilian, S.M.; Nouri, A. Aqueous free-radical polymerization of PEGMEMA macromer: Kinetic studies via an on-line $^1$H NMR technique. *Iran. Polym. J. Engl. Ed.* **2012**, *21*, 683–688. [CrossRef]
37. Fukuda, T.; Inagaki, H. Free-Radical Copolymerization. 3. Determination of rate constants of propagation and termination for the styrene/methyl methacrylate system. A critical test of terminal-model kinetics 1. *Macromolecules* **1985**, *18*, 17–26. [CrossRef]
38. Coote, M.L.; Davis, T.P. Mechanism of the propagation step in free-radical copolymerisation. *Prog. Polym. Sci.* **1999**, *24*, 1217–1251. [CrossRef]
39. Olaj, O.F.; Schnöll-Bitai, I.; Kremminger, P. Evaluation of individual rate constants from the chain-length distribution of polymer samples prepared by intermittent (rotating sector) photopolymerization-2. The copolymerization system styrene-methyl methacrylate. *Eur. Polym. J.* **1989**, *25*, 535–541. [CrossRef]
40. De La Cal, J.C.; Leiza, J.R.; Asúa, J.M. Estimation of reactivity ratios using emulsion copolymerization data. *J. Polym. Sci. Part A Polym. Chem.* **1991**, *29*, 155–167. [CrossRef]
41. García, G. Síntesis de Floculantes Catiónicos en Reactores Continuos. Ph.D. Thesis, University of the Basque Country (UPV/EHU), Donostia-San Sebastián, Spain, 2009.
42. Beuermann, S.; Buback, M.; Hesse, P.; Hutchinson, R.A.; Kukučková, S.; Lacík, I. Termination kinetics of the free-radical polymerization of nonionized methacrylic acid in aqueous solution. *Macromolecules* **2008**, *41*, 3513–3520. [CrossRef]
43. Riahinezhad, M.; Mcmanus, N.; Penlidis, A. Effect of monomer concentration and pH on reaction kinetics and copolymer microstructure of Acrylamide/Acrylic acid copolymer. *Macromol. React. Eng.* **2015**, *9*, 100–113. [CrossRef]
44. Paril, A.; Alb, A.M.; Giz, A.T.; Çatagil-Giz, H. Effect of medium pH on the reactivity ratios in acrylamide acrylic acid copolymerization. *J. Appl. Polym. Sci.* **2007**, *103*, 968–974. [CrossRef]
45. Ponratnam, S.; Kapur, S.L. Reactivity ratios of ionizing monomers in aqueous solution. Copolymerization of acrylic and methacrylic acids with acrylamide. *Makromol. Chem.* **1977**, *178*, 1029–1038. [CrossRef]
46. Cabaness, W.R.; Lin, T.Y.-C.; Párkányi, C. Effect on the pH on the reactivity ratios in the copolymerization of acrylic acid and acrylamide. *J. Polym. Sci. Part A Polym. Chem.* **1971**, *9*, 2155–2170. [CrossRef]

*processes*

**MDPI**

*Article*

# Polymerization Kinetics of Poly(2-Hydroxyethyl Methacrylate) Hydrogels and Nanocomposite Materials

**Dimitris S. Achilias * and Panoraia I. Siafaka**

Department of Chemistry, Aristotle University of Thessaloniki, Thessaloniki 54124, Greece; siafpan@gmail.com
* Correspondence: axilias@chem.auth.gr; Tel.: +30-2310-99-7822

Academic Editor: Alexander Penlidis
Received: 15 March 2017; Accepted: 19 April 2017; Published: 24 April 2017

**Abstract:** Hydrogels based on poly(2-hydroxyethyl methacrylate) (PHEMA) are a very important class of biomaterials with several applications mainly in tissue engineering and contacts lenses. Although the polymerization kinetics of HEMA have been investigated in the literature, the development of a model, accounting for both the chemical reaction mechanism and diffusion-controlled phenomena and valid over the whole conversion range, has not appeared so far. Moreover, research on the synthesis of nanocomposite materials based on a polymer matrix has grown rapidly recently because of the improved mechanical, thermal and physical properties provided by the polymer. In this framework, the objective of this research is two-fold: to provide a kinetic model for the polymerization of HEMA with accurate estimations of the kinetic and diffusional parameters employed and to investigate the effect of adding various types and amounts of nano-additives to the polymerization rate. In the first part, experimental data are provided from Differential Scanning Calorimetry (DSC) measurements on the variation of the reaction rate with time at several polymerization temperatures. These data are used to accurately evaluate the kinetic rate constants and diffusion-controlled parameters. In the second part, nanocomposites of PHEMA are formed, and the in situ bulk radical polymerization kinetics is investigated with DSC. It was found that the inclusion of nano-montmorillonite results in a slight enhancement of the polymerization rate, while the inverse holds when adding nano-silica. These results are interpreted in terms of noncovalent interactions, such as hydrogen bonding between the monomer and polymer or the nano-additive. X-Ray Diffraction (XRD) and Fourier Transform Infra-Red (FTIR) measurements were carried out to verify the results.

**Keywords:** polymerization kinetics; HEMA; nanocomposites; in situ polymerization; nano-clays; nano-silica

## 1. Introduction

Hydrogels are biomaterials with properties such as hydrophilicity, biocompatibility and high water absorption by swelling without dissolving, which have attracted great interest from several researchers worldwide [1]. They can be of natural origin (e.g., hyaluronic acid, chitosan, etc.) or synthetic (e.g., poly(ethylene glycol) (PEG), poly(lactic acid) (PLA), etc.). 2-hydroxyethyl methacrylate (HEMA) is a hydrophilic monomer used to prepare such synthetic polymeric (i.e., PHEMA) hydrogels. The polymer, PHEMA, due to its excellent biocompatibility and physicochemical properties, similar to those of living tissues, is widely used in many biomedical applications, such as drug-delivery systems, cell carriers, tissue engineering, contacts lenses, regenerative medicine, etc. [2–4]. Other applications of PHEMA include those presented by Kharismadewi et al. [5], as an adsorbent for the removal of the methylene blue cationic dye from an aqueous solution and Moradi et al. [6] for the adsorption of $Cu^{2+}$ and $Pb^{2+}$ ions from aqueous single solutions [6].

The great interest in the polymerization of HEMA is based on the combination of methacrylate groups in its structure, resulting in relatively easy radical reactions, with the hydroxyl groups, that provide hydrophilicity. Thus, PHEMA allows obtaining tissue in-growth due to high permeability to small molecules and the soft consistency, which minimizes mechanical frictional irritation to surrounding tissues. The polymer (PHEMA) is a glassy amorphous material with low water absorption, high adhesion to glass and glass transition temperature ranging between 50 and 90 °C depending on the conditions of the polymerization process [7].

During the last decade, research on the synthesis of nanocomposite materials based on a polymer matrix has grown rapidly because of the improved mechanical, thermal and physical properties that are provided by the polymer. Several types of nano-fillers have been used, such as nano-clays, nano-silica, carbon nanotubes, fullerenes, graphene, etc. In this research, we investigated the formation of PHEMA-based nanocomposites with either a nano-organomodified montmorillonite (OMMT) or nano-silica. Incorporation of MMT is well known to improve the mechanical properties of the polymer matrix [8,9], whereas adding silica to PHEMA results in bioactive materials that keep the swelling properties of neat polymer; thus, they assure both morphological and bioactive characteristics [7]. Nanocomposites can be prepared by various methods, such as in situ polymerization, melt intercalation/exfoliation and solution casting. The in situ polymerization method was used here in order to obtain nanocomposites with a uniform dispersion of the filler in the polymer matrix.

The kinetics of radical polymerization has been extensively studied for a long time. Differential scanning calorimetry (DSC) has long been used for monitoring polymerization reactions for several systems. It offers the advantage of continuous recording of the variation of the reaction rate with time through the measurements of the amount of heat released, since additional reactions are exothermic. This technique has been also used for recording the polymerization kinetics of HEMA and its copolymers with some dimethacrylate monomers (such as ethylene glycol dimethacrylate (EGDMA) or diethylene glycol dimethacrylate (DEGDMA)) [10–14]. Modeling of polymerization kinetics has been carried out using semi-empirical models, such as those developed for resins-curing [11] or based on the isoconversional principle [10]. A pioneering work on the polymerization kinetics of PHEMA crosslinked with EGDMA was presented thirty years ago by Mikos and Peppas [15]. Later on, the group of Bowman was the only one to develop kinetic models for the polymerization kinetics of HEMA based on the reaction mechanism [16,17]. A method for determining the kinetic parameters was also proposed [16]. In our previous investigation, we also used DSC measurements and compared the results obtained by a simple mechanistic model with those from isoconversional analysis [18]. Only the low degrees of conversion were investigated to avoid the diffusion-controlled phenomena.

This research consists of two parts: In the first part, a detailed kinetic model is proposed for the polymerization kinetics of HEMA in bulk based on DSC measurements of the polymerization rate and the reaction mechanism, including also the effect of diffusion-controlled phenomena on the termination and propagation reactions. Polymerizations at different temperatures are carried out, and all kinetic parameters used are either based on accurate independent experimental measurements or estimated in this work. The aim of this part was to provide accurate kinetic data concerning the radical polymerization of HEMA over the whole conversion range. In the second part, the effect of adding different types and amounts of nano-particles on the polymerization kinetics is investigated. The in situ bulk radical polymerization is examined with two different types of nano-additives: an organo-modified montmorillonite, which is a 2D nanoclay, and a nano-silica, which can be considered a 3D nanoparticle. The polymerization kinetics is studied with DSC operating isothermally at different temperatures. Depending on specific interactions, a different effect on the polymerization kinetics was observed from those two nano-additives.

## 2. Materials and Methods

### 2.1. Materials

The monomers used were composed of 2-hydroxyethyl methacrylate (HEMA), purchased from Aldrich with purity ≥99%. Before any use, it was passed at least twice through a disposable inhibitor-remover packed column, supplied from Aldrich, in order to remove the inhibitor included. The free radical initiator used, i.e., benzoyl peroxide (BPO) with a purity >97%, was provided by Fluka AG and purified by fractional recrystallization twice from methanol (purchased from Merck). For the preparation of the nanocomposites, two nanofillers were used: the first was a commercially-available organically-modified, with a quaternary ammonium salt (i.e., dimethyl hydrogenated tallow), montmorillonite clay, under the trade name Cloisite 15A, (provided by Southern Clay Products Inc., Gonzales, TX, USA). The chemical structure of the ammonium salt was, $N^+R_2(CH_3)_2$ where R is the hydrogenated tallow (~65% C18, ~30% C16, ~5% C14), and its cationic exchange capacity (CEC) was 125 meq/100 g clay. Typical physical properties, according to the manufacturer, include: size, as measured by a transmission electron microscope for a PA6 nanocomposite, 75–150 nm × 1 nm; surface area 750 $m^2$/g when exfoliated.

The second nanoparticle used was hydrophilic fumed silica under the trade name Aerosil 200. The average primary particle size was 12 nm, the specific surface area 200 $m^2$/g and the $SiO_2$ content greater than 99.8% (from Evonik Resource Efficiency GmbH). All other chemicals used were of reagent grade.

### 2.2. Polymerization Kinetics

Polymerization experiments were performed using the DSC Diamond (Perkin-Elmer, Waltham, MA, USA). For the temperature and enthalpy calibration of the instrument, indium was used. Polymerizations were run isothermally at temperatures of 52, 60, 72 and 82 °C. Although a significant amount of heat is produced during the reaction, especially in the autoacceleration region, the equipment set up is so to maintain the reaction temperature constant (within ± 0.01 °C) during the whole conversion range. The liquid mixture of the monomer with the initiator at an initial concentration of 0.03 mol/L was placed into aluminum Perkin-Elmer pans, accurately weighted (approximately 10 mg), sealed and positioned into the appropriate holder of the instrument. In order to have an estimation of the overall reaction rate $(dx/dt)$, the amount of heat released $(d(\Delta H)/dt)$, the reaction exotherm in normalized values (W/g) was continuously recorded as a function of time. Then, the reaction rate is estimated from:

$$\frac{dx}{dt} = \frac{1}{\Delta H_T} \frac{d(\Delta H)}{dt} \tag{1}$$

where $x$ denotes fractional conversion and $\Delta H_T$ total reaction enthalpy.

By integrating the area between the DSC thermogram and the baseline established after extrapolation from the trace produced when polymerization has been completed (no change in the heat produced), the degree of conversion can be calculated. In order to determine the total reaction enthalpy, a dynamic experiment followed, where samples were heated from the polymerization temperature to 180 °C at a rate of 10 K $min^{-1}$. The sum of enthalpies of the isothermal $(\Delta H_i)$ plus the dynamic $(\Delta H_d)$ experiment was the total reaction enthalpy. The values thus estimated were always near 422 J/g, similar to the theoretical one obtained by dividing the standard heat of polymerization of a methacrylate double bond (i.e., 54.9 kJ/mol) over the monomer molecular weight (i.e., 130). The ultimate monomer conversion was estimated by the quotient $\Delta H_i/\Delta H_T$. In order to check for possible monomer evaporation during the reaction, the pans were weighed again after the end of the polymerization. A negligible monomer loss (less than 0.2 mg) was observed only in a few experiments.

It should be noticed that at low polymerization temperatures (i.e., 50 or 60 °C), a small inhibition time was observed. This is due to the non-adequate removal of oxygen, dissolved in the monomer during the initial mixing, which was performed in air.

All of the experiments were performed at least twice, and the best results are included in the Results section.

## 2.3. Synthesis of Nanocomposites

For the synthesis of the nanocomposite materials, initially, the appropriate amount (1, 3 or 5 wt %) of the nano-particles (i.e., OMMT or silica) was dispersed in the monomer, using magnetic and ultrasound agitation. In the final homogeneous suspension, the initiator, BPO at a concentration 0.03 mol/L, was added, and the mixture was degassed by passing nitrogen and immediately used. The procedure followed for the measurement of the reaction rate with time was exactly the same as that reported previously.

## 2.4. Measurements

X-ray diffraction (XRD) patterns of the materials were obtained using an X-ray diffractometer (3003 TT, Rich. Seifert, Ahrenburg, Germany) equipped with a CuKa generator ($\lambda$ = 0.1540 nm). Scans were taken in the diffraction angle range $2\theta$ = $1°$–$10°$.

Details of the chemical structure of neat PHEMA and its nanocomposites were identified by recording their IR spectra. The particular FTIR instrument used was the Spectrum 1 spectrophotometer from Perkin-Elmer. The spectra were recorded over the range 4000–500 cm$^{-1}$ at a resolution of 1 cm$^{-1}$, and 32 scans were averaged to reduce noise.

## 3. Results

### 3.1. Polymerization Kinetics of HEMA

Bulk free-radical polymerization kinetics of PHEMA has been investigated previously by our group at reaction temperatures ranging from 50–80 °C [18,19]. Similar, though slightly different reaction temperatures were also used here in order to check previous results and have appropriate reaction rate data. The variation of heat released with time at temperatures of 52, 60, 72 and 82 °C is illustrated in Figure 1a. From these data, the reaction rate was estimated according to Equation (1). By integrating this equation, the monomer double bond conversion was calculated and is plotted in Figure 1b at all reaction temperatures studied.

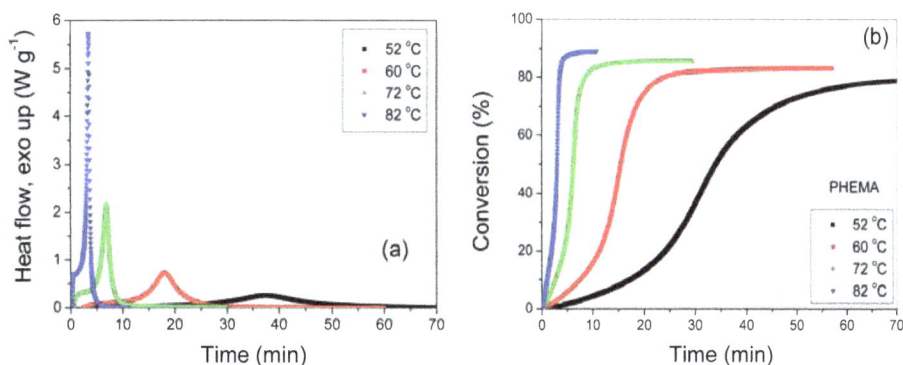

**Figure 1.** Variation of the amount of heat released (**a**) and conversion (**b**) with time during the bulk radical polymerization of HEMA at several constant temperatures.

The curves shown in Figure 1 exhibit the typical characteristics of a radical polymerization reaction with diffusion-controlled phenomena affecting the reaction rate. The feature points are briefly analyzed below. As is well known, the radical polymerization mechanism includes mainly three steps (i.e., decomposition of an initiator to primary radicals, reaction of these radicals with monomer molecules to form macro-radicals, propagation of the macro-radicals by reacting with several monomer molecules and, as the final point, termination of these macroradicals for the formation of the final polymer). At the early stages of polymerization (low conversions), 'classical' free-radical kinetics apply, with a purely chemical control of the reaction [20]. This is denoted by an almost constant polymerization rate and an almost linear dependence of monomer conversion with time. After a certain point in the region of 7%–12% conversion, an increase in the reaction rate takes place accompanied by an increase in the conversion values. This is the well-known gel-effect or autoacceleration phenomenon, which is attributed to the effect of diffusion-controlled phenomena mainly on the termination reaction. Accordingly, in reactions carried out without any solvent, in bulk, as the concentration of the macromolecules increases, the diffusion of the macro-radicals in space in order to find one another and react is significantly reduced. This results in a local increase in concentration, leading to increased reaction with monomer molecules and, hence, to an increased polymerization rate [20]. At higher degrees of conversion, the macro-radical chains are even more restricted in their movement, and their center-of-mass diffusion becomes very slow. However, termination of these macro-radicals is continued at a smaller rate, by means of their implicit movement caused by the addition of monomer molecules at the chain end. This diffusion mechanism is the so-called 'residual termination' or 'reaction diffusion'. The higher the propagation reaction rate, the more likely is the reaction-diffusion to be rate determining. Afterwards, the polymerization rate falls significantly and tends asymptotically to zero. This stage corresponds to the well-known glass-effect. This is attributed to the effect of diffusion-controlled phenomena on the propagation reaction since at such high monomer conversions, even the small monomer molecules are hindered in their movement to find a macro-radical and react [20]. All of these phenomena are quantified next.

As was reported previously, the free-radical polymerization mechanism constitutes mainly three steps: initiation (with an initiator decomposition kinetic rate constant, $k_d$, and initiator efficiency, $f$), propagation (with propagation rate constant $k_p$) and termination (with rate constant $k_t$). After a number of assumptions, including the steady-state approximation (rate of change of radical concentration with time equal to zero), the long chain hypothesis (consumption of monomer only in the propagation reactions) and negligible chain transfer to the monomer rate constant compared to propagation, the polymerization rate, $R_p$, is expressed as a function of conversion, $X$, by:

$$R_p = \frac{dX}{dt} = k_p \left(\frac{fk_d}{k_t}\right)^{1/2} [I]^{1/2}(1-X) \cong k_{eff}(1-X) \quad \text{with} \quad k_{eff} = k_p \left(\frac{fk_d}{k_t}\right)^{1/2} [I]^{1/2} \tag{2}$$

In order to estimate the overall kinetic rate constant, $k_{eff}$, Equation (2) can be integrated assuming that the initiator concentration, $[I]$, the initiator efficiency, $f$, and all kinetic rate constants are constant.

$$-\ln(1-X) = k_{eff}t \tag{3}$$

It should be noted that the assumptions used to result in Equation (3) are valid only at low degrees of monomer conversion. Then, from a plot of $-\ln(1-X)$ vs. $t$, the slope of the initial linear part is directly equal to $k_{eff}$. Such plots at conversion values in the range of 1%–7% have been created and illustrated in Figure 2a. The data followed very good straight lines at high temperatures (i.e., 72 and 82 °C), whereas a slight curvature appeared at the lower temperature of 52 °C (the correlation coefficient, $R^2$, ranged from 0.992–0.999). Then, from the $k_{eff}$ values measured at different temperatures, the overall activation energy of the polymerization rate, $E_{eff}$, can be estimated from the slope of $\ln(k_{eff})$ versus $1/T$ assuming an Arrhenius-type expression. The individual activation energies of the elementary reactions, i.e., propagation ($E_p$), initiation ($E_i$) and termination ($E_t$), are correlated to $E_{eff}$ according to the following equation, extracted from Equation (2):

$$E_{eff} = E_p + 1/2(E_i - E_t) \tag{4}$$

The Arrhenius-type plot is illustrated in Figure 2b. As can be seen, all data follow a very good straight line, and the slope estimated provides an activation energy equal to 90.6 ± 1.1 kJ/mol ($R^2$ = 0.999). This value is close to the literature value of PHEMA found in our previous publication, i.e., 89 ± 3.1 kJ/mol ($R^2$ = 0.997) [18], and also near to that of PMMA, i.e., 84 kJ/mol [21]. It should be noted that lower values have been reported in the literature (i.e., 56.7 [10]; 63.6 [12]; 73.2 kJ/mol [11]), though they have been estimated by integral data on non-isothermal polymerization that have been proven to be not strictly correct. In addition, the commercially-available monomer, HEMA, used by different groups of authors is not pure and includes inhibitors or other products, which affect the polymerization kinetics.

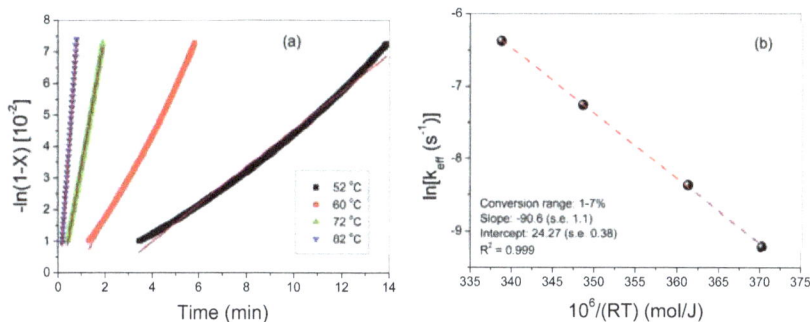

**Figure 2.** Estimation of the effective kinetic rate constant, $k_{eff}$, using experimental data in the conversion region 1–7% at several temperatures (**a**) and the Arrhenius-type plot to calculate the overall (effective) activation energy of HEMA polymerization (**b**).

For PHEMA, the propagation activation energy has been estimated by Buback et al. using the pulsed laser polymerization (PLP)/size-exclusion chromatography (SEC) technique and found equal to 21.9 ± 1.5 kJ/mol [22]. Moreover, for the BPO initiator, the decomposition activation energy is equal to 143 kJ/mol [23]. Then, using the value estimated for the overall (effective) activation energy (i.e., 90.6 kJ/mol) and Equation (4), the activation energy of the termination reaction can be estimated. This was found to be 5.6 kJ/mol, very close to the value proposed for MMA polymerization (i.e., 5.89 kJ/mol [24]).

Furthermore, in order to provide values for the individual kinetic rate constants of the HEMA polymerization, we used for the initiator BPO the values reported in [23], i.e., $k_d$ = 5 × $10^{16}$ exp(−143,000/RT) s$^{-1}$ and $f$ = 0.5. Then, from Equation (2) and assuming that $[I] \cong [I]_0$, the values of $(k_{p0}/\sqrt{k_{t0}})$ can be estimated and included in Table 1. The assumption of a nearly constant initiator concentration is valid since in the time horizon where the values of $k_{eff}$ were estimated, the ratio of $[I]/[I]_0$ was never lower than 0.998. For the propagation rate constant, that reported by Buback [22], i.e., ln ($k_p$ (L/mol/s)) = 16.0 − 2634/$T$ (K), can be used. In such a way, the termination rate constant can be estimated, and values at the temperatures investigated are included in Table 1.

**Table 1.** Values of kinetic and diffusion-controlled parameters obtained in this study during bulk polymerization of HEMA with the benzoyl peroxide (BPO) initiator.

| $T$ (°C) | $k_{eff}$ (min$^{-1}$) | $k_{p0}/\sqrt{k_{t0}}$ (L/mol/s)$^{1/2}$ | $k_{p0}$ [22] (L/mol/s) | $k_{t0}$ (L/mol/s) | $R$ (L/mol) | $A_p$ | $f_{cp}$ | $A_t$ | $f_{ct}$ |
|---|---|---|---|---|---|---|---|---|---|
| 52 | 0.00597 | 1.1278 | 2685 | 5.667E6 | 3.20 | 0.595 | 0.0500 | 1.40 | 0.083 |
| 60 | 0.01400 | 1.4014 | 3262 | 5.78E6 | 2.01 | 0.64 | 0.0485 | 1.35 | 0.084 |
| 72 | 0.04235 | 1.7257 | 4295 | 6.195E6 | 1.34 | 0.85 | 0.0481 | 1.28 | 0.0876 |
| 82 | 0.10222 | 2.0641 | 5326 | 6.657E6 | 0.91 | 1.14 | 0.0445 | 1.39 | 0.0858 |

In order to simulate the experimental data over the whole conversion range, using Equation (2), the next step is to use appropriate equations accounting for the variation of the termination and propagation rate constants during polymerization. Though a number of sophisticated models have been developed [20], we followed here the one proposed by Bowman et al. [16], for the following reasons: it is rather simple and allows estimation of all parameters directly from the simulation of experimental data without integrating the reaction rate; it was originally developed for methacrylate polymerization, such as HEMA; and it is based on DSC data on the reaction rate, as those reported here. In brief, the effect of diffusion-controlled phenomena on the termination and propagation rate constants is taken into consideration using the following equations [16]:

$$\frac{1}{k_p} = \frac{1}{k_{p0}} + \frac{1}{k_{p0}\exp\left[-A_p\left(1/V_f - 1/V_{f,cp}\right)\right]} \tag{5}$$

$$\frac{1}{k_t} = \frac{1}{k_{t0}} + \frac{1}{k_{t,res} + k_{t0}\exp\left[-A_t\left(1/V_f - 1/V_{f,ct}\right)\right]} \tag{6}$$

where $k_{t,res}$ denotes the contribution of the reaction-diffusion term on the termination rate constant, estimated from:

$$k_{t,res} = Rk_p[M] \tag{7}$$

$V_f$ is the fractional free volume of the system, related to the fractional free volumes of monomer ($V_{f,m}$) and polymer ($V_{f,p}$) from:

$$V_f = V_{f,m}\phi_m + V_{f,p}(1 - \phi_m) \tag{8}$$

with:

$$\phi_m = \frac{1 - X}{1 - X + X(\rho_m/\rho_p)} \tag{9}$$

$$V_{f,m} = 0.025 + \alpha_m(T - T_{g,m}) \tag{10}$$

$$V_{f,p} = 0.025 + \alpha_p(T - T_{g,p}) \tag{11}$$

$\phi_m$ is the volume fraction of the monomer; $\alpha$ denotes the coefficient of expansion; $T_g$ the glass transition temperature; and $\rho$ density. Subscripts $m$ and $p$ are associated with the monomer and polymer, respectively.

In the above set of equations, the thermal expansion coefficients, glass transition temperatures and densities of monomer and polymer were taken from Goodner et al. [16], as $\alpha_m = 0.0005\,°C^{-1}$, $\alpha_p = 0.000075\,°C^{-1}$, $T_{g,m} = -60\,°C$, $T_{g,p} = 55\,°C$, $\rho_m = 1.073\,g/cm^3$, $\rho_p = 1.15\,g/cm^3$. The value of the polymer glass transition temperature has been also experimentally measured by Bolbukh et al. [7], and the monomer density has been calculated by Buback and Kurz [22]. The initial concentration of the monomer can then be estimated as $[M]_0 = \rho_m/MW_m = 1.073 \times 1000/130 = 8.25\,mol/L$. The parameters of the diffusion-controlled model that have to be evaluated are $A_p$, $V_{f,cp}$, $A_t$, $V_{f,ct}$ and $R$. To determine these parameters, we followed a slightly modified procedure from that described by Goodner et al. [16]. Accordingly, every set of these parameters can be estimated from linear plots, if we focus only on the particular region of the diffusion-controlled phenomena on the termination and propagation rate constants. As was mentioned above, during polymerization, four regions can be identified. At the early stages (less than 10%), no diffusional limitations on either termination or propagation reactions occur. The second region is that of the gel-effect region, where strong diffusional limitations are present on the termination rate constant, which decreases by several orders of magnitude, while $k_p$ remains constant. The latter also holds in the third region, where the decrease in $k_t$ slows down due to the so-called reaction diffusion controlled termination. This means that although macroradicals cannot easily move in space, they can implicitly move by the addition of monomer molecules. In the fourth region, propagation becomes also diffusion controlled, and $k_p$ significantly reduces with conversion.

Initially, in order to estimate the parameter, $R$, introduced to account for the reaction-diffusion controlled termination, it is assumed that the particular term dominates $k_t$ in this region (near 40–50% conversion), and thus, $k_t$ can be set equal to $k_{t,res}$, i.e., $k_t = k_{t,res} = R\,k_p\,[M]$. Under this condition, the polymerization rate, Equation (2), is expressed as:

$$R_p = \frac{dX}{dt} = k_p \left( \frac{fk_d[I]}{Rk_p[M]} \right)^{1/2} (1 - X) = \left( \frac{k_p f k_d[I](1 - X)}{R[M]_0} \right)^{1/2} \tag{12}$$

The initiator concentration was assumed approximately equal to its initial value, $[I] \sim [I]_0$, since during the whole polymerization time and for all temperatures investigated, the initiator consumption never exceeds 1% (calculated according to its decomposition rate constant). In the reaction-diffusion region, propagation is not diffusion-controlled and $k_p$ is approximated by $k_{p0}$. Then, by rearranging Equation (12), the term, $R$, can be estimated from:

$$R = \left( \frac{k_{p0} f k_d[I](1 - X)}{R_p^2 [M]_0} \right) \tag{13}$$

By plotting the right-hand side of Equation (13) as a function of the conversion, a straight horizontal line should be obtained. This has been done for all temperatures investigated, and at the particular region where this holds, the parameter $R$ was evaluated and reported in Table 1.

Following, the diffusion-controlled parameters for the propagation reaction (i.e., $A_p$ and $V_{f,cp}$) can be estimated from the results obtained in the fourth region, where the termination reaction is still reaction-diffusion controlled, but the propagation rate constant is also diffusion-controlled. Then, squaring Equation (12), rearranging and taking the natural logarithm leads to:

$$\ln \left[ \frac{k_{p0} f k_d[I](1 - X)}{R_p^2 R[M]_0} - 1 \right] = A_p \frac{1}{V_f} - \frac{A_p}{V_{f,cp}} \tag{14}$$

Thus, plotting the left-hand side of Equation (14) vs. $1/V_f$ will provide $A_p$ from the slope and $V_{f,cp}$ from the intercept.

Finally, the gel-effect region can be analyzed similarly. In this region, $k_p$ can again be approximated by $k_{p0}$. During autoacceleration, termination is governed by center-of-mass diffusional limitations, while the reaction-diffusion is still negligible. Under these conditions, the expression for $k_t$ is reduced to:

$$k_t = \frac{k_{t0}}{1 + \exp\left[ A_t \left( 1/V_f - 1/V_{f,ct} \right) \right]} \tag{15}$$

Again, squaring Equation (2), rearranging and taking the natural logarithm leads to:

$$\ln \left[ \frac{k_{t0} R_p^2}{k_{p0}^2 f k_d[I](1 - X)^2} - 1 \right] = A_t \frac{1}{V_f} - \frac{A_t}{V_{f,ct}} \tag{16}$$

By plotting the left-hand side of Equation (16) vs. $1/V_f$, the last two parameters, $A_t$ and $V_{f,ct}$, can be determined.

Using the above procedure, all of the parameters were estimated, and results at different polymerization temperatures are included in Table 1.

It is seen that, as the polymerization temperature increases, the reaction-diffusion parameter, $R$, decreases, whereas $A_p$ increases. In contrast, parameters, $V_{f,cp}$, $V_{f,ct}$ and $A_t$ are slightly affected by temperature. Thus, it seems that increased reaction temperatures lead to a higher mobility of the monomer molecules, and diffusion controlled phenomena affect the propagation reaction at higher conversions. In contrast, higher reaction temperature results in lower termination rate constants in the reaction-diffusion region. This means that the increased temperatures provide greater mobility to the macroradicals to move in space before freezing and terminate only by the implicit addition of monomer molecules. It should be noted here that the estimated values of $R$ lie in between the values of 0.703 and 9.353 calculated for rigid or totally flexible chains [25]. Moreover, Goodner et al. [16] provided a value for $R$ equal to four, which follows our estimations quite well if we extrapolate our vales to a temperature near 35 °C, where these authors carried out their experiments.

Using the parameters reported in Table 1, the termination and propagation rate constants can be estimated at different temperatures, and results appear in Figure 3. It can be observed that $k_p$ maintains a constant value up to 55–75% depending on the temperature. Afterwards, it drops approximately 1.5 orders of magnitude due to the effect of diffusion-controlled phenomena. The $k_t$ curve shows a more complex evolution. It drops quickly from its initial value, as growing macro-radicals experience diffusion limitations and autoacceleration occurs. After a reduction by two orders of magnitude and a conversion approximately equal to 50%, reaction-diffusion termination starts to dominate, and the decrease in $k_t$ is much more gradual. This slow decrease over the conversion range from 45–65% is due to the fact that the reaction-diffusion termination rate constant, $k_{t,res}$, is proportional to the monomer concentration, $[M]$, denoted by the unreacted double-bonds' concentration. Since polymerization proceeds, the latter is still decreasing. When the glass-effect appears and $k_p$ starts to drop, $k_t$ again decreases through the reaction-diffusion proportionality. Moreover, from Figure 3, it can be observed that at lower temperatures, where both the small molecule and the macro-radical mobility is lower, the effect of diffusion-controlled phenomena is more pronounced in both $k_t$ and $k_p$.

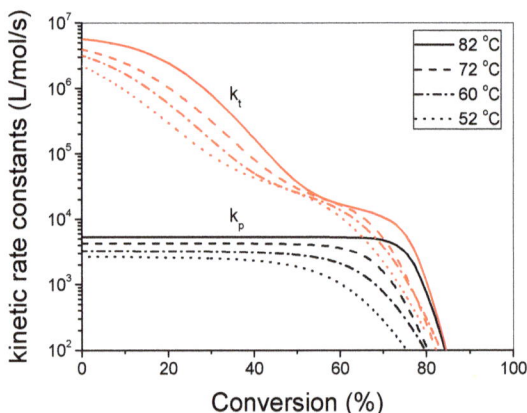

**Figure 3.** Variation of the termination and propagation rate constants estimated from Equations (5) and (6) with conversion at several reaction temperatures, using the parameters reported in Table 1.

Finally, as can be seen in Figure 4, using this simplified model, an extremely good simulation of the experimental data at all different temperatures was obtained. Both the initial rate and the location and value of the maximum rate of polymerization are reproduced to within a few percent in all cases. Furthermore, the simulation captures the decrease in the rate almost exactly. What is important to note from the results shown in Figure 4 is that all parameters used were determined from the previous polymerization study, and the fitting of the reaction rate did not employ any additional adjustable parameter.

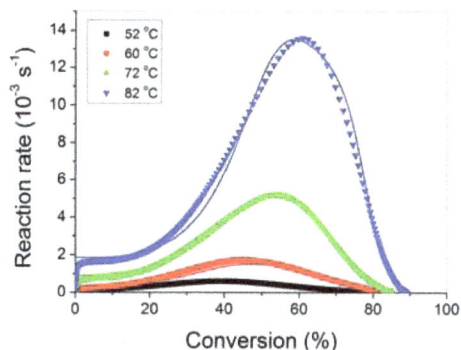

**Figure 4.** Comparison of the experimental and simulated rate curves for HEMA polymerization at different temperatures.

## 3.2. Polymerization Kinetics during the Formation of PHEMA/OMMT Nanocomposites

Subsequently, the effect of adding a commercially-available nano-OMMT, namely Cloisite 15A, on the polymerization kinetics of HEMA at several temperatures was investigated. Nanocomposites of HEMA with different amounts of nano-montmorillonite were prepared.

The morphology of the nanocomposites was examined using X-ray analysis. The XRD diffractograms of the nanocomposites with 1, 3 and 5 wt % OMMT (i.e., Cloisite 15A), as well as of the pristine nano-clay appear in Figure 5. For the commercial OMMT, a large peak was measured at 2.99° denoting a *d*-spacing of 2.97 nm, close to the value reported by the manufacturer, i.e., 3.15 nm. All nanocomposites presented clear, but lower in intention peaks at $2\theta$ angles equal to 2.07, 2.21 and 2.34° for the materials with 1, 3 and 5 wt % OMMT, respectively. These correspond to $d_{001}$-spacing of 4.29, 4.02 and 3.79 nm, respectively (Figure 5). All of these values were higher compared to the pristine nano-clay. This shift suggests an increase in the basal spacing of the silicate platelets, which is attributed to the penetration of the macromolecular chains into the clay platelets. From these observations and considering that the measured peaks are rather weak and broad, it can be said that the morphology of the nanocomposites produced was mainly intercalated and partially exfoliated.

The effect of the amount of nanofiller on the variation of the polymerization rate and double bond conversion with time at two constant temperatures appears in Figures 6 and 7, respectively. The presence of the nano-clay was found to enhance polymerization kinetics leading to higher conversion values at a specific reaction time.

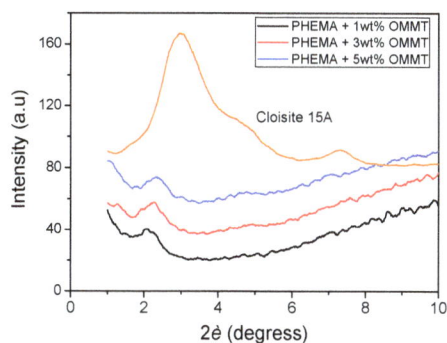

**Figure 5.** XRD diffraction patterns of the organomodified montmorillonite (OMMT) (Cloisite 15A) used and the PHEMA/OMMT nanocomposites.

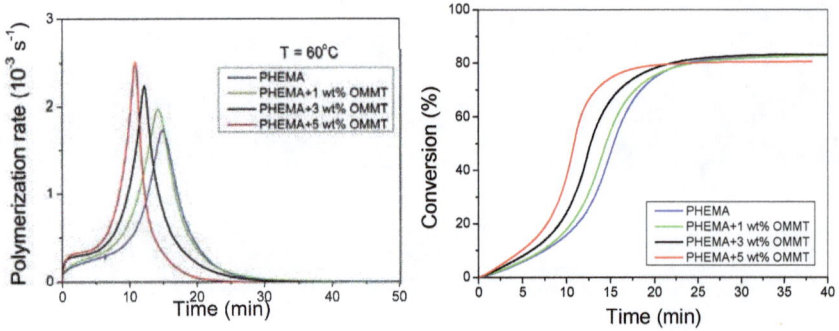

**Figure 6.** Effect of the amount of nano-OMMT on the variation of polymerization rate (**a**) and conversion (**b**) with time during polymerization of HEMA at 60 °C.

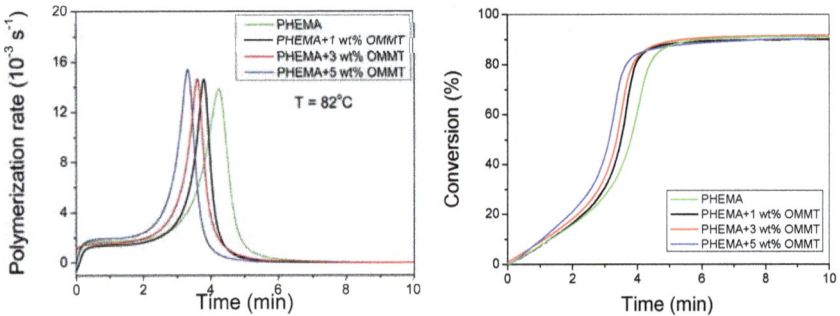

**Figure 7.** Effect of the amount of nano-OMMT on the variation of polymerization rate (**a**) and conversion (**b**) with time during polymerization of HEMA at 82 °C.

The effective rate constants were estimated at both 60 and 82 °C according to the procedure described in the previous section. The relative amounts of $k_{eff}$ compared to that of neat PHEMA, appear in Figure 8. It was observed that, at relatively low temperatures (i.e., 60 °C), the relative effective rate constant significantly increases with the amount of the nano-OMMT, whereas at higher temperatures, a much lower increase is clear.

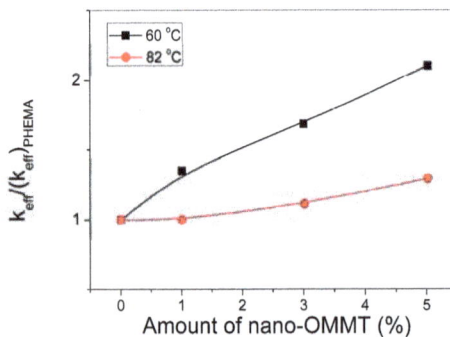

**Figure 8.** Relative effective kinetic rate constants obtained during polymerization of PHEMA/n-OMMT nanocomposites at two temperatures as a function of the amount of n-OMMT.

Furthermore, the overall activation energy of the polymerization was calculated according to the procedure described in Section 3.1, and a value equal to 76.5 kJ mol$^{-1}$ for the nanocomposite with 5 wt % nano-clay was obtained. This is lower compared to the corresponding neat PHEMA (i.e., 90.6 kJ mol$^{-1}$). In general, one would assume higher activation energy of the nanocomposites due to the extra barriers that the nano-filler adds into the polymerizing mixture. The reason for the enhanced polymerization rate can be found if one examines polymerization at a microscopic level, presented in the Discussion section.

### 3.3. Polymerization Kinetics during the Formation of PHEMA/Silica Nanocomposites

In this section, the effect of adding different amounts of nano-silica in the polymerization of HEMA is investigated. The results of the variation of the polymerization rate and conversion with time at 82 °C are included in Figure 9. In contrast to the nano-OMMT, adding nano-silica seems to retard the polymerization rate and shift the conversion vs. time curves to higher reaction times. In order to have safe results, the experiments were repeated also at lower reaction temperatures. Results are illustrated in Figure 10. It is seen that at all different reaction temperatures, the effect of nano-silica on the polymerization rate is the same. The reaction is retarded, and the conversion curves are shifted to higher times with increasing amounts of the nano-silica added.

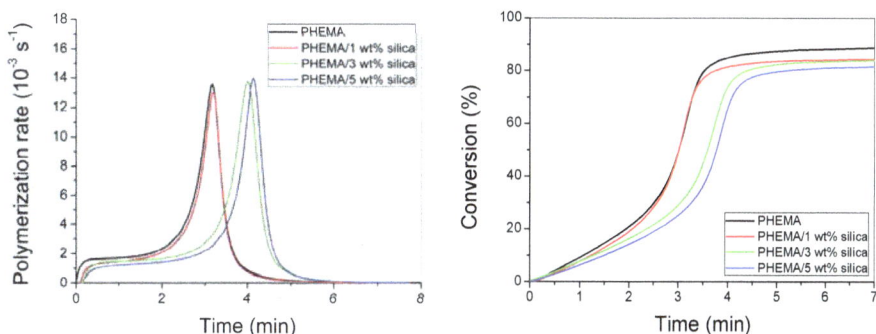

**Figure 9.** Effect of the amount of nano-silica on the variation of polymerization rate (**a**) and conversion (**b**) with time during polymerization of HEMA at 82 °C.

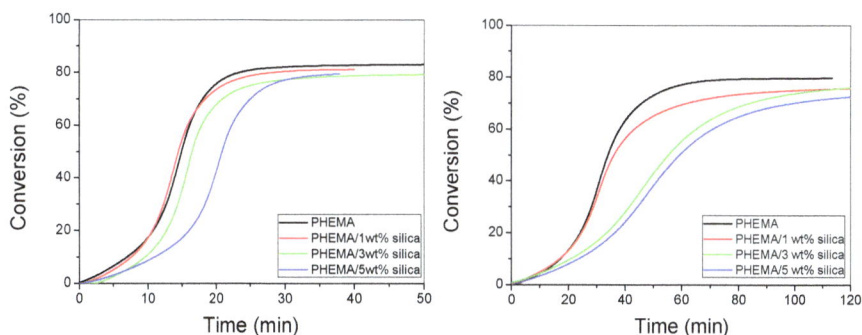

**Figure 10.** Effect of the amount of nano-silica on the variation of conversion with time during polymerization of HEMA at 60 (**a**) and 52 °C (**b**).

Furthermore, the effective rate constants were estimated at all four temperatures, according to the procedure described in the previous section. The relative amounts of $k_{eff}$ compared to that of neat

PHEMA appear in Figure 11. It can be noticed that, at all reaction temperatures, the relative effective rate constant significantly decreases with the amount of the nano-silica, reaching almost 70% of the initial value when adding 5% of the nano-additive.

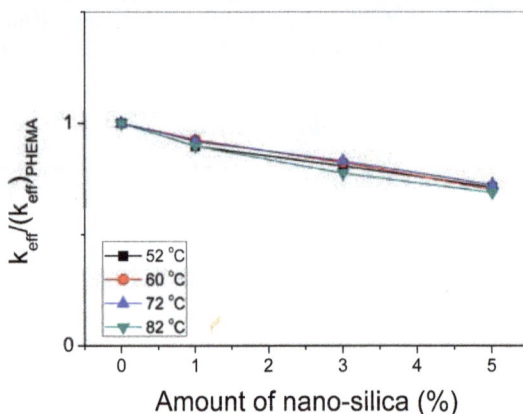

**Figure 11.** Relative effective kinetic rate constants obtained during polymerization of PHEMA/nano-silica nanocomposites at temperatures of 52, 60, 72 and 82 °C as a function of the amount of the additive.

Finally, in order to estimate the effect of the nano-silica on the overall activation energy of the polymerization, Arrhenius-type plots were constructed at all amounts of the nano-additive and illustrated in Figure 12. Initially, it can be seen that the effective kinetic rate constant of the nanocomposites is always slightly lower compared to neat PHEMA at all temperatures and decreased with the amount of the nano-silica. Moreover, the activation energies estimated were $90.6 \pm 2.4$, $90.8 \pm 3.9$ and $91.2 \pm 3.2$ kJ/mol for the nanocomposites with 1, 3 and 5 wt % nano-silica, respectively. These values are similar to neat PHEMA (i.e., 90.6 kJ/mol) showing a slightly increasing trend with the amount of the nano-silica added.

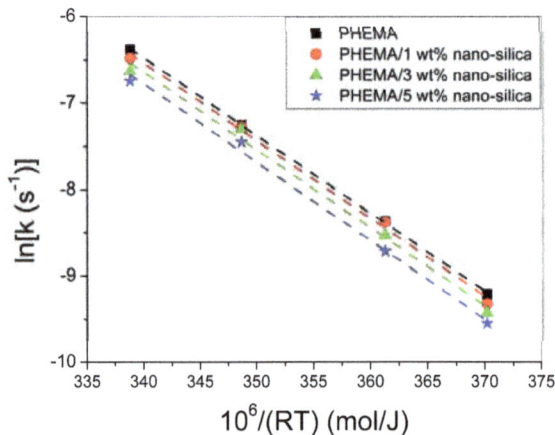

**Figure 12.** Arrhenius-type plots to calculate the overall (effective) activation energies of PHEMA/nano-silica nanocomposites.

## 4. Discussion

Although the polymerization kinetics of HEMA have been investigated in the literature [12–17], the development of a model, including both kinetic and diffusion-controlled parameters that have been estimated from experimental measurements, valid over the whole conversion range, has not appeared so far. In this framework, in the first part of this research, an accurate, rather simple, kinetic model for the polymerization of HEMA was provided, based on experimental DSC measurements. The variation of the polymerization rate with time at several isothermal reaction temperatures was presented, and kinetic rate constants and diffusion-controlled parameters were evaluated. It was found that using these parameters the evolution of the experimental polymerization rate with time can be simulated very well. All parameters estimated at different reaction temperatures are included in Table 1. These could be used in the development of more complex models for the simulation of the polymerization kinetics of HEMA and its copolymers.

In the second part, nanocomposites of PHEMA with several relative amounts of nano-clay and nano-silica were prepared in order to investigate the effect of the nano-additive on the polymerization kinetics of HEMA. The in situ bulk radical polymerization technique was investigated. Isothermal experimental data were obtained from DSC measurements. It was found that the inclusion of the nano-montmorillonite results in a slight enhancement of the polymerization rate, while the inverse holds when adding nano-silica.

The interpretation of these results can be carried out in terms of specific interactions and particularly the formation of intra- and inter-chain hydrogen bonds between the monomer and the polymer molecules. The monomer, 2-hydroxyethyl methacrylate, contains one hydroxyl (–OH) and one carbonyl (C=O) group on its molecule. The C=O group acts only as the proton acceptor, while the OH group acts as both the proton donor and acceptor [26]. Hydrogen bonding between the monomer hydroxyl group and carbonyl oxygen atom strengthens the positive partial charges at the carbonyl C atom and at the double bond, as shown schematically in Scheme 1, leading to a significant charge transfer in the transition state of propagation [27]. In the polymer, PHEMA, both $OH \cdots OH$ and $C = O \cdots HO$ types of hydrogen-bonds can occur (Scheme 1). Not only the dimer structure ($OH \cdots OH \cdots$), but also the aggregate structure ($\cdots OH \cdots OH \cdots OH \cdots$) have been found in many systems, including liquid alcohols and solid polymers [26]. It has been found that 53.7% of the OH group on the PHEMA side chain terminal contributes to the $OH \cdots OH$ type of hydrogen-bond, while the remaining 47.3% are engaged in the $OH \cdots O = C$ type of hydrogen bond, at ambient temperature [26].

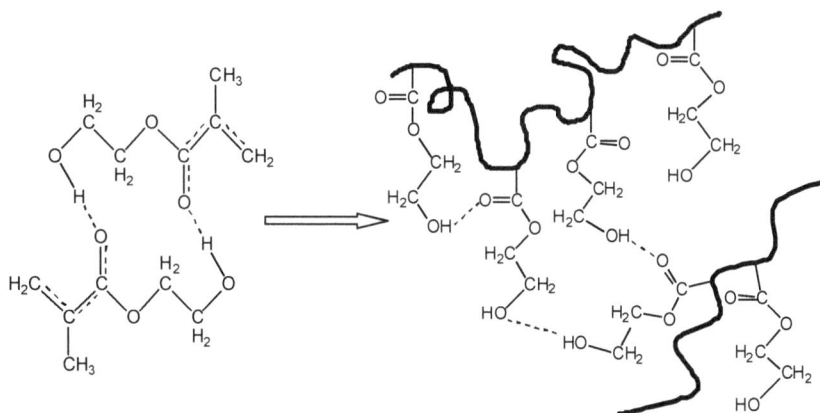

**Scheme 1.** Schematic illustration of the polymerization of HEMA to PHEMA and the intra- and inter-chain hydrogen bonds formed.

When the nano-clay, OMMT, is added to the system, the clay platelets are partially exfoliated due to the ultrasound agitation and polymerization, by the insertion of the macromolecular chains in between the clay galleries, as was observed from XRD measurements. Then, the HEMA-HEMA interactions with the hydrogen bonding are disrupted, by the existence of the clay platelets inserted in between the monomer molecules and macromolecular chains, as shown schematically in Scheme 2. This could result in more reactive monomer molecules during polymerization that could facilitate the reaction rate, resulting in higher kinetic rate constants and less overall activation energy. The disruption of hydrogen bonding between HEMA molecules is thus consistent with the increased rate of monomer addition to the macroradicals compared to that in neat polymerization. Moreover, at high temperatures, monomer molecules and macroradicals have enough mobility, and their movement is not affected much by the presence of the side hydrogen bonds. However, at lower temperatures, hydrogen bonding significantly decreases the reactivity of radicals and monomer molecules, and as a result, the presence of nano-clays, which disrupt those bonds, contributes much to a higher polymerization rate (Figure 7).

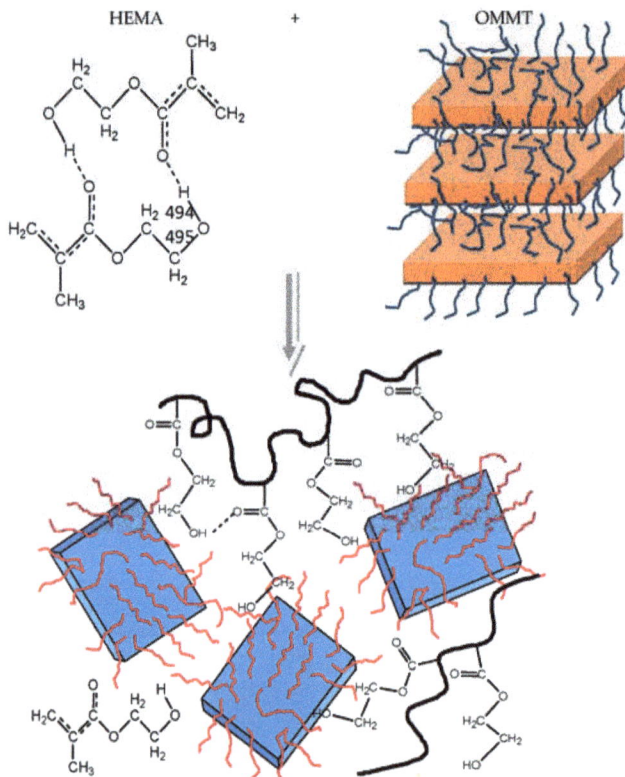

**Scheme 2.** Schematic illustration of the polymerization of HEMA to PHEMA in the presence of nano-OMMT.

In contrast, polymerization of HEMA in the presence of nano-silica resulted in retarded polymerization rates and lower effective kinetic rate constants compared to neat PHEMA. It seems that the presence of hydroxyl groups in the surface of the nano-silica results in the formation of hydrogen bonds between the polymer and the nano-additive, resulting in lower reactivity of the monomer. Intra-molecular HEMA-HEMA hydrogen bonding interactions are disturbed and replaced by inter-molecular hydrogen bonds between the hydroxyls in the surface of silica with carbonyls and/or hydroxyls present in the PHEMA macromolecular chain, as shown schematically in Scheme 3. This gives rise to a reduction in the polymerization rate compared to neat HEMA, more pronounced as the amount of nano-silica is increased. It should be noted here that similar results have been observed during polymerization of HEMA in polar solvents, such as DMF [28].

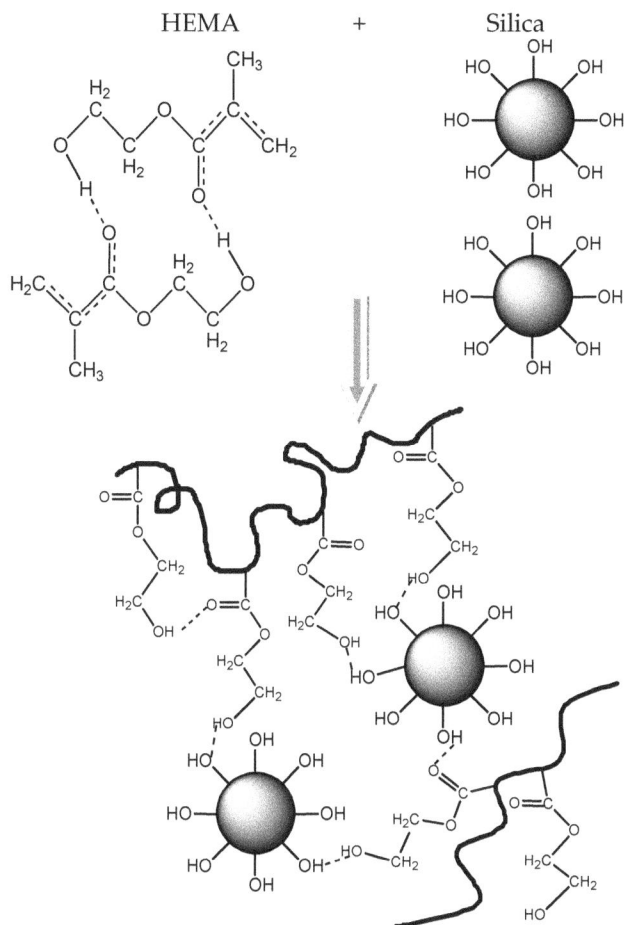

**Scheme 3.** Schematic illustration of the polymerization of HEMA to PHEMA in the presence of nano-silica.

The above assumptions, concerning hydrogen bonding of PHEMA during polymerization, were partially verified by FTIR measurements. The FTIR spectra of neat PHEMA and the nanocomposites were recorded, and details in the regions of interest are shown in Figure 13. Particularly, Figure 13a shows the IR spectra in the O–H stretching region, whereas Figure 13b shows the corresponding IR spectra in the C=O stretching region.

Several contributions from 3100–3700 cm$^{-1}$ are identified in the O–H stretching region. According to Morita et al. [26], the band at 3536 cm$^{-1}$ is attributed to hydrogen-bonded hydroxyl groups, whereas at 3624–3660 cm$^{-1}$ to free OH. According to Figure 13a, signals at the latter region were not identified in any material investigated. Therefore, it seems that there are not many free hydroxyls, i.e., the OH groups are not donating hydrogen bonds. Moreover, a dominant peak at 3536 cm$^{-1}$ was recorded for neat PHEMA, meaning the existence of hydrogen-bonded hydroxyls. In the PHEMA/SiO$_2$ nanocomposites, this peak was shifted to lower wavenumber at 3430 cm$^{-1}$, providing evidence of the gradual association of the OH$\cdots$OH type of hydrogen bonds with the addition of the nanosilica.

In the C=O stretching region, two contributions around 1730 and 1637 cm$^{-1}$ were identified. Again, these bands are assigned to free C=O groups and hydrogen-bonded carbonyl groups, respectively [26]. The peak position and width of the band around 1730 cm$^{-1}$ slightly changed with the addition of the nanofiller, whereas for the band around 1637 cm$^{-1}$, the position again slightly changed, although its width varied significantly. Particularly, the peak at 1637 cm$^{-1}$ almost disappeared at the PHEMA/OMMT nanocomposite, whereas it became very broad and intensive in the PHEMA/silica materials. Therefore, it seems that only a small number of hydrogen bonded C=O appears in neat PHEMA, which is negligible in PHEMA/OMMT nanocomposites. In contrast, in the C=O stretching region, it was found that the association of the C = O$\cdots$HO-type of hydrogen bond occurs to a large extent when silica is used.

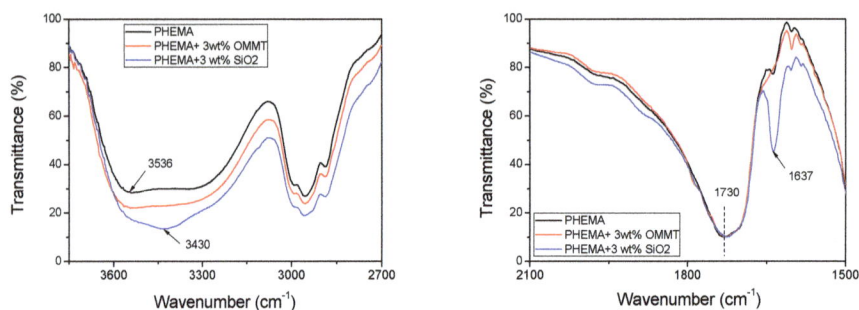

**Figure 13.** Details of the FTIR spectra of neat PHEMA and PHEMA nanocomposites with 3 wt % OMMT or 3 wt % silica in the region 3750–2700 cm$^{-1}$ (**a**) and 2100–1500 cm$^{-1}$ (**b**).

## 5. Conclusions

In this study, the polymerization kinetics of PHEMA was studied both experimentally using DSC measurements and theoretically. A simplified kinetic model was developed taking into account both the chemical reactions and diffusion-controlled phenomena and valid over the whole conversion range. Kinetic and diffusion-controlled parameters were evaluated, and the model was able to simulate successfully the variation of the polymerization rate with time. The main conclusion drawn from the study on the formation of nanocomposite materials based on PHEMA is that the polymerization rate can be either enhanced or decreased depending on the type and the amount of the nanofiller used. Thus, the effective overall kinetic rate constant was found to increase with the amount of the nano-filler when a nano-clay was used, whereas the inverse was observed when nano-silica was used. It seems that nano-compounds having surface functional groups, such as hydroxyls in the case of silica, may form hydrogen bonds with the carbonyls or hydroxyl groups of the polymer, resulting somehow in the retardation of the reaction. In contrast, the addition of clay platelets may result in the dissociation of hydrogen bonds existing in the polymer molecules, resulting thus in slightly higher reaction rates.

**Author Contributions:** D.S.A. and P.I.S. conceived of and designed the experiments. P.I.S. performed the experiments. D.S.A. wrote the paper and performed the simulation analysis.

**Conflicts of Interest:** The authors declare no conflict of interest.

## References

1.  Nogueira, N.; Conde, O.; Minones, M.; Trillo, J.M.; Minones, J.R. Characterization of poly(2-hydroxyethyl methacrylate) contact lens using the Langmuir monolayer technique. *J. Colloid Interface Sci.* **2012**, *385*, 202–210. [CrossRef] [PubMed]

2.  Montheard, J.P.; Chatzopoulos, M.; Chappard, D. 2-Hydroxyethyl methacrylate (HEMA): Chemical properties and applications in biomedical fields. *J. Macromol. Sci. C Polym. Rev.* **1992**, *32*, 1–34. [CrossRef]

3.  Gupta, M.K.; Bajpai, J.; Bajpai, A.K. Optimizing the release process and modelling of in vitro release data of *cis*-dichlorodiamminoplatinum (II) encapsulated into poly(2-hydroxyethyl methacrylate) nanocarriers. *Mater. Sci. Eng. C* **2016**, *58*, 852–862. [CrossRef] [PubMed]

4.  Costantini, A.; Luciani, G.; Annunziata, G.; Silvestri, B.; Branda, F. Swelling properties and bioactivity of silica gel/pHEMA Nanocomposites. *J. Mater. Sci. Mater. Med.* **2006**, *17*, 319–325. [CrossRef] [PubMed]

5.  Kharismadewi, D.; Haldorai, Y.; Nguyen, V.H.; Tuma, D.; Shim, J.-J. Synthesis of graphene oxide-poly(2-hydroxyethyl methacrylate) composite by dispersion polymerization in supercritical $CO_2$: Adsorption behavior for the removal of organic dye. *Compos. Interfaces* **2016**, *23*, 7. [CrossRef]

6.  Moradi, O.; Aghaie, M.; Zare, K.; Monajjemi, M.; Aghaie, H. The study of adsorption characteristics $Cu^{2+}$ and $Pb^{2+}$ ions onto PHEMA and P(MMA-HEMA) surfaces from aqueous single solution. *J. Hazard. Mater.* **2009**, *170*, 673–679. [CrossRef] [PubMed]

7.  Bolbukh, Y.; Klonos, P.; Roumpos, K.; Chatzidogiannaki, V.; Tertykh, V.; Pissis, P. Glass transition and hydration properties of polyhydroxyethylmethacrylate filled with modified silica nanoparticles. *J. Therm. Anal. Calorim.* **2016**, *125*, 1387–1398. [CrossRef]

8.  Nikolaidis, A.K.; Achilias, D.S.; Karayannidis, G.P. Synthesis and characterization of PMMA/organomodified montmorillonite nanocomposites prepared by in situ bulk polymerization. *Ind. Eng. Chem. Res.* **2011**, *50*, 571–579. [CrossRef]

9.  Achilias, D.S.; Siafaka, P.I.; Nikolaidis, A.K. Polymerization kinetics and thermal properties of poly(alkyl methacrylate)/organomodifiend montmorillonite nanocomposites. *Polym. Int.* **2012**, *61*, 1510–1518. [CrossRef]

10. Passos, M.F.; Dias, D.R.C.; Bastos, G.N.T.; Jardini, A.L.; Benatti, A.C.B.; Dias, C.G.B.T.; Maciel Filho, R. PHEMA Hydrogels. *J. Therm. Anal. Calorim.* **2016**, *125*, 361–368. [CrossRef]

11. Ning, L.; Xu, N.; Xiao, C.; Wang, R.; Liu, Y. Analysis for the reaction of hydroxyethyl methacrylate/benzoyl peroxide/polyethacrylate through DSC and viscosity changing and their resultants as oil absorbent. *J. Macromol. Sci. Part A Pure Appl. Chem.* **2015**, *52*, 1017–1027. [CrossRef]

12. Huang, C.-W.; Sun, Y.-M.; Huang, W.-F. Curing kinetics of the synthesis of poly(2-hydroxyethyl methacrylate) (PHEMA) with ethylene glycol dimethacrylate (EGDMA) as a crosslinking agent. *J. Polym. Sci. A Polym. Chem.* **1997**, *35*, 1873–1889. [CrossRef]

13. Kaddami, H.; Gerard, J.F.; Hajji, P.; Pascault, J.P. Silica-filled poly(hema) from hema/grafted SiO₂ nanoparticles: Polymerization kinetics and rheological changes. *J. Appl. Polym. Sci.* **1999**, *73*, 2701–2713. [CrossRef]

14. Li, L.; Lee, L.J. Photopolymerization of HEMA/DEGDMA hydrogels in solution. *Polymer* **2005**, *46*, 11540–11547. [CrossRef]

15. Mikos, A.G.; Peppas, N.A. A model for prediction of the structural characteristics of EGDMA crosslinked PHEMA microparticles produced by suspension copolymerization/crosslinking. *J. Controll. Release* **1987**, *5*, 53–62. [CrossRef]

16. Goodner, M.D.; Lee, H.R.; Bowman, C.N. Method for determining the inetic parameters in diffusion-controlled free-radical homopolymerizations. *Ind. Eng. Chem. Res.* **1997**, *36*, 1247–1252. [CrossRef]

17. Hacioglu, B.; Berchtold, K.A.; Lovell, L.G.; Nie, J.; Bowman, C.N. Polymerization kinetics of HEMA/DEGDMA: using changes in initiation and chain transfer rates to explore the effects of chain-length-dependent termination. *Biomaterials* **2002**, *23*, 4057–4064. [CrossRef]

18. Achilias, D.S. Investigation of the radical polymerization kinetics using DSC and mechanistic or isoconversional methods. *J. Therm. Anal. Calorim.* **2014**, *116*, 1379–1386. [CrossRef]

19. Siafaka, P.; Achilias, D.S. Polymerization kinetics and thermal degradation of poly(2-hydroxyethyl methacylate)/organomodified montmorillonite nanocomposites prepared by in situ bulk polymerization. *Macromol. Symp.* **2013**, *331–332*, 166–172. [CrossRef]

20. Achilias, D.S. A review of modeling of diffusion controlled polymerization reactions. *Macromol. Theory Simul.* **2007**, *16*, 319–347. [CrossRef]
21. Achilias, D.S.; Verros, G.D. Modeling of Diffusion-Controlled Reactions in Free Radical Solution and Bulk Polymerization: Model Validation by DSC Experiments. *J. Appl. Polym. Sci.* **2010**, *116*, 1842–1856. [CrossRef]
22. Buback, M.; Kurz, C.H. Free-radical propagation coefficients for cyclohexyl methacrylate, glycidyl methacrylate and 2-hydroxyethyl methacrylate homopolymerizations. *Macromol. Chem. Phys.* **1998**, *199*, 2301–2310. [CrossRef]
23. Siddiqui, M.N.; Redhwi, H.H.; Vakalopoulou, E.; Tsagkalias, I.; Ioannidou, M.D.; Achilias, D.S. Synthesis, characterization and reaction kinetics of PMMA/silver nanocomposites prepared via in situ radical polymerization. *Eur. Polym. J.* **2015**, *72*, 256–269. [CrossRef]
24. Zoller, A.; Gigmes, D.; Guillaneuf, Y. Simulation of radical polymerization of methyl methacrylate at room temperature using a tertiary amine/BPO initiating system. *Polym. Chem.* **2015**, *6*, 5719–5727. [CrossRef]
25. Achilias, D.S.; Kiparissides, C. Development of a general mathematical framework for modeling of diffusion-controlled free-radical polymerization reactions. *Macromolecules* **1992**, *25*, 3739–3750. [CrossRef]
26. Morita, S. Hydrogen-bonds structure in poly(2-hydroxyethyl methacrylate) studied by temperature-dependent infrared spectroscopy. *Front. Chem.* **2014**, *2*, 1–5. [CrossRef] [PubMed]
27. Liang, K.; Hutchinson, R.A. Solvent Effects on Free-Radical Copolymerization Propagation Kinetics of Styrene and Methacrylates. *Macromolecules* **2010**, *43*, 6311–6320. [CrossRef]
28. Furuncuoglu Ozaltin, T.; Dereli, B.; Karahan, O.; Salman, S.; Aviyente, V. Solvent effects on free-radical copolymerization of styrene and 2-hydroxyethyl methacrylate: A DFT study. *New J. Chem.* **2014**, *38*, 170–178. [CrossRef]

*processes*

MDPI

*Article*

# Aqueous Free-Radical Polymerization of Non-Ionized and Fully Ionized Methacrylic Acid

Eric Jean Fischer, Giuseppe Storti and Danilo Cuccato *

Institute for Chemical and Bioengineering, ETH Zurich, Vladimir-Prelog-Weg 1, 8093 Zurich, Switzerland; eric.fischer@chem.ethz.ch (E.J.F.); giuseppe.storti@chem.ethz.ch (G.S.)
* Correspondence: danilo.cuccato@chem.ethz.ch; Tel.: +41-44-632-66-60

Academic Editor: Alexander Penlidis
Received: 16 February 2017; Accepted: 24 April 2017; Published: 27 April 2017

**Abstract:** Water-soluble, carboxylic acid monomers are known to exhibit peculiar kinetics when polymerized in aqueous solution. Namely, their free-radical polymerization rate is affected by several parameters such as monomer concentration, ionic strength, and pH. Focusing on methacrylic acid (MAA), even though this monomer has been largely addressed, a systematic investigation of the effects of the above-mentioned parameters on its polymerization rate is missing, in particular in the fully ionized case. In this work, the kinetics of non-ionized and fully ionized MAA are characterized by in-situ nuclear magnetic resonance (NMR). Such accurate monitoring of the reaction rate enables the identification of relevant but substantially different effects of the monomer and electrolyte concentration on polymerization rate in the two ionization cases. For non-ionized MAA, the development of a kinetic model based on literature rate coefficients allows us to nicely simulate the experimental data of conversion versus time at a high monomer concentration. For fully ionized MAA, a novel propagation rate law accounting for the electrostatic interactions is proposed: the corresponding model is capable of predicting reasonably well the electrolyte concentration effect on polymerization rate. Nevertheless, further kinetic information in a wider range of monomer concentrations would be welcome to increase the reliability of the model predictions.

**Keywords:** methacrylic acid; free radical polymerization; modeling; propagation; termination; electrostatic interactions; electrostatic screening; kinetics; NMR

---

## 1. Introduction

Water soluble polymers are very attractive materials which have applications in many different fields: they are particularly employed in the manufacturing of pharmaceuticals and cosmetics, in enhanced oil separation and water purification processes, and as additives for thickening, flocculation, coating, etc. [1–3]. The peculiar properties of such polymers come from their building blocks, namely water soluble vinyl monomers exhibiting polarizable, ionizable, or charged moieties. Typical examples of such monomers are acrylic and methacrylic acid as well as their esters, acrylamides and vinyl amides, and ammonium salts. The presence of charges and polarized groups in such monomers makes the corresponding polymers suitable for establishing electrostatic interactions between the polymer chains as well as between the polymer and its environment. As a consequence, the solution and surface properties of the polymer as well as the viscoelastic behavior of the material are affected by such interactions, determining the peculiar features of water-soluble polymers which make them so interesting and versatile [4,5].

In addition to the effect on the material properties, these interactions involving monomer and polymer moieties have an impact on the reaction kinetics during aqueous radical polymerization, which is the usual method of synthesis of water-soluble polymers. The presence of charges or dipoles can induce interactions of various nature between the reacting species, and they can in turn substantially affect the observed reaction order of some relevant kinetic steps with respect to the reactants [6]. Monomer-monomer association [7], hindrance of the internal motions of reaction complexes due to intermolecular interactions [8], and electrostatic forces producing diffusion limitations [9–11] are some of the relevant phenomena which may affect the kinetic behavior of water-soluble, ionized, or ionizable monomers, especially with respect to propagation reactions.

A general effect of reduction of the propagation rate coefficient upon increasing monomer concentration has been detailed by studies on various water-soluble monomers, and it has been explained in terms of "fluidization" of the transition state structure due to its interactions with the surrounding species, in particular hydrogen bonding involving water molecules [8,12,13]. The electrostatic effects on the propagation kinetics of ionized or ionizable monomers have been discussed in terms of a reduced diffusion of the monomer to the radical site of an active chain when both are similarly charged, as a result of the repulsion forces. In this view, the sensitivity of the propagation kinetics to the initial monomer concentration is explained by the phenomenon of electrostatic screening. Increasing the concentration of electrolyte in a solution containing charged species reduces the strength of the electrostatic interactions among the charges by screening of the repulsive forces. In the case of polymerization of ionized monomers, the monomer itself acts as an electrolyte in solution: accordingly, increasing the concentration of ionized monomer enhances its propagation kinetics [6,11,14]. The most recent studies on the effect of charge interactions on the kinetics of water-soluble, ionized, or ionizable monomers have been largely focused on copolymer composition: in this way, it is possible to isolate the contribution of propagation reactions and specifically address the sensitivity of propagation reactivity ratios upon changes in ionic strength, monomer concentration and ionization, pH, etc. [10,11,15,16]. On the other hand, the investigation of such effects on the polymerization rate is of great interest due to its implications on the final polymer properties (e.g., molar masses).

In this context, methacrylic compounds offer the advantage of not involving secondary reactions, i.e., backbiting, which can strongly affect the polymerization rate and make it difficult to focus on the effect of charge interactions only [17–19]. The kinetics of methacrylic acid (MAA) has been the subject of several studies by pulsed-laser polymerization (PLP), which have been focused especially on the estimation of propagation [12,20] and termination [21,22] rate coefficients. In addition, some attempts of characterizing and modeling the polymerization kinetics of MAA have been carried out recently, in some cases together with the analysis of the molar masses and of the effect of chain transfer reactions [23–25]. Nevertheless, the majority of the existing studies are focused on non-ionized MAA and limited to medium-large monomer concentrations. Experimental and modeling studies on fully ionized MAA are missing in the literature, with the notable exception of PLP analysis [20]. Moreover, specific studies on the influence of non-monomeric electrolyte addition on the polymerization kinetics cannot be found.

For all these reasons, in this work we focus on MAA with the aim of elucidating the effects of monomer concentration and ionization on the rate of polymerization. Namely, the time evolution of conversion is characterized by in-situ NMR for both non-ionized and fully ionized acids, with focus on the impact of the monomer concentration and ionic strength on the kinetic behavior. Experimental reactions are carried out in the medium-low monomer concentration range (i.e., 1 to 10 wt % of initial monomer). In combination with fundamental kinetic modeling, substantial differences between the polymerization behaviors of the two ionization forms of MAA are revealed. Finally, the addition of NaCl to the fully ionized system is applied to shed light on the effect of the electrostatic interactions on the reaction kinetics.

## 2. Materials and Methods

The polymerization reactions are carried out in deuterium oxide ($D_2O$, D, 99%, 99.5% chemical purity, Cambridge Isotope Laboratories, Inc., Tewksbury, MA, USA) using NMR tubes (5 mm NMR tube, Type 5UP (Ultra Precision), 178 mm, ARMAR AG, 5312 Döttingen, Switzerland) as reactors, which are directly inserted in an operating NMR spectrometer (UltraShield 500 MHz/54 mm magnet system, Bruker Inc., Billerica, MA, USA). For each experiment, the monomer MAA (99.5%, extra pure, stabilized with ca. 250 ppm 4-methoxyphenol (MEHQ), Acros Organics, 2440 Geel, Belgium) is dissolved in $D_2O$. If the reaction is to be carried out at full monomer ionization, i.e., at the initial degree of acid dissociation $\alpha = 1$, a solution of NaOH (pellets, analytical reagent grade, Fisher Scientific Ltd., Loughborough, LE11 5RG, UK) is added. If the ionic strength is to be altered, pure NaCl (EMSURE, ACS, ISO, Reag. Ph Eur, for analysis, Merck KGaA, 64271 Darmstadt, Germany) is added. As the last step, a solution of the radical initiator 2,2′-azobis(2-methylpropionamidine) dihydrochloride (V-50, 98%, Acros Organics, 2440 Geel, Belgium) in $D_2O$ is added. All the reagents are used as received, and the V-50 solution is renewed once a week to avoid any degradation. The reaction mixture is degassed with $N_2$ for 5 min at room temperature, as cooling with ice water leads to partial monomer precipitation. About 0.6 mL of the reaction mixture is transferred into the NMR tube, which is purged with $N_2$ during and after the transfer in order to avoid any contamination with oxygen. Once closed, the NMR tube is immediately put into the magnet at room temperature, where the reaction is monitored by in-situ NMR. It takes about 10 min to reach a constant reaction temperature of 50 °C. During this relatively short time of temperature adjustment, the inhibitor and possible impurities associated to the monomer are consumed, as no polymer formation is detected by NMR. The monomer conversion is measured as a function of the polymerization time by a series of $^1H$ NMR acquisitions. Since the relaxation time T1 of MAA is ca. 5 s, the instrument is set to wait ca. 25 s between each scan. As each data point consists of an average of 4 scans and considering the acquisition time, a data point is generated every 2.5 min. This resolution is enough to follow the evolution of the reaction. The conversion is evaluated as

$$\chi(t) = 1 - \frac{A(t)}{A_0} \tag{1}$$

where $A(t)$ is the sum of the areas of the two hydrogen atoms at the double-bond carbon ($sp^2$ hybridized) and $A_0$ is the area of the same hydrogens at time zero, as detailed in the Supporting Information (cf. Figures S1 and S2). More details about the in-situ NMR procedure adopted are given elsewhere [11].

In the experiments, the initial monomer concentration is varied from low to medium values (1 to 10 wt %). In order to mimic the ionic strength of higher monomer concentration, sodium chloride (NaCl) is added. For example, to obtain the ionic strength of the 10 wt % MAA reference (1.16 mol/kg), the reaction mixture contains 1 wt % MAA and 6.1 wt % NaCl. MAA concentrations in $D_2O$ higher than 10 wt % are avoided due to solubility limitations at room temperature, which hinders an accurate transfer of the mixture into the NMR tube. The initial V-50 concentration is kept constant at a level high enough to ensure adequate kinetic rates, i.e., 0.02 wt % for the reactions at $\alpha = 0$ and 0.10 wt % for the reactions at $\alpha = 1$ (which are considerably slower). In each reaction, a certain inhibition time is observed, which is due to the presence of MEHQ in the monomer. During this time, polymer formation is instead detected by NMR, contrary to what was observed throughout the temperature adjustment. The rate of polymer formation is very slow at the beginning of the reaction and then increases with time as the inhibitor is consumed. As a consequence, the derivative of the conversion-time curve reaches its maximum value only with some delay, when the inhibitor is completely depleted. Therefore, the time axis of the conversion versus time plots is adjusted to remove this effect: a delay time is considered for each experiment so that the tangent to the conversion-time curve at its maximum slope runs through the origin. It is worth noting that the observed inhibition time increases with monomer content (e.g., from ca. 20 min at 1 wt % MAA to ca. 70 min at 10 wt % MAA), in agreement with a larger inhibitor content.

## 3. Results

### 3.1. Non-Ionized System

Reactions are carried out at 50 °C with 1, 5, and 10 wt % of initial monomer content ($w_{M,0}$, weight fraction) and a constant amount of V-50 (0.02 wt %). With "non-ionized MAA" we mean the situation of natural dissociation of the monomer in the absence of any added base. The conversion versus time plots are shown in Figure 1. At 10 wt % of initial monomer, the slope of the conversion-time curve has reached its maximum value (i.e., the inhibitor is completely consumed) and remains constant until ca. 40% of conversion. After about 60 min, an increase of the slope is noticeable, which is a sign of diffusion limitations to termination (gel effect). This is fully consistent with the experimental results of Buback et al. [23], who ran their reactions at larger initiator concentrations (5 mmol/L, equivalent to ca. 0.12 wt %). At lower monomer concentrations, the initial polymerization rate is faster than at 10 wt %, in agreement with the increase in the propagation rate coefficient as a function of $w_{M,0}$, revealed by PLP studies [12]. It is also worth noting that at 5 wt % and 1 wt % of MAA, the autoacceleration of the reaction due to gel effect is less relevant: this behavior is consistent with the lower viscosity of the reacted solution observed in the NMR tube in the low monomer concentration cases compared to the situation at 10 wt % of initial MAA. Furthermore, it appears that at high conversion, the polymerization rate is slowing down: such an effect is particularly pronounced at 1 wt % of MAA, where full conversion is reached only hours later. The reproducibility of the obtained conversion versus time curves has been tested for the reaction at 1 wt % and 10 wt % of initial monomer. As shown in Figure S3 in the Supporting Information, at 1 wt %, the repeated experiment exhibits a slightly smoother behavior at conversions above 60%, i.e., showing a slightly less pronounced autoacceleration behavior. For the repetition of the reaction with 10 wt % MAA (Figure S4), the difference in the slope above 40% conversion is a bit larger. Since this discrepancy (of similar size) at the same reaction conditions has already been reported in the literature [23], we consider this accuracy to be sufficient.

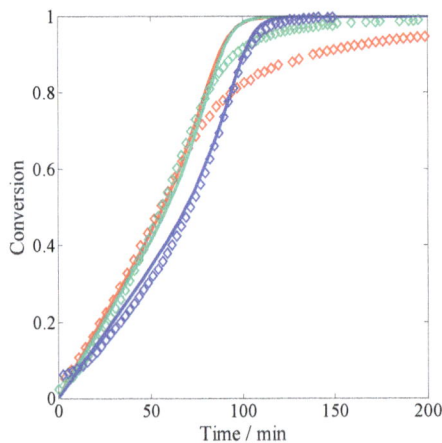

**Figure 1.** Monomer conversion versus time profiles of the radical batch polymerization of non-ionized methacrylic acid (MAA) in aqueous solution at 50 °C, 0.02 wt % of 2,2′-azobis(2-methylpropionamidine) dihydrochloride (V-50) initiator, and initial monomer concentration equal to 1 wt % (red), 5 wt % (green), and 10 wt % (blue): comparison between experimental data by in-situ nuclear magnetic resonance (NMR) (diamonds) and simulated curves (lines).

A kinetic model suitable for evaluating the reaction rate and number average molecular weight of the active chains has been developed. The most conventional free-radical kinetic scheme is considered, with all the reactions listed in Table 1. The initiator $I_2$ decomposes to form two radicals $I^{\cdot}$ with rate coefficient $k_d$. Assuming that all terminations are accounted for by an initiator efficiency factor $f$, and applying the quasi-steady-state assumption for the radical fragments $I^{\cdot}$, the rate of propagation of $I^{\cdot}$ is equal to the rate of initiation (i.e., the first term in the monomer balance equation in Table 2). The radicals of any chain length $R_n^{\cdot}$ undergo propagation with a monomer unit $M$ (rate coefficient $k_p$), and termination by combination ($k_{tc}$) and by disproportionation ($k_{td}$). Chain transfer to the monomer is also considered ($k_{ctm}$). The population balance equations (PBEs) are given in Table 2, where the definitions of the zero- and first-order moments of the chain length distribution of the active chains ($\lambda_0$ and $\lambda_1$, respectively) and of their number average molecular weight ($M_n^R$) are as follows:

$$\lambda_0 = \sum_{n=1}^{\infty} R_n^{\cdot} \tag{2}$$

$$\lambda_1 = \sum_{n=1}^{\infty} n R_n^{\cdot} \tag{3}$$

$$M_n^R = MW_{MAA} \frac{\lambda_1}{\lambda_0} \tag{4}$$

where $MW_{MAA}$ is the molar mass of MAA, thus the average chain length is defined in g mol$^{-1}$. The reaction rate coefficients are defined in Table 3, where $\chi$ is the conversion and $P$ and $T$ are set to 1.013 bar and 323 K, respectively. In the definition of the composite model for the termination rate coefficient, a critical average chain length $M_{n,C}^R = 68\ MW_{MAA}$ is considered, according to Wittenberg et al. [25]. Since we do not include the population balance equations for the dead chains, we cannot compute the weight-average molecular weight required to calculate the viscosity parameter $C_\eta$ as detailed by Wittenberg et al. [25]; nevertheless, we obtained a satisfactory overlap of our model with the $k_t$ versus conversion profiles at 30 wt % of MAA reported in the reference by setting $C_\eta$ to 9. In the model equations, a constant density of 1.1 g mL$^{-1}$ is assumed for $D_2O$ solutions.

**Table 1.** Reaction scheme used in the modeling of non-ionized methacrylic acid (MAA) polymerization.

| Reaction | Scheme |
|---|---|
| Initiator decomposition | $I_2 \xrightarrow{k_d} 2I^{\cdot}$ |
| Propagation of initiator fragment | $I^{\cdot} + M \xrightarrow{k_p} R_1^R$ |
| Propagation of active chain | $R_n^{\cdot} + M \xrightarrow{k_p} R_{n+1}^{\cdot}$ |
| Chain transfer to monomer | $R_n^{\cdot} + M \xrightarrow{k_{ctm}} P_n + R_1^{\cdot}$ |
| Termination by combination | $R_n^{\cdot} + R_m^{\cdot} \xrightarrow{k_{tc}} P_{n+m}$ |
| Termination by disproportionation | $R_n^{\cdot} + R_m^{\cdot} \xrightarrow{k_{td}} P_n + P_m$ |

**Table 2.** Population balance equations on the involved species $I_2$, $M$ and the zero and first order moments of the active chain distribution, $\lambda_0$ and $\lambda_1$.

| Reaction |
|---|
| $\dfrac{dI_2}{dt} = -k_d I_2$ |
| $\dfrac{dM}{dt} = -2fk_d I_2 - k_p M \sum_{m=1}^{\infty} R_m - k_{ctm} M \sum_{m=1}^{\infty} R_m$ |
| $\dfrac{d\lambda_0}{dt} = 2fk_d I_2 - 2(k_{td} + k_{tc})\lambda_0^2$ |
| $\dfrac{d\lambda_1}{dt} = 2fk_d I_2 + k_p M \lambda_0 - k_{ctm} M(\lambda_1 - \lambda_0) - 2(k_{td} + k_{tc})\lambda_1 \lambda_0$ |

**Table 3.** Reaction rate coefficients used in the modeling of non-ionized MAA polymerization.

| Reaction |
| --- |
| **Initiation [23,25]** |
| $k_d/s^{-1} = 2.24 \cdot 10^{15} exp\left(-\dfrac{1.52 \cdot 10^4}{T}\right)$ <br><br> $f = 0.8$ |
| **Propagation [21]** |
| $k_p/(L\ mol^{-1}s^{-1})$ <br><br> $= 4.1 \cdot 10^6 exp\left(-\dfrac{1.88 \cdot 10^3}{T}\right)$ <br><br> $\left\{0.08 + 0.92 exp\left[\dfrac{-5.3 w_{M,0}(1-\chi)}{1-\chi w_{M,0}}\right]\right\} exp\left\{\dfrac{P}{T}\left[0.096 + \dfrac{0.11 w_{M,0}(1-\chi)}{1-\chi w_{M,0}}\right]\right\}$ |
| **Termination [22,23,25]** |
| $k_t^{1,1}/(L\ mol^{-1}s^{-1}) = 2.29 \cdot 10^{12} exp\left(-\dfrac{2.64 \cdot 10^3}{T}\right)$ <br><br> $k_t^{CLD}/(L\ mol^{-1}s^{-1}) = \begin{cases} k_t^{1,1}\left(M_n^R\right)^{-0.61} & @\ M_n^R \leq M_{n,c}^R \\ k_t^{1,1}\left(M_{n,c}^R\right)^{-0.444}\left(M_n^R\right)^{-0.166} & @\ M_n^R > M_{n,c}^R \end{cases}$ <br><br> $k_t/(L\ mol^{-1}s^{-1}) = k_t^{CLD}\left[0.96 + 0.04 exp(\chi C_\eta)\right]^{-1} + C_{RD} w_{M,0}(1-\chi)k_p$ <br><br> $k_{tc}/(L\ mol^{-1}s^{-1}) = 0.2 k_t$ <br><br> $k_{td}/(L\ mol^{-1}s^{-1}) = 0.8 k_t$ |
| **Chain transfer to monomer [23]** |
| $k_{ctm}/(L\ mol^{-1}s^{-1}) = 5.37 \cdot 10^{-5} k_p$ |

Once all parameter values are set, the model is used to predict the kinetic profiles for our experiments at 1 wt %, 5 wt %, and 10 wt % of MAA. The resulting curves are shown in Figure 1: a good agreement between the experimental data and the model predictions is found, with the initial slope perfectly reproduced at all the concentrations. At 10 wt % of initial monomer, both the simulated curve and the experimental data show an increase in polymerization rate due to the gel effect; moreover, they are nicely superimposed. On the other hand, at 1 wt % and 5 wt % the model predicts a slight increase in reaction rate, which is not observed experimentally. This discrepancy could reflect an increase in termination or a decrease in the propagation rate not accounted for by the model.

To better elucidate the possible reason for this disagreement, the experimental data have been processed as follows: (i) the local slope $\Delta\chi/\Delta t$ is calculated at each time step; (ii) the product between the propagation rate coefficient and the actual radical concentration, $\lambda_0$, is evaluated as:

$$k_p\lambda_0 = \frac{\Delta\chi/\Delta t}{1-\chi} \tag{5}$$

(iii) using the $k_p$ value defined as in Table 3, the radical concentration is then obtained; and (iv) the apparent rate coefficient of termination is finally estimated as:

$$k_t^{exp} = \frac{f k_d I_2}{\lambda_0^2} \tag{6}$$

These resulting $k_t^{exp}$ values are depicted in Figure 2: the increase in $k_t$ reflecting the slowdown mentioned above becomes relevant at residual monomer concentrations below 0.5 wt %, i.e., at large polymer contents. Generally, a decrease (instead of an increase) in termination rate towards higher conversion is expected, thus the observed deceleration in the conversion versus time profiles should be imputed to a decrease in the propagation rate coefficient. Since $k_t$ is chain-length dependent (cf. Table 3), there is also the possibility that at high monomer conversion, the termination rate coefficient increases due to reduced monomer concentration leading to the production of shorter chains. However, from Figure 2 such an increase in the literature $k_t$ is observed only at very large conversion (i.e., above 95%): before that point, the termination rate coefficient decreases monotonously with conversion.

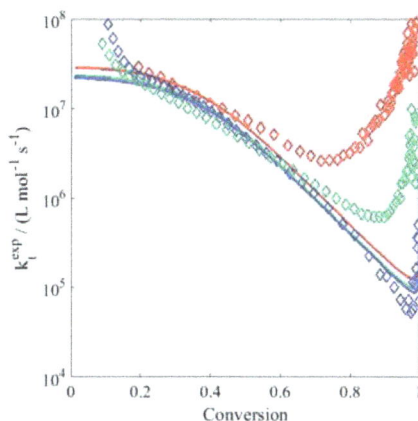

**Figure 2.** Termination rate coefficient as a function of the monomer conversion calculated from the propagation rate coefficient reported in Table 3 and the experimental data of conversion versus time of non-ionized MAA polymerization (diamonds) at an initial monomer concentration equal to 1 wt % (red), 5 wt % (green), and 10 wt % (blue): comparison with the simulated $k_t$ using the model equations (lines).

Therefore, a similar procedure as above has been applied to calculate the propagation rate coefficient deduced from experiments, this time by using the termination rate coefficient as in Table 3. Since $k_t$ is function of $k_p$, the procedure is not as straightforward: at each $(t, \chi)$ step, the set of PBEs in Table 2 is solved for different values of $k_p$ while minimizing the error between the experimental and calculated conversion. Figure 3 shows the resulting $k_p^{exp}$ curves as a function of conversion: all the curves exhibit a strong decrease at high conversion, more or less when the residual MAA concentration falls below 0.5 wt %. A similar behavior has been previously reported for the polymerization of acrylic acid (AA): a decrease of the apparent $k_p$ at concentrations below 3 wt % has been noticed and imputed to differences in the local and overall monomer concentrations in solution, which may be a result of preferential solvation [26]. This decrease explains the decreasing reaction rate at high conversion shown in Figure 1. Apart from this discrepancy, the $k_p$ predicted by the model equations provides a reasonable description of $k_p^{exp}$ as shown by Figure 3. Therefore, we can conclude that the developed model is suitable for reproducing the experimental behavior of the system at medium-high monomer concentration and to qualitatively capture the trends of the propagation and termination rate coefficients as a function of conversion. The residual weakness in predicting the polymerization rate at lower monomer concentration is imputed to inaccurate $k_p$ evaluation at high conversion.

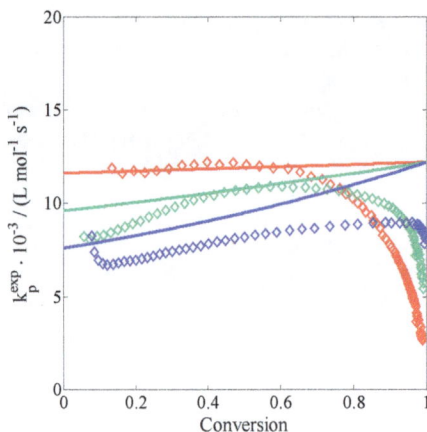

**Figure 3.** Propagation rate coefficient as a function of the monomer conversion calculated from the termination rate coefficient reported in Table 3 and the experimental data of conversion versus time of non-ionized MAA polymerization (diamonds) at an initial monomer concentration equal to 1 wt % (red), 5 wt % (green), and 10 wt % (blue): comparison with the simulated $k_p$ using the model equations (lines).

## 3.2. Fully Ionized System

Three different sets of experiments are carried out for the polymerization of fully ionized MAA at 50 °C and 0.10 wt % V-50: the first set at different values of initial monomer concentration; the second set at different values of initial monomer concentration while keeping the ionic strength constant at an ionic strength equivalent to a monomer concentration of 10 wt %; and the third set at different values of ionic strength while keeping $w_{M,0}$ constant at 5 wt %. In the second and third cases, the ionic strength is adjusted by the addition of NaCl. An overview over all these polymerization reactions is provided in Table 4. A remarkably good reproducibility of the experiments in the fully ionized case has been observed for the reactions at 1 wt % and 10 wt % initial MAA, as reported in Figures S5 and S6 of the Supporting Information.

**Table 4.** List of all polymerization reactions of fully ionized MAA (alpha = 1) in aqueous solution at 50 °C and 0.1 wt % of 2,2′-azobis(2-methylpropionamidine) dihydrochloride (V-50) initiator.

| Reaction No. | Initial Monomer Concentration $w_{M,0}$ (wt %) | Ionic Strength Corresponding to (wt %) | Ionic Strength (kg mol$^{-1}$) |
|:---:|:---:|:---:|:---:|
| 1 | 1 | 1 | 0.12 |
| 2 | 2.5 | 2.5 | 0.29 |
| 3 | 5 | 5 | 0.58 |
| 4 | 10 | 10 | 1.16 |
| 5 | 1 | 5 | 0.58 |
| 6 | 1 | 10 | 1.16 |
| 7 | 5 | 10 | 1.16 |
| 8 | 5 | 15 | 1.74 |
| 9 | 5 | 20 | 2.32 |
| 10 | 5 | 30 | 3.48 |

Let us consider the experiments at different monomer concentrations. When comparing the resulting conversion versus time curves shown in Figure 4 to those reported above for the non-ionized case (Figure 1), the initial reaction rates are much slower at full MAA ionization: this feature can be explained by the electrostatic repulsion between the charges of the ionized reactants in solution.

An increase in the polymerization rate is observed when moving from 1 wt % of MAA to the higher concentrations, whereas between 2.5 wt % and 10 wt % the kinetics appears to be unaffected by the monomer concentration. The former effect corresponds to an increase in the propagation kinetics due to an increase in the solution ionic strength, in agreement with the so-called phenomenon of electrostatic screening: namely, higher initial concentration of the monomer corresponds to higher concentration of the electrolyte in the system [6,11]. The latter effect is more surprising, and reveals either a saturation of the electrostatic screening already at 2.5 wt % of MAA or a competition between electrostatic and non-electrostatic effects of monomer concentration on propagation kinetics. Another counterbalancing effect to the electrostatic-driven increase in $k_p$ could be represented by the known dependence of termination kinetics of fully ionized MAA on monomer concentration [27].

The impact of monomer concentration on the reacting system at constant ionic strength (equivalent to 10 wt % MAA) and 0.10 wt % of initiator is shown in Figure 5. The idea behind this set of experiments is to reproduce the non-electrostatic influence of the monomer concentration, keeping the ionic strength constant. As shown in the figure, the initial reaction rate does not change with the monomer concentration. Only at a conversion higher than 50% does the curve at low monomer concentration seem to become slower than those at higher monomer concentration, where the propagation rate seems to be fully independent of the non-electrostatic (i.e., intrinsic) reactivity.

The impact of NaCl addition on the reacting system at constant monomer concentration is shown in Figure 6. The aim is to reproduce the ionic strength of higher monomer concentrations while keeping constant the initial monomer concentration, in order to separate the effect of the charges from the mixed effects associated to an increase in monomer content (i.e., electrostatic and non-electrostatic ones). As shown in the figure, the initial reaction rate increases continuously with increasing amounts of added salt. Additionally, there is no visible effect of autoacceleration on the rate of polymerization in this case.

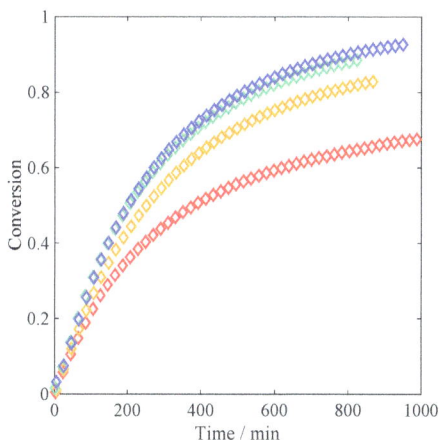

**Figure 4.** Monomer conversion versus time profiles of the radical batch polymerization of fully ionized MAA in aqueous solution at 50 °C, 0.1 wt % of V-50 initiator, and initial monomer concentration equal to 1 wt % (red), 2.5 wt % (yellow), 5 wt % (green), and 10 wt % (blue).

Since the propagation reaction seems to be affected by the electrolyte concentration in the entire explored range (i.e., up to 30 wt % of equivalent MAA), this same reaction is expected to never become fully kinetically-controlled: namely, the effects of electrostatic interactions and electrostatic screening on the propagation kinetics are likely to play a role at all explored values of initial monomer concentration, and some kind of "electrostatic saturation" at higher monomer content (i.e., above 2.5 wt % of MAA) cannot be assumed.

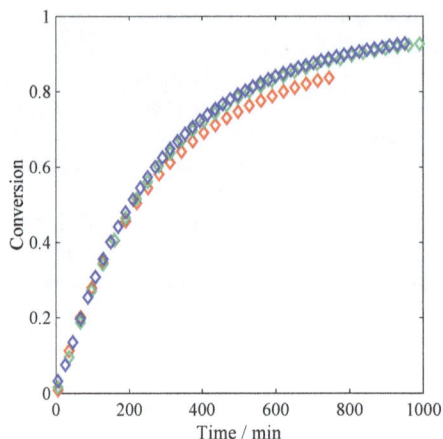

**Figure 5.** Monomer conversion versus time profiles of the radical batch polymerization of fully ionized MAA in aqueous solution at 50 °C, 0.1 wt % of V-50 initiator and constant ionic strength equivalent to 10 wt % MAA by addition of NaCl. Variation of monomer concentration: 1 wt % (red), 5 wt % (green), and 10 wt % (blue, no NaCl addition).

Aiming to develop a kinetic model for the polymerization of fully ionized MAA, the reaction scheme and population balances in Tables 1 and 2 are again considered, whereas the validity of the reaction rate coefficients in Table 3 is critically discussed hereinafter.

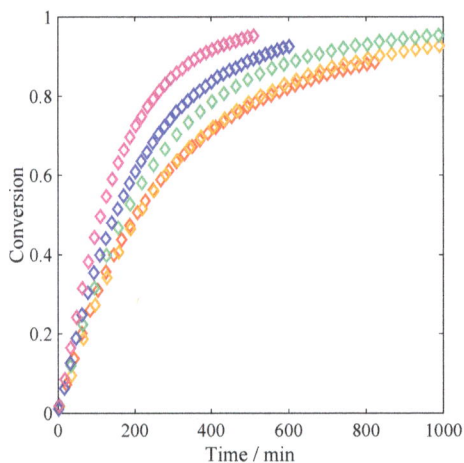

**Figure 6.** Monomer conversion versus time profiles of the radical batch polymerization of fully ionized MAA in aqueous solution at 50 °C, 0.1 wt % of V-50 initiator, initial monomer concentration equal to 5 wt %, without addition of salt (red), salt addition to obtain an ionic strength equivalent to a monomer concentration of 10 wt % (yellow), 15 wt % (green), 20 wt % (blue), and 30 wt % (purple).

The propagation rate coefficient proposed by Beuermann et al. [21] has been improved by Lacik et al. to take into account the effect of acid ionization [20]. However, as this equation has been proposed for the specific range of initial monomer concentration 5 wt % $\leq w_{M,0} \leq$ 40 wt %, it cannot be used for our experiments at 1 wt % and 2.5 wt % of initial monomer content, where the calculated $k_p$ would be negative. Furthermore, this equation cannot be used to simulate the effect of ionic strength on the propagation kinetics coming from NaCl addition. For these reasons, we propose a rate law of propagation of fully ionized MAA composed of an intrinsic kinetics term and a diffusion term, the latter being driven by electrostatic interactions and thus a function of the electrolyte concentration, similar to the equation developed in a previous work [11]:

$$k_p = \left( \frac{1}{k_{p,i}^0 \exp(-Bw_{M,0})} + \frac{1}{k_{D,0}C_E^\beta} \right)^{-1} \tag{7}$$

where $k_{p,i}^0$, $k_{D,0}$, $\beta$, and $B$ are fitting parameters, while $C_E$ is the electrolyte concentration in mol kg$^{-1}$, which is a function of $w_{M,0}$:

$$C_E = 2 \left( \frac{w_{S,0}}{MW_{NaCl}} + \frac{w_{M,0}\alpha}{MW_{MAA}} \right) \tag{8}$$

where $w_{S,0}$ and $MW_{NaCl}$ are the initial weight fraction and molar mass of NaCl, respectively.

The first part of Equation (7) introduces the non-electrostatic dependence on the monomer concentration, similar to what has been done by Beuermann et al. for non-ionized MAA [21]. However, it should be noted that such dependence can be in principle different for the fully ionized case: for this reason, the parameter $B$ is estimated as well. For simplicity, the conversion-dependent term of the exponent $(1 - \chi)/(1 - w_{M,0}\chi)$ has been omitted. The parameter $k_{p,i}^0$ identifies the intrinsic kinetics rate coefficient of fully ionized MAA, thus without any electrostatic interaction effect. The electrostatic and non-electrostatic dependence of $k_p$ on $w_{M,0}$ for the fully ionized case, defining respectively an increase and a decrease of the rate coefficient as the initial monomer concentration increases, are expected to compensate each other at higher monomer concentrations, thus reproducing the experimental behavior observed in Figure 4.

In order to verify if the proposed rate law for $k_p$ is reasonable, the propagation rate coefficient is evaluated as a function of conversion using the same procedure described for the non-ionized case, namely using the experimental data reported in Figures 4 and 5. As for $k_t$, the rate coefficient from a recent publication by Kattner et al. [27] has been adopted: the authors proposed an expression for the termination rate coefficient at monomer concentrations of 5 and 10 wt % as shown in Table 5.

**Table 5.** Termination rate coefficients for the fully ionized MAA [27]. For 10 wt %, $M_{n,C}^R$ is equal to 80 $MW_{MAA}$.

| Ionic Strength Equivalent to $w_{M,0}$/wt % | $k_t^{1,1}/\left( \text{L mol}^{-1}\text{s}^{-1} \right)$ | $k_t/\left( \text{L mol}^{-1}\text{s}^{-1} \right)$ |
|---|---|---|
| 5 | $1.97{\cdot}10^8 \exp\left( -\dfrac{999}{T} \right)$ | $k_t^{1,1} M_n^{R\,-0.59}$ |
| 10 | $7.24{\cdot}10^8 \exp\left( -\dfrac{1049}{T} \right)$ | $\begin{cases} k_t^{1,1} M_n^{R\,-0.59} & @\ M_n^R \leq M_{n,C}^R \\ k_t^{1,1} M_{n,C}^{R\,-0.41} M_n^{R\,-0.18} & @\ M_n^R > M_{n,C}^R \end{cases}$ |

Similar to the propagation rate, it can be assumed that the termination reaction is also made of two contributions, one related to an intrinsic kinetics and the other driven by electrostatic interactions. Due to the limited amount of available data, we need to assume that only one of these two contributions majorly affects the termination rate. Since bimolecular termination of growing radicals is a reaction with a low activation energy barrier (mostly controlled by diffusion), we assume that the change in ionic strength upon change in monomer concentration is principally responsible for the $k_t$ variation as a function of $w_{M,0}$. This is considered more realistic than an effect of monomer concentration on the fluidity of the transition state, as previously considered in the non-electrostatic effect of monomer concentration on the propagation kinetics of other water soluble compounds [8]. Therefore, we will focus on those experiments with an ionic strength (obtained by a certain amount of monomer concentration or addition of NaCl) equivalent to a monomer concentration of 5 wt % or 10 wt % when considering the $k_t$ expressions reported in Table 5.

Equivalent to the procedure for the non-ionized case (cf. Figure 3), the experimental propagation rates are shown in Figures 7 and 8 for ionic strengths equal to 5 wt % and 10 wt %, respectively. The propagation rate coefficient of 5 wt % MAA without addition of NaCl can be compared with the experimental results of Lacik et al. [20]: at our reaction temperature (50 °C), the Lacik equation predicts a propagation rate of about 770 L mol$^{-1}$ s$^{-1}$, whereas our experimentally deduced propagation rate is about five times smaller. At monomer concentrations of 5 wt % and 10 wt %, the propagation rate increases at low conversion, which is probably due to the competition between propagation and radical scavenging by the inhibitor. Above 50% conversion, a general decrease is visible. Apart from these minor variations with conversion, the results reveal that the reaction rate increases with ionic strength. In fact, for any monomer concentration, the propagation rate at an ionic strength of 10 wt % monomer concentration is 2–3 times faster than at 5 wt %. At constant ionic strength, a reduction of monomer concentration leads to a slight increase of the propagation rate coefficient. However, this effect is less relevant than the change in ionic strength. The observed competition between the influence of monomer concentration and ionic strength confirms the suitability of the propagation rate law in Equation (7) to describe at least qualitatively the kinetic behavior of the system.

The four parameters $k_{p,i}^0$, $k_{D,0}$, $\beta$, and $B$ are estimated by fitting the model equations to the experimental data of conversion versus time for the fully ionized case. The ratio of the parameters $k_{p,i}^0 \exp(-Bw_{M,0})$ and $k_{D,0}C_E^\beta$ is expected to be close to unity: this way, the diffusion limitation to $k_p$ due to the electrostatic interactions should be effective inside the entire range of ionic strength values explored, as deduced from the experimental results. It should be noted that $k_{D,0}$ must be expressed in L mol$^{-1}$ s$^{-1}$ (kg mol$^{-1}$)$^\beta$ in agreement with the definition of $C_E$ in the model. The exponent $\beta$ is dimensionless and it is expected to range between 0.1 and 10, while $B$ is a measure of the influence of initial monomer concentration on the kinetics. As the value for non-ionized MAA has been determined as 5.3 [21], a similar value (between 1 and 10) is expected.

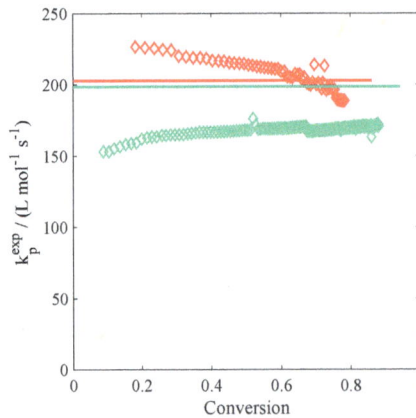

**Figure 7.** Propagation rate coefficient as a function of the monomer conversion calculated from the termination rate coefficient in Table 5 and the experimental data of conversion versus time of fully ionized MAA polymerization at constant ionic strength equivalent to 5 wt % MAA by addition of NaCl. Variation of monomer concentration 1 wt % (red) and 5 wt % (green, no NaCl addition): comparison with the simulated $k_p$ using the model equations (lines).

The parameters are estimated by minimizing the square of the error between experimental and simulated conversion versus time data. This minimization is carried out by a genetic algorithm which is implemented using Matlab. An optimization run was conducted to evaluate the four mentioned adjustable parameters and leads to the results and parameter values reported in Figures 9–11 and Table 6, respectively. Figure 9 compares the influence of monomer concentration without salt on conversion: while the curve at 10 wt % is perfectly reproduced, at 5 wt % monomer, the model predicts a reaction rate which is slightly faster than that at 10 wt %, but while looking at the experimental data the rates should be exactly the same (i.e., the electrostatic and intrinsic kinetics effects on the propagation and termination rate coefficients should be perfectly balanced). Towards higher conversion, the model correctly predicts a decrease of the reaction rate at 5 wt %.

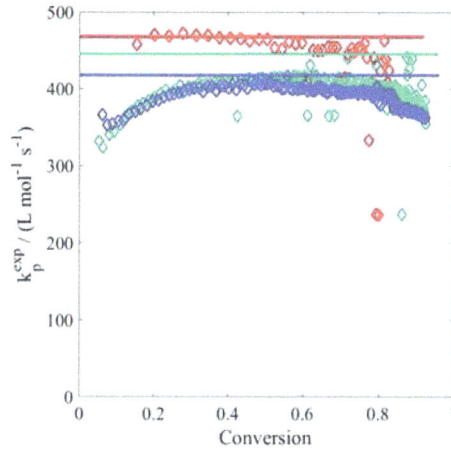

**Figure 8.** Propagation rate coefficient as a function of the monomer conversion calculated from the termination rate coefficient in Table 5 and the experimental data of conversion versus time of fully ionized MAA polymerization at constant ionic strength equivalent to 10 wt % MAA by addition of NaCl. Variation of monomer concentration 1 wt % (red), 5 wt % (green), and 10 wt % (blue, no NaCl addition): comparison with the simulated $k_p$ using the model equations (lines).

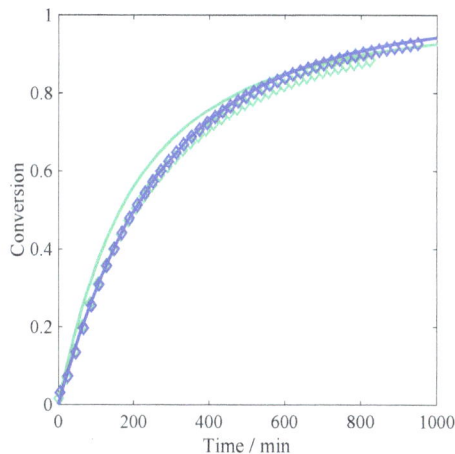

**Figure 9.** Monomer conversion versus time profiles of the radical batch polymerization of fully ionized MAA in aqueous solution at 50 °C, 0.1 wt % of V-50 initiator, and initial monomer concentration equal to 5 wt % (green) and 10 wt % (blue): comparison between experimental data by in-situ NMR (diamonds) and simulated curves (lines, using the parameter values of Table 6).

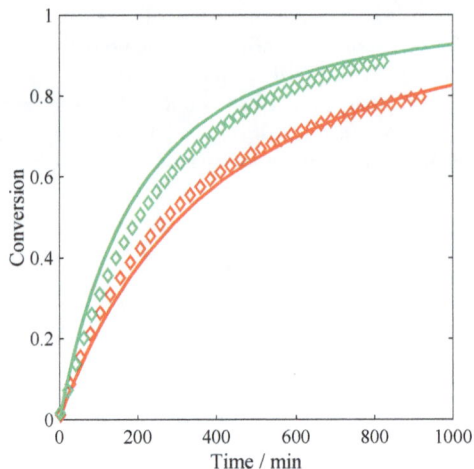

**Figure 10.** Monomer conversion versus time profiles of the radical batch polymerization of fully ionized MAA in aqueous solution at 50 °C, 0.1 wt % of V-50 initiator and constant ionic strength equivalent to 5 wt % MAA by the addition of NaCl. Variation of monomer concentration 1 wt % (red) and 5 wt % (green, no NaCl addition): comparison between experimental data by in-situ NMR (diamonds) and simulated curves (lines, using parameter values of Table 6).

The influence of monomer concentration at an ionic strength equivalent to 5 wt % is shown in Figure 10. Qualitatively, the initial slopes at varying amounts of MAA are well reproduced, although the impact of monomer concentration on the initial rate is slightly too strong. The decrease in the reaction rate at conversions above 50% is very well described for both monomer concentrations.

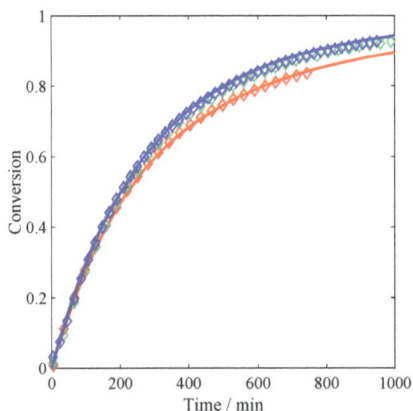

**Figure 11.** Monomer conversion versus time profiles of the radical batch polymerization of fully ionized MAA in aqueous solution at 50 °C, 0.1 wt % of V-50 initiator and constant ionic strength equivalent to 10 wt % MAA by addition of NaCl. Variation of monomer concentration 1 wt % (red), 5 wt % (green), and 10 wt % (blue, no NaCl addition): comparison between experimental data by in-situ NMR (diamonds) and simulated curves (lines, using the parameter values of Table 6).

**Table 6.** Estimated parameter values of the parameter optimization carried out by fitting the model equations to the experimental data of conversion versus time (cf. Figures 9–11) for the fully ionized MAA case.

| Parameter | Value |
|---|---|
| $k_{p,i}^0 / (\text{L mol}^{-1}\,\text{s}^{-1})$ | 758 |
| $k_{D,0} / (\text{L mol}^{-1}\,\text{s}^{-1}\,(\text{kg mol}^{-1})^\beta)$ | 201 |
| $\beta$ | 2.18 |
| $B$ | 1.92 |

A similar comparison is shown in Figure 11, where the ionic strength is set equivalent to 10 wt % MAA. As already mentioned in the discussion of Figure 5, the initial slope is the same for all MAA concentrations, which is well reproduced by the model, as well as the behavior at higher conversion. Since the model is able to correctly simulate the reduction at 1 wt %, this decrease of the reaction rate cannot be imputed to smaller local concentrations around the polymer chain. It is rather due to a conversion-dependent increase of the termination rate coefficient (caused by a decrease of $M_n^R$). Note that an effect of local concentration can still be present when considering the reaction at 1 wt % without salt addition, but since we lack an appropriate description of the termination rate at this ionic strength, we are unable to predict this situation.

The parameter values are provided in Table 6: the kinetic rates $k_{p,i}^0$ and $k_{D,0}$ are in the same order of magnitude and similar to the kinetic parameter obtained by Lacík et al. [20], whereas $B$ is slightly smaller than two. It is therefore clearly smaller than the 5.3 obtained by Beuermann et al. [21], but since the reactivity of the ionized MAA is much smaller, it might be reasonable to assume that the effect of monomer concentration is also reduced. The parameter $\beta$ is slightly larger than two, which is large enough to alter $k_{D,0}c_E^\beta$ in a way to qualitatively reproduce the experimental behavior at increasing ionic strength. The overall good balance between $k_{p,i}^0 \exp(-Bw_{M,0})$ and $k_{D,0}c_E^\beta$ leads to a good description of the system: at low ionic strength, the propagation reaction is diffusion limited, whereas at ionic strengths larger than 5 wt % (either by addition of NaCl or MAA), the propagation seems to become more reaction controlled, although never completely (as shown by the experimental results in Figure 6). A comparison between the intrinsic kinetics and electrostatic contributions to the propagation rate coefficient for the considered reactions involving fully ionized MAA is given by Figure S7 and Table S2 in the Supporting Information. Moreover, the sensitivity of the model performances upon limited variations of the model parameter values is reported in Table S3 of the Supporting Information, indicating that small variations of each parameter (i.e., within 10 wt %) induce a very large change in the total error between experiments and simulations.

Eventually, in order to better describe what happens at ionic strengths below 5 wt % and above 10 wt %, a more generalized description of the termination rate is required. This knowledge is especially needed to elucidate the reason for the drastically reduced reaction rate at 1 wt % MAA without salt addition.

As a final remark, the knowledge of molecular weights would allow us to improve the reliability of the developed kinetic models, especially for the fully ionized case. Nevertheless, our choice of focusing on time-conversion data only in the analysis and modeling of the system kinetics was motivated by the following reasons: (i) the estimation of accurate MWD for MAA and generally for aqueous polymers is not trivial and requires ad hoc experimental setup and procedures; (ii) the accuracy and reliability of the kinetic data obtained by in-situ NMR is much higher compared to our current capability of measuring MWD; (iii) the availability of independent values of the termination rate coefficients from PLP of non-ionized MAA and, for specific reaction conditions, also for fully ionized MAA makes our approach reasonable for determining the propagation kinetics by focusing only on the measured polymerization rates.

## 4. Conclusions

Free radical aqueous polymerization reactions of non-ionized and fully ionized MAA were carried out in $D_2O$ by in-situ NMR, with a focus on the effect of the monomer concentration (varied between 1 wt % and 10 wt %) on the polymerization rate. The experimental results revealed substantial differences in the behavior of the two systems. The degree of ionization of the monomer plays a major role in determining the overall reaction rate as well as its sensitivity to changes in monomer and electrolyte concentration.

In the non-ionized case, we were able to develop a kinetic model based on literature rate coefficients which is capable of nicely predicting the experimental data at the initial monomer concentration of 10 wt %. At lower monomer concentration and high conversion (ca. 0.5 wt % residual monomer concentration), a decrease in the reaction rate is observed, which cannot be explained by the current version of the model. Similar to what has also been previously highlighted for AA polymerization, interactions between MAA monomer molecules and their environment at high dilution are assumed to be responsible for this kinetic feature not contemplated by the currently developed polymerization models.

In the case of the fully ionized monomer, we improved the model for the non-ionized case by introducing a new rate law for propagation: namely, both electrostatic and non-electrostatic effects of monomer concentration on the reaction kinetics are explicitly considered. Due to the limited amount of termination data, we focused this development on the cases with an ionic strength relative to MAA concentrations 5 wt % and 10 wt %. The modeling of higher and lower monomer concentrations (or equivalent ionic strengths) would require the knowledge of either molecular weights or termination rate coefficients—both of which are very hard to determine. A reasonable agreement between model simulations and experimental results was obtained by considering together the effects of monomer concentration and salt addition on the polymerization kinetics. In particular, the model is capable of reproducing the effect of increasing reaction rate upon increasing the initial ionic strength.

A better characterization of the rate coefficients of propagation and termination of fully ionized MAA is supposed to substantially improve the elucidation of their combined impact on the polymerization rate, especially in view of the investigation of the effects of monomer and electrolyte concentration. With this respect, a promising method to access on-line the propagation and termination rate is represented by in-situ diffusion-ordered spectroscopy (DOSY) NMR [28], which might allow for measuring at the same time the residual monomer concentration and molecular weight. Furthermore, relevant information about the solution behavior of polar and ionizable molecules such as the MAA monomer and polymer as well as their interactions with the reaction environment (i.e., water and electrolytes) could be obtained by the application of computational chemistry, namely molecular dynamics simulations. Both strategies are currently under development in our research group.

**Supplementary Materials:** The following are available online at at www.mdpi.com/2227-9717/5/2/23/s1, Figure S1: Details about the monomer peak area selected for the calculation of the monomer conversion from the 1H NMR analysis for the polymerization of non-ionized MAA. Figure S2: Details about the monomer peak area selected for the calculation of the monomer conversion from the 1H NMR analysis for the polymerization of fully ionized MAA. Figure S3: Reproducibility of the monomer conversion versus time profiles of the radical batch polymerization of non-ionized MAA in aqueous solution at 50 °C, 0.02 wt % of V-50 initiator and 1 wt % MAA. Figure S4: Reproducibility of the monomer conversion versus time profiles of the radical batch polymerization of non-ionized MAA in aqueous solution at 50 °C, 0.02 wt % of V-50 initiator and 10 wt % MAA. Figure S5: Reproducibility of the monomer conversion versus time profiles of the radical batch polymerization of fully ionized MAA in aqueous solution at 50 °C, 0.1 wt % of V-50 initiator and 10 wt % MAA. Figure S6: Reproducibility of the monomer conversion versus time profiles of the radical batch polymerization of fully ionized MAA in aqueous solution at 50 °C, 0.1 wt % of V-50 initiator and 1 wt % MAA. Figure S7: Representation of the intrinsic and electrostatic contributions to the overall propagation reaction rate, based on the parameter set provided in Table 6. Table S1: Full list of reactions with non-ionized (0.02 wt % initiator V-50) and fully ionized MAA (0.1 wt % initiator V-50) including repeated experiments. Table S2: Comparison of intrinsic vs. electrostatic contributions to the propagation rate. Table S3: Sensitivity analysis on the parameter values. The total error (mean-square error) between the experiments at $\alpha = 1$ and simulations is calculated upon a 10% increase of each model parameter with respect to its optimized value reported in Table 6. The error corresponding to the set of unchanged parameters is reported as a reference (bottom line in the table).

**Acknowledgments:** The financial support from the Swiss National Science Foundation with Grant Number 200021.153403/1 is gratefully acknowledged.

**Author Contributions:** All three coauthors E.J.F., G.S. and D.C. designed and conceived the experiments, which were performed by E.J.F., whereas the resulting data was analyzed by E.J.F. and D.C. The paper was written by E.J.F. and D.C.

**Conflicts of Interest:** The authors declare no conflict of interest.

# References

1. Nemec, J.W.; Bauer, J.W. *Encyclopaedia of Polymer Science and Engineering*; Wiley: New York, NY, USA; Volume I, 1988.
2. Goin, J. Water Soluble Polymers. In *CEH Marketing Research Report 582.0000 D-E*; SRI International: Menlo Park, CA, USA, 1991.
3. Kumar, J.; Nalwa, H.S.; MacDiarmid, A.G. *Handbook of Polyelectrolytes and Their Applications*; Tripathy, S.K., Ed.; ASP: Stevenson Ranch, CA, USA, 2002.
4. Oosawa, F. *Polyelectrolytes*; Marcel Dekker: New York, NY, USA, 1971.
5. Dobrynin, A.V.; Rubinstein, M. Theory of Polyelectrolytes in Solutions and at Surfaces. *Prog. Polym. Sci.* **2005**, *30*, 1049–1118. [CrossRef]
6. Losada, R.; Wandrey, C. Non-Ideal Polymerization Kinetics of a Cationic Double Charged Acryl Monomer and Solution Behavior of the Resulting Polyelectrolytes. *Macromol. Rapid Commun.* **2008**, *29*, 252–257. [CrossRef]
7. Hahn, M.; Jaeger, W.; Wandrey, C.; Reinisch, G. Zur Kinetik der Radikalischen Polymerisation von Dimethyl-diallyl-ammoniumchlorid. IV. Mechanismus von Start-und Abbruchreaktion mit Persulfat als Initiator. *Acta Polym.* **1984**, *35*, 350–358. [CrossRef]
8. Buback, M. Solvent Effects on Acrylate $k_p$ in Organic Media: A Response. *Macromol. Rapid Commun.* **2015**, *36*, 1979–1983. [CrossRef] [PubMed]
9. Kurenkov, V.F.; Nadezhdin, I.N.; Antonovich, O.A.; Lobanov, F.I. Phase Separation in Aqueous Solutions of Binary Copolymers of Acrylamide with Sodium 2-Acrylamido-2-Methylpropanesulfonate and Sodium Acrylate. *Russ. J. Appl. Chem.* **2004**, *29*, 804–808. [CrossRef]
10. Riahinezhad, M.; Kazemi, N.; McManus, N.; Penlidis, A. Effect of Ionic Strength on the Reactivity Ratios of Acrylamide/Acrylic Acid (sodium acrylate) Copolymerization. *J. Appl. Polym. Sci.* **2014**, *131*, 40949. [CrossRef]
11. Cuccato, D.; Storti, G.; Morbidelli, M. Experimental and Modeling Study of Acrylamide Copolymerization with Quaternary Ammonium Salt in Aqueous Solution. *Macromolecules* **2015**, *48*, 5076–5087. [CrossRef]
12. Beuermann, S.; Buback, M.; Hesse, P.; Lacik, I. Free-Radical Propagation Rate Coefficient of Nonionized Methacrylic Acid in Aqueous Solution from low Monomer Concentrations to Bulk Polymerization. *Macromolecules* **2006**, *39*, 184–193. [CrossRef]
13. Lacík, I.; Chovancova, A.; Uhelska, L.; Preusser, C.; Hutchinson, R.A.; Buback, M. PLP-SEC Studies into the Propagation Rate Coefficient of Acrylamide Radical Polymerization in Aqueous Solution. *Macromolecules* **2016**, *49*, 3244–3253. [CrossRef]
14. Drawe, P.; Buback, M.; Lacik, I. Radical Polymerization of Alkali Acrylates in Aqueous Solution. *Macromol. Chem. Phys.* **2015**, *216*, 1333–1340. [CrossRef]
15. Rintoul, I.; Wandrey, C. Polymerization of Ionic Monomers in Polar Solvents: Kinetics and Mechanism of the Free Radical Copolymerization of Acrylamide/Acrylic Acid. *Polymer* **2005**, *46*, 4525–4532. [CrossRef]
16. Preusser, C.; Ezenwajiaku, I.H.; Hutchinson, R.A. The Combined Influence of Monomer Concentration and Ionization on Acrylamide/Acrylic Acid Composition in Aqueous Solution Radical Batch Copolymerization. *Macromolecules* **2016**, *49*, 4746–4756. [CrossRef]
17. Junkers, T.; Barner-Kowollik, C. The Role of Mid-Chain Radicals in Acrylate Free Radical Polymerization: Branching and Scission. *J. Polym. Sci. A* **2008**, *46*, 7585–7605. [CrossRef]
18. Preusser, C.; Chovancova, A.; Lacik, I.; Hutchinson, R.A. Modeling the Radical Batch Homopolymerization of Acrylamide in Aqueous Solution. *Macromol. React. Eng.* **2016**, *10*, 490–501. [CrossRef]
19. Wittenberg, N.F.G.; Preusser, C.; Kattner, H.; Stach, M.; Lacik, I.; Hutchinson, R.A.; Buback, M. Modeling Acrylic Acid Radical Polymerization in Aqueous Solution. *Macromol. React. Eng.* **2016**, *10*, 95–107. [CrossRef]

20. Lacik, I.; Ucnova, L.; Kukuckova, S.; Buback, M.; Hesse, P.; Beuermann, S. Propagation Rate Coefficient of Free-Radical Polymerization of Partially and Fully Ionized Methacrylic Acid in Aqueous Solution. *Macromolecules* **2009**, *42*, 7753–7761. [CrossRef]

21. Beuermann, S.; Buback, M.; Hesse, P.; Hutchinson, R.A.; Kukucova, S.; Lacik, I. Termination Kinetics of the Free-Radical Polymerization of Nonionized Methacrylic Acid in Aqueous Solution. *Macromolecules* **2008**, *41*, 3513–3520. [CrossRef]

22. Barth, J.; Buback, M. SP-PLP-EPR Study into the Termination Kinetics of Methacrylic Acid Radical Polymerization in Aqueous Solution. *Macromolecules* **2011**, *44*, 1292–1297. [CrossRef]

23. Buback, M.; Hesse, P.; Hutchinson, R.A.; Kasak, P.; Lacik, I.; Stach, M.; Utz, I. Kinetics and Modeling of Free-Radical Batch Polymerization of Nonionized Methacrylic Acid in Aqueous Solution. *Ind. Eng. Chem. Res.* **2008**, *47*, 8197–8204. [CrossRef]

24. Wittenberg, N.F.G.; Buback, M.; Stach, M.; Lacik, I. Chain Transfer to 2-Mercaptoethanol in Methacrylic Acid Polymerization in Aqueous Solution. *Macromol. Chem. Phys.* **2012**, *213*, 2653–2658. [CrossRef]

25. Wittenberg, N.F.G.; Buback, M.; Hutchinson, R.A. Kinetics and Modeling of Methacrylic Acid Radical Polymerization in Aqueous Solution. *Macromol. React. Eng.* **2013**, *7*, 267–276. [CrossRef]

26. Lacik, I.; Beuermann, S.; Buback, M. PLP-SEC Study into Free-Radical Propagation Rate of Nonionized Acrylic Acid in Aqueous Solution. *Macromolecules* **2003**, *36*, 9355–9363. [CrossRef]

27. Kattner, H.; Drawe, P.; Buback, M. Chain-Length-Dependent Termination of Sodium Methacrylate Polymerization in Aqueous Solution Studied by SP-PLP-EPR. *Macromolecules* **2017**, *50*, 1386–1393. [CrossRef]

28. Rosenboom, J.-G.; Roo, J.D.; Storti, G.; Morbidelli, M. Diffusion (DOSY) 1H NMR as an Alternative Method for Molecular Weight Determination of Poly(ethylene furanoate) (PEF) Polyesters. *Macromol. Chem. Phys.* **2017**, *218*, 1600436. [CrossRef]

*processes*

[MDPI]

*Review*

# Applications of Water-Soluble Polymers in Turbulent Drag Reduction

**Wen Jiao Han, Yu Zhen Dong and Hyoung Jin Choi \***

Department of Polymer Science and Engineering, Inha University, Incheon 22212, Korea;
22151728@inha.edu (W.J.H.); 22152270@inha.edu (Y.Z.D.)
\* Correspondence: hjchoi@inha.ac.kr; Tel.: +82-32-860-7486

Academic Editor: Alexander Penlidis
Received: 28 February 2017; Accepted: 26 April 2017; Published: 4 May 2017

**Abstract:** Water-soluble polymers with high molecular weights are known to decrease the frictional drag in turbulent flow very effectively at concentrations of tens or hundreds of ppm. This drag reduction efficiency of water-soluble polymers is well known to be closely associated with the flow conditions and rheological, physical, and/or chemical characteristics of the polymers added. Among the many promising polymers introduced in the past several decades, this review focuses on recent progress in the drag reduction capability of various water-soluble macromolecules in turbulent flow including both synthetic and natural polymers such as poly(ethylene oxide), poly(acrylic acid), polyacrylamide, poly(*N*-vinyl formamide), gums, and DNA. The polymeric species, experimental parameters, and numerical analysis of these water-soluble polymers in turbulent drag reduction are highlighted, along with several existing and potential applications. The proposed drag reduction mechanisms are also discussed based on recent experimental and numerical researches. This article will be helpful to the readers to understand better the complex behaviors of a turbulent flow with various water-soluble polymeric additives regarding experimental conditions, drag reduction mechanisms, and related applications.

**Keywords:** drag reduction; aqueous polymer; turbulent flow; poly(ethylene oxide); polyacrylamide; gum

## 1. Introduction

A liquid flow passing through pipelines generally experiences chaotic changes in the flow velocity and pressure in the process of overcoming frictional losses. These changes depend on the fluid characteristics such as the velocity, viscosity, and density and on the flow geometry, including the pipe diameter and length. This flow regime with a sufficiently high Reynolds number is commonly called turbulent flow [1] and is characterized by a less orderly flow phase in which the eddy currents of the fluid elements can result in chaotic lateral mixing [2]. For turbulent flow in pipelines, as well as in open and external flow, the friction can be reduced by introducing a minor quantity of a flexible, linear polymer with a high molecular weight into the flow [3]. This behavior is termed the polymer-induced turbulent drag reduction (DR) effect [4,5] which is an active friction reduction mechanism compared to the prevailing passive friction reduction. Thus, DR techniques have been applied in various industrial and engineering arenas, such as oil pipelines, open channels, marine applications, agricultural field irrigation, fire hoses, and biomedical systems [6–8].

In general, different types of additives, including solid particles, surfactants, and polymers have been used to induce the turbulent DR phenomenon. Several surfactants have been applied as drag-reducing agents because of their self-assembly characteristics and ability to form thread-like micelles in fluids [9–11]. Self-assembly of surfactants is a physicochemical process; the self-assembly properties depend on the surfactant type, concentration, and temperature, etc. Because of these characteristics, surfactant molecules can be rapidly recombined under high shear, allowing the surfactants to resist mechanical degradation and become more stable in the flow, leading to highly efficient DR [12–14]. Nonetheless, one of the disadvantages of using surfactants is that, compared to water-soluble polymeric drag reducers, a rather high concentration is required to initiate the DR. Thus, high-molecular-weight polymers are considered as more efficient drag reducing agents when dissolved in a suitable solvent system of either aqueous or non-aqueous media [15,16]. Due to the particular viscoelastic properties, polymeric additives strongly influence the turbulent flow characteristics, even when used in small quantities. High levels of DR can be obtained when polymers are added to the turbulent flow [17]. It is also possible to customize polymers to harness the specific characteristics of different polymers. The selection of polymeric additives for a potential application is related to the unique properties of the polymer, including the polymer structure, molecular weight, and fluid characteristics [18]. The energy loss caused by the turbulent flow increases operating costs; thus, the DR capability of polymeric additives has been widely studied based on the recognized economical benefits.

When a polymer is dissolved in a suitable solvent, its DR effect can be maximized under the right conditions. Although several water-soluble polymers successfully reduced the drag in the turbulent flow of aqueous systems, oil-soluble polymers such as polyisobutylene and polystyrene are known to demonstrate high DR efficiency when dissolved in organic solvents or oils [19]. Therefore, these polymers are very important in engineering areas, including crude oil pipeline transport.

Over several decades, many kinds of natural and synthetic polymers have been discussed in isolation as promising turbulent DR additives, whereas there are few general discussions and summaries covering polymer combinations. In addition, the role of DNA on turbulent DR is also focused. In this short review, we focus on the recent progress in research on the DR of water-soluble polymers in turbulent flow, with special emphasis on the polymer species, experimental parameters, and applications, along with the fundamental characteristics of these polymers. Synthetic and natural water-soluble drag reducing polymers for various potential agricultural and industrial applications, including field irrigation, slurry transport, fracking, enhanced oil recovery, and biomedical applications [20,21], are covered.

## 2. Mechanism of Turbulent Drag Reduction with Polymers

Based on the extensive industrial and engineering applications of DR, this phenomenon has been the topic of numerous theoretical and experimental studies for more than sixty years [22–24]. However, the DR mechanism still requires further investigation. Herein, we summarize the qualitative explanations of polymer-induced DR. Some mechanisms focusing on viscoelasticity, vortex stretching, non-isotropic properties, molecular stretching of polymers, and laminarization of turbulent flow will be mainly discussed along with the behaviors of polymeric additives in a turbulent flow such as polymer extension, extensional viscosity of polymers, mechanical degradation, and polymer relaxation.

Astarita [25] proposed a general explanation of the DR phenomenon based on the viscoelasticity of the turbulent DR solution, suggesting that the energy dissipation in a turbulent flow can be used to explain the DR phenomenon by investigation of the energy dissipation of viscoelastic liquids.

On the other hand, Gadd [26] suggested that the DR phenomenon induced by the addition of miniscule amounts of polymers (as little as tens of ppm) to the turbulent flow might be related to the anomalous performance of the fluid particles during rapid deformation. In this scenario, vortex stretching arises near the wall, and extensive elongation of the polymeric molecules may take place, causing large elongational forces in the direction of the streamlines. Finally, this system resists the vortex stretching, resulting in the DR. The fluid with added polymer molecules has a longer relaxation time, and large polymeric chains with a longer relaxation time would likely induce large eddies with a lower stretching rate, leading to the rapid decay of the eddies.

For the mechanism of non-isotropic properties, it is known that polymeric chains will migrate in the solution influenced by the shear flow, hence the structure and the viscosity of solution will be changed. Since the shear rate is directionally dependant, the structure and the viscosity of the solution will be directionally dependant, in other words is non-isotropic. These non-isotropic properties were considered to be sources of turbulent structure changes and the DR phenomenon [27].

Molecular stretching of the polymers was also assumed to be responsible for the DR. According to this assumption, the macromolecules added to the turbulent solution increase the resistance to extensional flow, which is associated with shear hardening behavior, impeding turbulent bursts near the wall. Tulin [28] concluded that an increase in the laminar sub-layer thickness caused greater turbulence dissipation, accounting for the DR. Furthermore, Lumley [29] proposed a mechanism for the DR phenomenon based on polymeric chain extension, in which extension of the molecules absorbs the energy from the turbulent eddies, giving rise to the DR. Min et al. [30] proposed that polymer additives store the elastic energy from the flow very near the wall as shown in Figure 1, in which when the relaxation time is long enough, this high elastic energy is transported to and dissipated into the buffer by near-wall vortical motion, resulting in significant drag reduction.

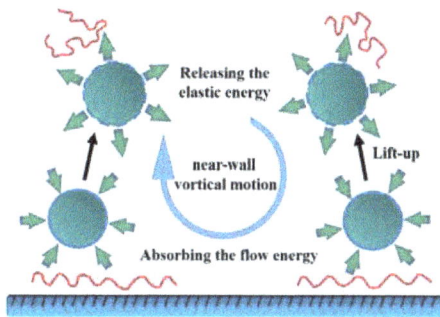

**Figure 1.** Schematic representation of polymer-induced turbulent drag reduction (DR) mechanism.

The idea of an extended laminar flow was also considered as an important mechanism for the generation of DR. The component of flow not in the main flow direction is futile to the main flow, which induces wasteful energy dissipation via turbulent eddies and leads to the increased frictional drag [31]. In other words, the laminarization of turbulent flow will reduce the dissipation of fluid energy, therefore contributing to the DR. The most low energy consumption of viscous fluid would occur in the corresponding laminar flow under no matter how big the Reynolds number is, in which the frictional drag becomes the minimum value, just enough to overcome the viscous friction. Hence, the maximum effect of DR can be obtained while the fluctuating velocity components in turbulent flow are completely suppressed. Meanwhile, the heat transfer in this laminar condition would be the physical minimum for a given flow rate. The DR effect range of polymeric additives is between the asymptotic polymer solution and turbulent Newtonian fluid [31].

On the other hand, many studies of the polymeric behavior in turbulent flow have been carried out. Armstrong and Jhon [32] explained the polymer conformation by empolying the bead-spring model by using the connector potential, in which the polymer was considered as a chain of idential beads linked together by arbitrary spring potential. They then established the self-consistent process based on the relationship between the friction factor and molecular dissipation. It was proposed that the effect of the stochastic velocity field on the molecule can be explained as a "renormalization" of the connector potential, where the dumbbell probability density was derived for an arbitrary connector potential. It was determined that for a Hookean potential, at some value of the turbulent strength, the probability density would have an infinite second moment. It was found that the renormalization of the connector potential between the beads will reduce the connector force, which makes the bead extend; that is to say, the polymer molecule expands in the turbulent flow.

Regarding the effect of the extensional viscosity on the DR, the two-dimensional grid turbulent flow produced from a vertically flowing soap film was evaluated by introducing polyethylene oxide

(PEO) using several extensional rates [33]. The mechanisms of energy transfer were found to be different in the normal and stream-wise directions. The critical polymer concentration for producing DR in a two-dimensional turbulent flow varied with the elongational rate. When the elongational rate was higher, the DR was efficient from a lower polymer concentration.

Moreover, compared to other drag reducing additives such as surfactants and particles, polymers under a turbulent flow face the issue of severe mechanical degradation. Brostow [34] reported that the energy supplied to a polymeric molecule in a turbulent flow leads to two processes, i.e., mechanical degradation and polymer relaxation. They proposed a model of the polymeric chain conformations based on statistical mechanics, and the effects on the DR efficiency and mechanical degradation of the polymer chains in a turbulent flow were established by considering the relationship between the extent of mechanical degradation of the polymer chains and the flow time. Recently, Pereira and Soares [35] further showed that the DR increased with time to reach a maximum, before decreasing due to chain degradation. They also presented the relative DR quantity, defined as the ratio of the current DR to the maximum DR obtained before degradation. They then suggested another decay function based on the relative drag reduction as a function of the molecular weight, polymer concentration, temperature, and Reynolds number.

In order to study the mechanism of DR by polymeric additives, some modern experimental and simulation techniques, such as direct numerical simulation, laser Doppler velocimetry, and particle image velocimetry, have been widely used for various measurements of the turbulence statistics. Den Toonder et al. [36] investigated the roles of stress anisotropy and elasticity by means of direct numerical simulation and laser Doppler velocimetry. They proposed that when polymers are extended, viscous anisotropic effects cause a change in the turbulence structure, and the entropy change results in drag reduction. White et al. [37] studied the turbulence structure in a drag-reduced flat-plate boundary layer flow by means of particle image velocimetry, providing information for further understanding the interaction between the polymer and the near-wall vortex. In addition, various research groups [38–42] performed a series of studies on the turbulent flow with polymer addition and clarified the velocity profile using particle image velocimetry and laser Doppler velocimetry.

Concurrently, the finite elastic non-linear extensibility-Peterlin (FENE-P) model as a kind of polymer model has been often adopted for DR studies. The FENE-P model is obtained by a pre-averaging approximation applied to a suspension of non interacting finitely extensible non-linear elastic (FENE) dumbbells, which accounts for the finite extensibility of the molecule [43]. Li et al. [44] carried out the direct numerical simulation of the forced homogeneous isotropic turbulence with/without polymer additives. They adopted the FENE-P model as the conformation tensor equation for the viscoelastic polymer solution.

Furthermore, Eshrati et al. [45] investigated the turbulent DR phenomenon in many oil-water flow systems with dissolved polymers. They used fuzzy logic [46] to study the DR phenomenon of water-soluble polymers in multiphase in-pipeline flow, and proposed that compared with the traditional logics, fuzzy logic is more useful for linking the multiple inputs to produce a single output. They also investigated the connection between DR and many parameters such as the polymer molecular weight, polymer concentration, charge density, mixture velocity, and oil fraction to estimate how fuzzy logic performs in prediction of the DR. The results showed that the effectiveness of the drag reducer increases with an increase in the velocity of the mixture; the polymer molecular weight and the increase in the oil fraction and the polymer charge density also affect the maximum drag reduction. From these results, a fuzzy logic model was developed to predict the DR, and the prediction was also verified through various results obtained with many polymers. Furthermore, dimensional analysis was carried out to find the best equation fitting the results. The result showed that a quadratic form was the most appropriate.

The two-dimensional (2-D) flows have also been adopted for DR studies [47–49]. Unlike the three-dimensional (3-D) flow, the flow velocity at each point in the 2-D flow is parallel to a fixed plane. Hidema et al. [50] pointed out that in the 3-D flow, it is difficult to avoid the effects of shear stress and to extract the extensional stress completely. In order to study the relationship between the extensional viscosity of polymers and turbulent DR, they used flowing soap films to generate 2-D turbulent flow to eliminate shear stress.

Therefore, it is highly possible that more than one kind of mechanism can lead to the DR phenomenon. Extensive research has led to the development of many proposed prediction models, but the accuracy of the mechanism of these predictions is still an open question; the above discussion is an attempt to provide a possible mechanism. In order to more accurately explain the mechanism of DR, further research is required.

## 3. Synthetic Water-Soluble Polymers

Synthetic water-soluble polymers usually have repeating units that can be synthesized by controlling the chemical structure, molecular weight, and number of functional groups on the polymeric backbone. These synthetic polymers, which possess a linear flexible chain and high molecular weight, have been extensively studied as drag reducing agents over the years. However, the drawbacks of chemical and mechanical degradation must be overcome. Table 1 summarizes the structures of four major synthetic water-soluble polymers that are used in DR research and are discussed in this review in detail.

**Table 1.** Structures of synthetic water-soluble polymers: Poly (ethylene oxide), Polyacrylamide, Poly (acrylic acid) and Poly(*N*-vinyl formamide).

| No. | Name | Chemical Structure |
|-----|------|-------------------|
| 1. | Poly (ethylene oxide) | |
| 2. | Polyacrylamide | |
| 3. | Poly (acrylic acid) | |
| 4. | Poly (*N*-vinyl formamide) | |

### 3.1. Poly(ethylene oxide)

Drag-reducing polymers generally have linear flexible chain structures with a very high average molecular weight. PEO, a polymer with ethylene oxide as the repeating unit, has been extensively adopted as an effective drag reducer in aqueous systems.

Virk et al. [51] studied the effects of solutions of five homologous PEOs with different molecular weights in distilled water on DR in a turbulent flow. They found that the strength of the DR produced by the homologous PEOs in a pipe flow is a universal function of the molecular weight, polymer concentration, and flow rate. The maximum DR efficiency was possibly limited by an asymptotic value that is independent of the pipe diameter and polymeric species. This experiment also presented the intrinsic concentration, $[C] = R_{Fmax}/[R]$ (where $[R]$ is the DR value of an initial increment of the polymeric chain), and demonstrated that the DR efficiency increased with a decrease in the diameter of the pipe and increasing polymer concentration [52].

Based on Virk's drag reduction equation, Little [53] presented a simplified equation from experimental analysis:

$$\frac{DR}{DR_{max}} = \frac{C}{[C]+C} \tag{1}$$

where $C$ is the concentration of the additive and $DR_{max}$ is the maximum DR value. From the concentration-dependence of the DR, the intrinsic concentration $[C]$ was found to be very useful for normalization of the DR values of different-molecular-weight components of a homologous series of polymers [54]. The relationship between the DR and the polymer properties, such as the molecular weight and concentration, was represented as Equation (2).

$$DR = \frac{C}{[A/(M-M_0)]+C} \times DR_{max} \tag{2}$$

Similarly, Choi et al. [55] investigated the effect of the concentration of dilute solutions of water-soluble PEO on turbulent DR in a rotating disk flow system. The results demonstrated that the DR efficiency of PEO increased with increasing concentration of PEO, to reach a critical concentration at which the DR was maximal. The DR efficiency then declined with a further increase in the polymer concentration. Figure 2 illustrates the dependence of the DR efficiency of PEO with different viscosity-average molecular weight ($Mv$) as a function of the additive concentration. The optimum critical polymer concentration decreased with an increase in the polymer viscosity-average molecular weight. On the other hand, the appearance of a maximum was reported to be due to the DR characteristics of the polymeric solute and the increased solution viscosity, wherein both become important at higher polymer concentration.

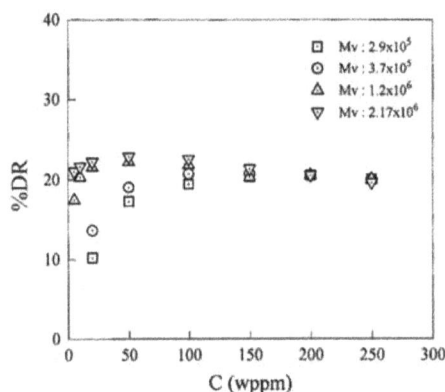

**Figure 2.** Drag reduction of PEO versus the concentration of polymers of various molecular weights at 2800 rpm, reprinted with permission from [55]. Copyright American Chemical Society, 1996.

Kim et al. [56] applied PEO as a potential drag reducer in seawater piping in an ocean thermal energy conversion (OTEC) process, and investigated the drag reducing characteristics and mechanical degradation of PEO with different molecular weights and concentrations. Figure 3 presents the stability and mechanical degradation of PEO with a weight-average molecular weight ($M_w$) of $5.0 \times 10^6$ g/mol in a turbulent flow. The drag reduction was initially time-dependent and then remained at the limiting value due to degradation of the polymer chains. With the degradation of the polymer chains, the drag reduction ability decreased significantly. The temperature-dependent DR efficiency was also investigated. As presented in Figure 4, although the initial percentage DR was the highest at a room temperature, the DR declined most rapidly at this temperature. A higher DR efficiency was obtained at a lower temperature than at a higher temperature over time. The mechanical degradation of PEO molecules in seawater was compared with that in the PEO-deionized water system, as shown in Figure 5. Equation (3) was used to plot the degradation behavior:

$$\frac{DR\,(t)}{DR_0} = \frac{1}{1 + W\left(1 - e^{-ht}\right)} \tag{3}$$

where DR*(t)* and $DR_0$ are the percentage drag reduction at time $t$ and at the beginning of flow, respectively. The term $h$ indicates the degradation velocity and $W$ indicates the shear stability. The results showed that PEO polymer chains in the deionized-water system undergo more degradation in the original state than in seawater; however, the shear stability in the aqueous system was higher than that in seawater. Based on this study, it was proved that the DR effect induced by PEO could be exploited in the OTEC system to reduce the pumping energy cost. Choi et al. [57] also investigated the drag reduction efficiency of PEO in synthetic seawater and found that a maximum DR of 30% was obtained near 50 wppm concentration for higher-molecular-weight PEO. They confirmed that the behavior of PEO in seawater was similar to that in the deionized-water system, which could be a possible indicator of the commercial viability of PEO.

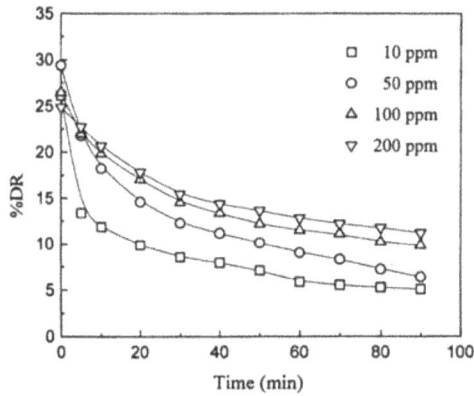

**Figure 3.** Percent drag reduction of PEO ($M_w = 5 \times 10^6$) as function of time for various polymer concentrations, reprinted with permission from [56]. Copyright The Society of Chemical Engineers, Japan, 1999.

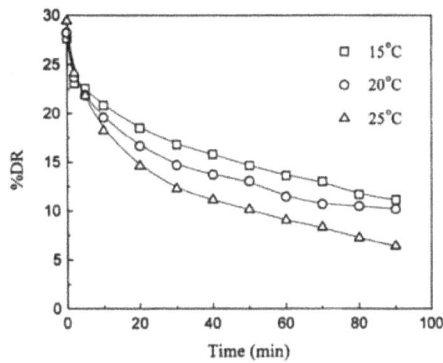

**Figure 4.** Percent drag reduction of PEO ($M_w = 5 \times 10^6$) as function of time for three different temperatures, reprinted with permission from [56]. Copyright The Society of Chemical Engineers, Japan, 1999.

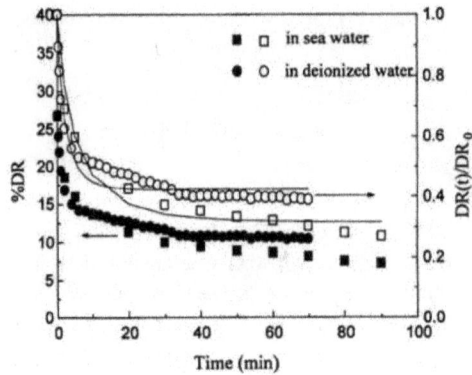

**Figure 5.** Comparison of drag reduction (solid symbol) and degradation process (open symbol) of 50 ppm PEO ($M_w = 4 \times 10^6$) in seawater and deionized water at same Reynolds number. Symbols represent experimental results (square for PEO in seawater and circle for PEO in deionized water); lines are obtained from equation [3], reprinted with permission from [56]. Copyright The Society of Chemical Engineers, Japan, 1999.

Recently, Lim et al. [58] used aqueous PEO solutions to explore the relationship between the drag reduction efficiency and the polymer molecular conformation. Note that the chain conformation may be affected by the solvent quality of the polymer solution. The specific temperature where the solvent is exactly poor enough to cancel the effect of excluded volume expansion of the dissolved polymeric chain is called the theta temperature ($T_\theta$). Thus, they measured the drag reduction efficiency of PEO aqueous solutions at various increments from $T_\theta$, and found that the drag reduction properties were sensitive to the conformation of the polymers, even at very low concentrations. In addition, they carried out temperature quenching experiments and proposed that the faster decrease in the drag reduction efficiency for polymers closer to $T_\theta$ is not due to mechanical molecular degradation, but rather, to shrinkage of the PEO molecules.

### 3.2. Polyacrylamide

Polyacrylamide (PAAM), a synthetic macromolecule formed from acrylamide subunits, has been extensively studied as the most commonly used drag reducing agent. One of the extensive usages of PAAM is in the flocculation of solid particles dispersed in a medium, with potential applications in water treatment and the papermaking process. Another common use is in the oil and gas industries. Enhanced oil recovery is an important example of PAAM application. A highly viscous aqueous solution can be obtained even with a low PAAM concentration. This technique could also be used to address (water) flooding problems.

The DR efficiency of PAAM has been more widely examined than that of other drag reducing polymeric agents. Sung et al. [59] investigated the DR efficiency of PAAM in a rotating disk flow system by comparison with that of PEO. The effect of temperature on the DR was evaluated at a polymer concentration of 50 wppm for both PAAM and PEO, as shown in Figure 6. The results showed that mechanical degradation of the PEO chains increased with temperature, whereas PAAM remained mechanically stable even at high temperature. Therefore, PEO was more susceptible to thermal and mechanical degradation, whereas PAAM exhibited relatively high shear resistance. Therefore, PAAM is prospectively an excellent DR additive for high-temperature and long-period transportation applications.

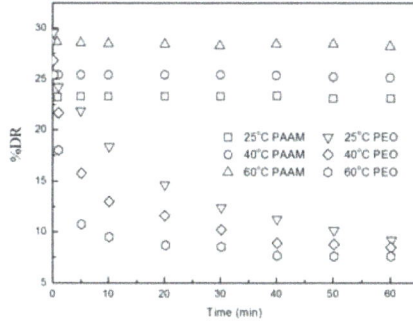

**Figure 6.** DR of both PAAM and PEO (50 wppm) at three different temperatures, reprinted with permission from [59]. Copyright Marcel Dekker, Inc., Japan, 2004.

Sandoval et al. [60] also compared the degradation phenomenon for three different aqueous solutions of PEO, PAAM, and xanthan gum (XG) using a pipe flow device. PEO and PAAM have flexible chains, whereas the chain structure of XG is rigid. Figure 7 presents a comparison of the DR efficiency of the flexible polymers versus the rigid polymer. PAAM was found to be as effective as PEO, whereas the DR efficiency of the PAAM solution declined to a lesser extent than that of the PEO solution, which suggested that polymeric chain scission was more significant for PEO. Moreover, the drag reduction achieved with rigid chain XG fell abruptly in the first step and remained constant, which is different from the cases with PEO and PAAM. Similarly, Pereira et al. [61] compared the drag reduction efficiency of PEO, PAAM, and XG over time by using a rotating cylindrical double-gap device. They also found that there was no decrease in the drag reduction at a high PAAM concentration, which suggested that the PAAM chain is mechanically stronger than that of PEO.

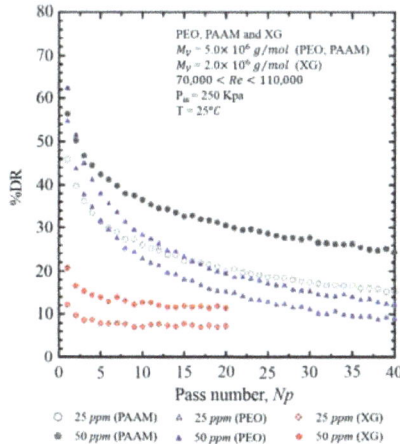

**Figure 7.** Comparison of DR between flexible polymers (PEO and PAM) and rigid polymer, XG, reprinted with permission from [60]. Copyright Engenharia Térmica (Thermal Engineering), 2015.

On the other hand, Lim et al. [62] examined the turbulent DR of λ-DNA and compared it to that of PAAM in continuous and stepwise mode using an in-house designed rotating disk apparatus. Figure 8 confirmed that the mechanical degradation behavior of linear flexible PAAM was very different from that of λ-DNA due to differences in the molecular weight, polydispersity, and flexibility. Although PAAM produced a much higher initial DR value, this drag reduction effect decreased to zero in 5 min.

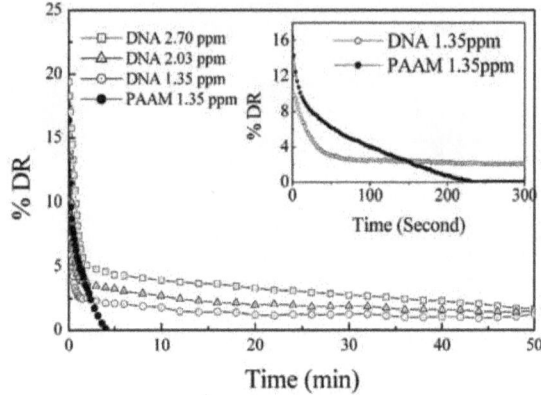

**Figure 8.** Comparison of λ-DNA percent DR (1.35, 2.03 and 2.70 wppm) with PAAM ($M_w$ = 18 × $10^6$ g/mol) on long-term scale at 1980 rpm and 25 °C. The inset represents the initial changes in the drag reducing efficiency for λ-DNA and PAAM at 1.35 wppm, reprinted with permission from [62]. Copyright American Chemical Society, 2015.

Zhang et al. [63] experimentally investigated heat transfer and frictional DR in the two-phase flow of air-water with and without PAAM additives in a horizontal circular tube. They showed that with the addition of PAAM, the heat transfer coefficients were reduced from 36.8% to 70.3%, and the pressure drop was reduced from 31.9% to 54.7% compared to those without the PAAM additive.

*3.3. Poly(acrylic acid)*

Poly(acrylic acid) (PAA) is another high-molecular-weight water-soluble polymer of acrylic acid, possessing a carboxyl group in each monomeric unit of the main chain. This polymer has a highly negative charge density in aqueous medium due to dissociation of the acid groups. Therefore, PAA is one of the most extensively used aqueous anionic polyelectrolytes, and its degree of ionization is dependent on the pH of the solution [64,65]. At a low pH, PAA is known to adopt a more compact conformation and can associate with various non-ionic polymers to form hydrogen-bonded inter-polymeric complexes due to the absence of negative charges in the backbone of PAA. In aqueous solutions, PAA can also form poly-complexes with oppositely charged polymers and surfactants.

Based on this interesting property of the ionic polymer PAA, Kim et al. [66] used a polymer–surfactant (SDS) complex system as a drag reducer and investigated the effects of the surfactant and pH on the DR efficiency of PAA in an external flow using a rotating disk system. A schematic diagram of the conformational changes in the molecular structure of PAA at various pH values is shown in Figure 9. As the pH increased, the PAA underwent a conformational transition from the compact helical structure to the highly extended rigid rod form, and its shear viscosity increased. Figure 10 illustrates the effect of pH on the turbulent DR efficiency of PAA at different pH values. The DR efficiency at pH 4 was found to be smaller than that at higher pH, indicating that the turbulent DR efficiency of the polymers in water is directly related to their chain conformations. They suggested that the extended conformation of PAA is more beneficial for drag reduction in a turbulent flow, compared with the compact helical form. Furthermore, as the SDS concentration (M = mol/L) increased, the drag reduction efficacy of PAA increased, as shown in Figure 11. This can be explained by the formation of an association complex between the polymer chains and surfactant, which enhanced the bonding force of the polymer molecules, resulting in extension of the polymer chains, especially at low pH. Figure 12 illustrates the conformational variation of the PAA molecule with addition of the surfactant SDS.

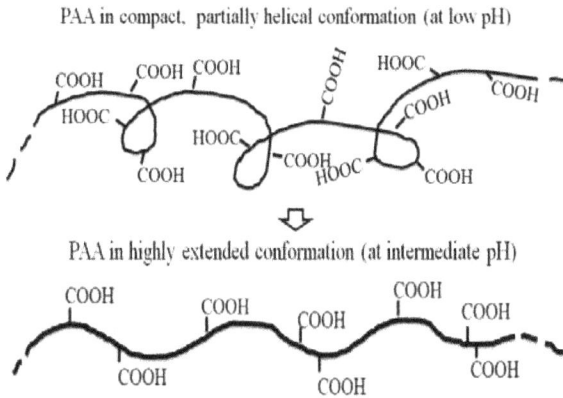

**Figure 9.** A schematic representation of the conformational variation in the PAA molecular at different pH levels, reprinted with permission from [66]. Copyright Elsevier, 2011.

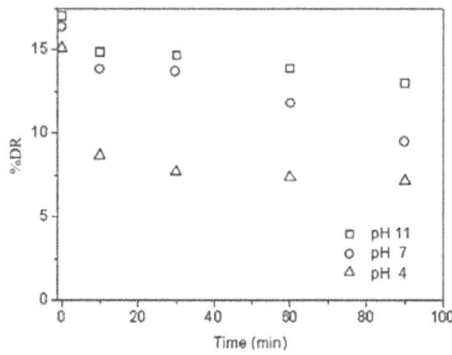

**Figure 10.** Effect of pH of PAA597 solution (100 ppm) on DR, reprinted with permission from [66]. Copyright Elsevier, 2011.

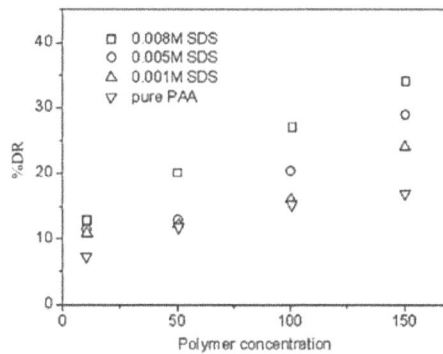

**Figure 11.** Effect of SDS concentrations as a function of polymer concentrations (pH 4), reprinted with permission from [66]. Copyright Elsevier, 2011.

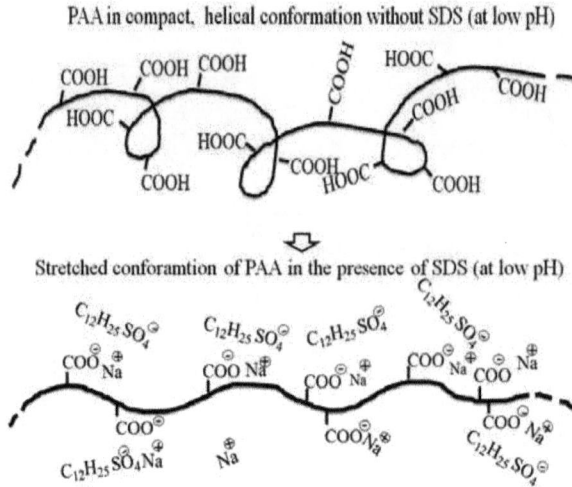

**Figure 12.** The variation in PAA conformation by adding surfactant SDS, reprinted with permission from [66]. Copyright Elsevier, 2011.

Kim et al. [67] studied the conformational changes of the PAA chains under high shear flow, and found a severe decrease in the turbulent DR (%DR) with increasing rotational speed of the disk and time. They also confirmed that these changes in the drag reduction are sensitive to external parameters such as the pH, polymer concentration, and PAA molecular weight. Such DR changes were explained as a direct consequence of inter-chain association via hydrogen bonding.

Recently, Zhang et al. [68,69] studied the turbulent DR efficiency of aqueous poly(acrylamide-*co*-acrylic acid) copolymers with various molecular parameters in a rotating disk flow system, and found a maximum DR of 45% at 50 wppm concentration. The influence of temperature on the drag reduction efficacy is presented in Figure 13. As the temperature increased, the initial drag reduction efficacy increased; however, the persistent drag reduction activity decreased over the course of the examination. This was attributed to the fact that the thermal motion of the molecules increased with increasing temperature under turbulent flow conditions, resulting in weakness of the solvated flow regions.

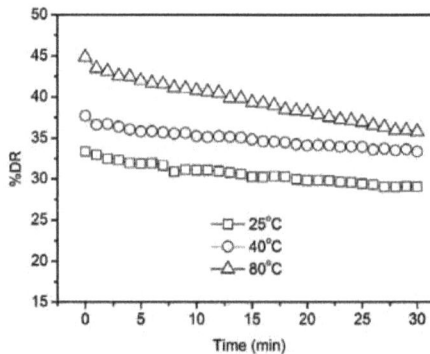

**Figure 13.** %DR versus time of copolymer solution ($M_w = 1.5 \times 10^7$ g/mole) for different temperatures (25 °C, 40 °C and 80 °C) at 1980 rpm, reprinted with permission from [68]. Copyright Springer-Verlag, 2011.

In addition, Cole et al. [70] synthesized 4-arm star PAA for use as a new water-soluble drag reducing agent by applying a Cu(0)-mediated polymerization technique. The results confirmed that 4-arm star PAA was effective as a drag reducing agent, with a drag reduction of 24.3%. Moreover, 4-arm star PAA showed superior mechanical stability compared with the commercial product Praestol, PEO, and linear PAA.

*3.4. Poly(N-vinyl formamide)*

With continued improvement of the methods for synthesis and purification of the hydrophilic cationic *N*-vinyl formamide monomer, water-soluble poly(*N*-vinyl formamide) (PNVF) and its copolymers have been adopted in industrial and biomedical applications. Due to the very high molecular weight and the structural similarity to PAAM, PNVF has been studied as a potential candidate as an effective drag reducing agent [71]. Furthermore, PNVF has been introduced as a replacement for acrylamide polymers for industrial application. The potential uses of PNVF and its hydrolyzed products have been demonstrated in the areas of papermaking, water treatment, coatings, rheology modifiers, and in the oil field industry [72].

Based on its very low toxicity and good solubility in water, Marhefka et al. [73] investigated the drag reduction efficacy, viscoelasticity, and mechanical degradation behavior of PNVF in a turbulent pipe flow, specifically for biomedical applications. They found that PNVF with a high molecular weight demonstrated DR characteristics, and its mechanical chain degradation was much lower than that of PEO. Besides this, in vivo analysis of PNVF suggested that it can be adopted as a drag reducing agent for clinical use. On the other hand, Mishra et al. [74] synthesized an aqueous graft copolymer of κ-carrageenan-g-*N*-vinyl formamide and studied its swelling behavior, metal ion sorption, flocculation, and anti-biodegradation characteristics. The results showed that this grafted polymer exhibits better water swelling behavior and resistance to biodegradation. Efficient flocculation capability could also be obtained, which may enable its application in the treatment of coal wastewater.

*3.5. Water-Soluble Copolymer*

Water-soluble copolymers are one of the fastest growing materials in industrial products. Due to their interesting characteristics that differ from those of homopolymers, water-soluble copolymers can be widely used in water treatment, coatings, personal care formulations, and commercial drag reducing agents.

McCormick and co-workers [75] extensively studied water-soluble copolymers. They synthesized four series of water-soluble acrylamide copolymers and investigated their corresponding DR effectiveness using a modified rotating disk in tube flow. The synthetic copolymers, including polyacrylamide-*co*-sodium acrylate (PAAM-*co*-NaA), polyacrylamide-*co*-(sodium2-(acrylamido)-2-methylpropanesulfonate) (PAAM-*co*-NaAMPS), polyacrylamide-*co*-(sodium3-(acrylamido)-3-methylbutanoate) (PAAM-*co*-NaAMB), and polyacrylamide-*co*-diacetone acrylamide (PAAM-*co*-DAAM), exhibited differences in terms of the hydrophobicity, charge density, and the degree of inter-molecular or intra-molecular association. The researchers presented a simple way of examining the DR effectiveness based on the DR efficacy as a function of the hydrodynamic volume of the polymer ([$\eta$]C), and demonstrated the dependence of the DR performance on the polymeric structure. They found that the incorporation of low quantities of charged co-monomers afforded copolymers with a high intrinsic viscosity and a high molecular weight with the greatest DR, even better than that of PEO or PAAM.

Moreover, Mumick et al. [76] synthesized a series of aqueous polyampholytes (containing both negative and positive charges in the same polymeric chain) based on acrylamide (AM), sodium 2-acrylamido-2-methylpropanesulfonate (NaAMPS), (2-acrylamido-2-methylpropyl)trimethylammonium chloride (AMPTAC), sodium 3-acrylamido-3-methylbutanoate (NaAMB), and 3-((2-acrylamido-2-methylpropyl)dimethylammonio)-1-propanesulfonate (AMPDAPS) and studied their chain confirmation and solvation in a turbulent flow. They reported that all of these macromolecules showed higher DR effectiveness and intrinsic viscosity with an increase in the ionic strength of the solvent. Among these synthesized copolymers, the betaine species exhibited the best DR characteristics, whereas the polyampholyte with a high charge density exhibited the lowest efficiency. Figure 14 illustrates the drag reduction characteristics of DAPS-10 (copolymers of AAM and AMPDAPS).

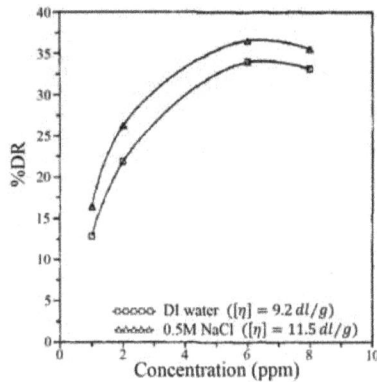

**Figure 14.** Percent DR versus concentration for DAPS-10 in deionized water and 0.5 M NaCl at wall shear stress $T_W = 1122$ dyn/cm$^2$ in the rotating-disk rheometer, reprinted with permission from [76]. Copyright American Chemical Society, 1994.

Recently, Brun et al. [77] investigated the DR and mechanical stability of poly(AAm-*co*-NaAMPS) solutions under shear in a turbulent pipe flow, and found that the NaAMPS portion in the main backbone of the drag reducer enhances the DR efficiency of PAAM. They also suggested that the presence of NaAMPS in the copolymer does not significantly influence the mechanical degradation of the polymeric chains. The lowest fanning friction factor value of poly(AAm-*co*-NaAMPS) was achieved with 15% NaAMPS, whereas the highest measured drag reduction of 75% was obtained for the 100 ppm solution. Further, the PAAM-PEO copolymer was also highly efficient for decreasing the pressure drop of the pipe flows. A DR efficiency of 33% was achieved with PAAM-PEO in water [78]. Reis et al. [79] carried out an experiment using PAAM-g-poly(propylene oxide) (PAAM-*g*-PPO) as a water-soluble drag reducer. The grafted amphiphilic copolymer PAAM-g-PPO produced a better DR effect than pure PAAM based on different parameters, including the average molecular weight, molecular chain length, intrinsic viscosity, and polymer size distribution. The amphiphilic graft copolymer was found to be more resistant to shear degradation and fluid recirculation than pure PAAM. Other derivatives, including copolymers of N-alkyl- and N-arylalkylacrylamides with AM [80], (acrylamido *tert*-butyl sulfonic acid)-acrylamide copolymer [81], polyacrylamide-grafted chitosan [82], and guar-*g*-polyacrylamide [83], etc., are also being studied as drag reducing agents.

## 4. Natural Water-Soluble Polymers

Polysaccharides, as an important class of natural polymers, have been studied as promising drag reducing agents due to their unique molecular structure, safety, shear-stability, low cost, biodegradability, and reproducibility (Table 2). They can be used to circumvent the safety and environmental influence issues associated with artificial polymers. However, their biodegradability may reduce their efficiency period and shelf life. To control this characteristic, many attempts (such as grafting, derivatization, and cross-linking) have been made to improve their properties to fit various applications, including polymer-induced DR.

*4.1. Guar Gum*

Guar gum (GG), which belongs to the hydrocolloid polysaccharide family, consists of mannose and galactose sugars, and is generally obtained from the endosperm of Cyamopsis tetragonolobus. It is mainly composed of the linear backbone chains of 1,4-linked β-D-mannopyranosyl units with short side-branches of 1,6-linked α-D-galactopyranosyl units. GG has been widely used in various industries, including food, hydraulic fracturing, suspending agents, agriculture, bioremediation, cosmetics, and pharmaceuticals.

Kim et al. [84] examined the DR behavior of GG with three different-molecular-weight fractions in water using a rotating disk apparatus. Because natural GG has a single average molecular weight, three different molecular weights of GGV, GG30 and GG60 were obtained by ultrasonic degradation for 0, 30 and 60 min, respectively. They tested the drag reduction efficacies of GG, and observed that GG is an effective aqueous drag reducing agent, and is more stable towards mechanical chain degradation than synthetic aqueous drag reducing agents like PEO. Figure 15 shows the results of mechanical degradation of these GG samples with the three molecular weights over 10 h at 1800 rpm. All of the GG solutions gave rise to a certain percentage drag reduction of between 62% and 80% of the initial DR efficiency.

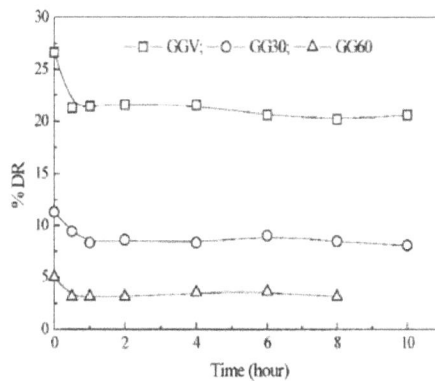

**Figure 15.** %DR versus time for a 100 wppm GG solution, reprinted with permission from [84]. Copyright John Wiley and Sons, 2002.

Chemical modification of GG has also been studied by many researchers as a method to overcome its inherent deficiencies, such as its facile biodegradability. Singh et al. [85] compared the DR efficiencies of pristine GG, purified GG, and grafted GG in water, as shown in Figure 16. They found that the grafted GG copolymer combined the robustness of GG and the efficiency of the synthetic PAAM polymer, resulting in enhanced drag reduction efficiency. Moreover, Deshmukh et al. [86] studied the grafting of polyacrylamide onto GG, and compared the graft polymer with commercial GG and purified GG. They found that the grafted GG and purified GG exhibited good biodegradation resistance and enhanced drag reduction efficiencies.

Behari et al. [87] prepared copolymers in which methacrylamide was grafted onto GG using a redox pair comprising potassium chromate/malonic acid and investigated the reaction conditions that could produce less of the homopolymer and generate more of the graft copolymer. They observed that the extent of grafting increased with an increase in the concentration of hydrogen ions or chromate ions. Furthermore, the grafting ratio and drag reduction efficiency increased with an increase in the malonic acid concentration from $3.5 \times 10^{-3}$ to $10 \times 10^{-3}$ mol·dm$^{-3}$.

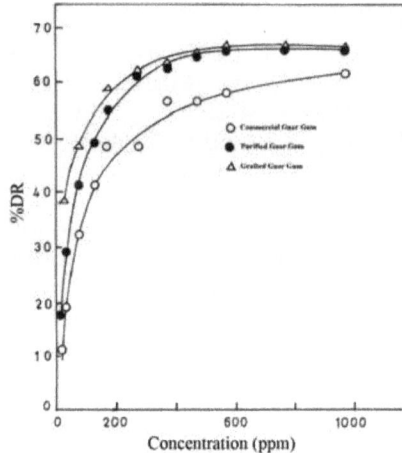

**Figure 16.** DR characteristics of commercial guar gum, purified guar gum, and grafted guar gum, reprinted with permission from [85]. Copyright De Gruyter, 2009.

### 4.2. Xanthan Gum

Xanthan gum (XG) is an acidic polymer that is secreted by Xanthomonas campestris bacterium. Its main structure consists of one D-glucopyranosyluronic unit, two D-glucopyranosyl units, and two D-mannopyranosyl units. Due to its ability for rheological control in aqueous systems, XG has been widely used in industries. In the case of agricultural applications, XG has been used to enhance the fluidity of fungicides and insecticides [88]. Moreover, XG can produce a high shear viscosity polymer solution at low concentrations or under shear forces, which makes it effective as a thickener and stabilizer in the food and petroleum industries, such as in cosmetics, toothpastes, oil drilling fluids, enhanced oil recovery, and pipeline cleaning [89].

The drag reduction capability of XG continues to be widely researched. Sohn et al. [90] evaluated the effect of various molecular parameters on the DR of XG, including the molecular weight, polymer concentration, temperature, solution ionic strength, and rotational speed of the disk. The results showed that the DR efficiency of XG is closely associated with various molecular parameters. Its higher shear-stability in both water and salt solutions compared to other flexible polymers was documented.

Hong et al. [91] studied the DR efficiency induced by different concentrations of XG in aqueous KCl solutions in a closed chamber using a rotating disc, and found that the mechanical degradation as a function of time decreased with increased KCl concentration. The anionic charges on XG allow interaction with the added salt ions to induce a conformational change of XG in solution, resulting in changes in the shear viscosity. The DR efficiency of XG as a function of the concentration in the initial period and after 40 min is illustrated in Figure 17. The DR increased with higher XG concentrations, and the DR efficiency of XG/KCl declined with increasing KCl concentration. This is because the polymeric chain conformation of XG tends to be more rigid, leading to lower sensitivity to the high shear conditions.

Recently, Andrade et al. [92] analyzed the drag reduction efficiency of PEO, PAAM, and XG by dissolving these three polymers in deionized water with and without synthetic sea salt. The effect of the salt concentration on the DR over time was studied by using a double-gap Couette-type rheometer device. In the presence of salt, the maximum DR efficiency was reduced for both the PEO and XG macromolecular solutions over time; however, the DR of the PAAM solutions did not change significantly. The time-dependent DR profiles, DR($t$), for XG in de-ionized water and synthetic saline solution are illustrated in Figure 18. The dramatic decrease in the efficiency is associated with the structural transition from helical to coiled with addition of the salt. For the 50 ppm solution, the maximum drag reduction decreased from 0.195 to 0.04.

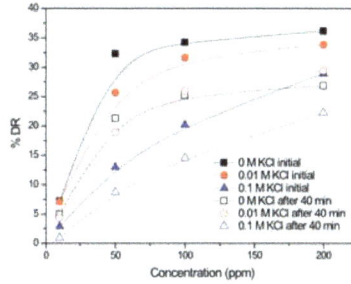

**Figure 17.** Concentration dependence of %DR with XG, reprinted with permission from [91]. Copyright Elsevier, 2015.

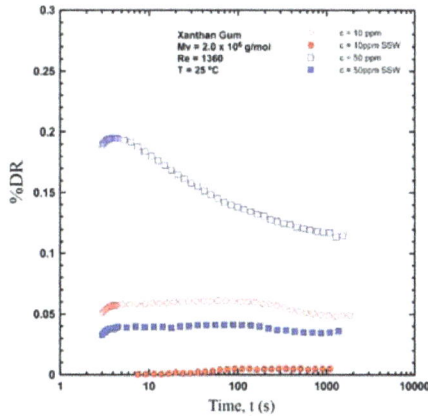

**Figure 18.** DR(*t*) for a range of concentrations of XG in synthetic seawater and deionized water reprinted with permission from [92]. Copyright American Society of Mechanical Engineers (ASME), 2016.

### 4.3. Carboxymethyl Cellulose

Cellulose, the most common biopolymer, is mainly found in the cell wall of plants. Cellulose possesses interesting properties such as a high mechanical strength and good thermal stability; however, the principal disadvantage is its insolubility in both aqueous and organic solvents, which limits its use in industry. Thus, chemical modification of the hydroxyl groups of cellulose may be used to confer greater solubility [93]. Carboxymethyl cellulose (CMC), which is a cellulose derivative possessing hydroxyl groups bound to carboxymethyl groups, shows good solubility in aqueous solvents. CMC is an important industrial polymer that is used in a number of applications such as DR, flocculation, detergents, foods, and oil-well drilling technology [94,95].

Deshmukh and co-workers [96] published a method of synthesizing CMC-based graft copolymers by grafting acrylamide chains onto the CMC backbone, and measured their DR efficacy, shear stability and biodegradability. The copolymers (CAm1, CAm2, CAm3 and CAm4) were synthesized with the same amount of CMC (1 g) and acrylamide (0.14 mol) with gradually increasing concentrations of ceric-ions (0.05, 0.10, 0.20, 0.30 mol). The increased ceric-ion concentrations resulted in an increase in the number of grafts, but led to shorter chains. The presence of the grafted polyacrylamide chains led to enhanced DR efficacy and good mechanical shear stability, and these factors were also found to be dependent on the number and length of the grafts. Figure 19 illustrates the drag reduction efficacies of unmodified CMC and its copolymers. As the polymer concentration increased, the drag reduction efficacies changed in two different ways. All of the graft copolymers were more effective than unmodified CMC, and a maximum DR of about 68% was obtained with a concentration of CAm1 of 75 ppm. These graft copolymer solutions exhibited significantly reduced biodegradability.

**Table 2.** Structures of natural water-soluble polymer: Guar Gum, Xanthan Gum, Carboxymethyl cellulose and DNA, reprinted with permission from [97]. Copyright Nature Publishing Group, 2007.

| Name | Chemical Structure | Name | Chemical Structure |
|------|--------------------|------|--------------------|
| Guar Gum | | Carboxymethyl cellulose | |
| Xanthan Gum | | DNA [97] | |

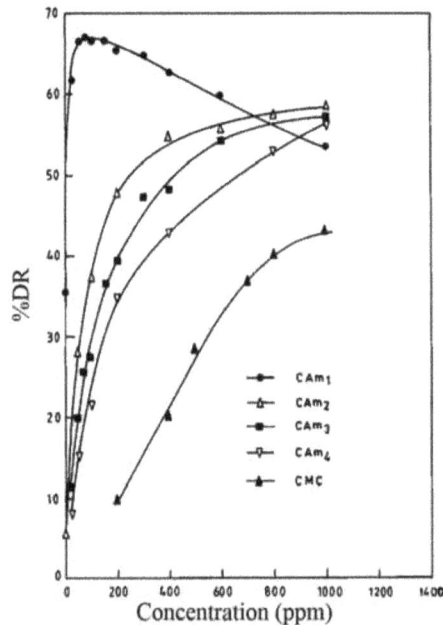

**Figure 19.** Drag reduction vs. polymer concentration for CMC and its graft copolymers, reprinted with permission from [96]. Copyright John Wiley and Sons, 1991.

Similarly, Biswal and Singh [94] synthesized six different CMC-*g*-PAAM copolymers with variation of the amount of monomer and catalyst, and found that these grafted copolymers exhibited significant flocculation and viscosifying characteristics.

*4.4. DNA*

DNA, a long polymer made from repeating nucleotide units, carries the genetic instructions for all living organisms and many viruses. Most DNA molecules are composed of two biopolymer strands that are coiled around each other by hydrogen bonds to form a double helix. The DNA structure is dynamic along the length and can be coiled into tight loops or other shapes [97].

It is well known that configurational changes of DNA can be induced by changing the pH and temperature. The DNA molecules can change from their double-stranded state to two single-strands under certain temperature or pH conditions. Based on the above property, Choi et al. [98] employed λ-DNA as a drag reducing agent to probe the mechanism of drag reduction and mechanical molecular degradation. They reported that λ-DNA is considered to be a better polymeric DR additive than the high-molecular-weight linear long-chain PEO. Figure 20 illustrates the DR efficiency of λ-DNA in comparison with that of PEO. The λ-DNA exhibited a high and stable percentage DR over time at a concentration of only 2.7 wppm. This DR behavior over a sufficiently long time is considered to be the main distinction compared with other high-molecular-weight, linear, flexible macromolecules. The mechanical degradation test showed that λ-DNA was always divided exactly into half, after which there was no further degradation. These results suggested that the mechanism of mechanical shear degradation of λ-DNA is quite different from that of traditional linear flexible macromolecules. Similarly, Lim et al. [99] compared the drag reduction efficiency of calf thymus DNA (CT-DNA) with that of λ-DNA, and found that CT-DNA has no drag reduction efficiency due to its rapid degradation.

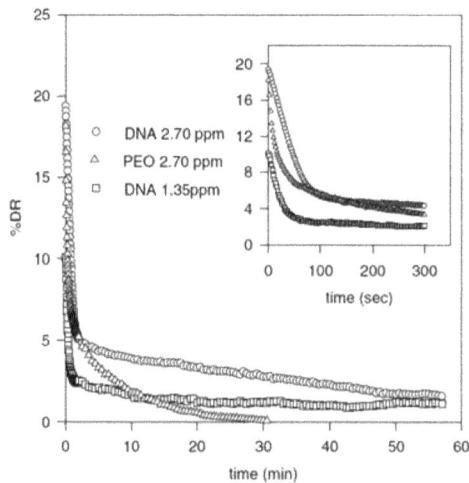

**Figure 20.** Time dependence of drag-reduction percentage for 1.35 and 2.70 wppm λ-DNA in buffer solution compared with PEO ($M_w = 5 \times 10^6$) at 1980 rpm ($Re = 1.2 \times 10^6$). Inset shows the same data at early times, reprinted with permission from [98]. Copyright American Physical Society, 2002.

Subsequently, Lim et al. [100] examined the coil-globule transition of monodisperse λ-DNA in an external flow of turbulence, thereby studying the effects of structure on drag reduction. They added spermidine (SPD) as an electrostatic condensing agent to generate the negatively charged λ-DNA in the turbulent flow, and compared its drag reducing phenomena with that of simple λ-DNA. As shown in Figure 21, in the absence of SPD, a relatively high drag reduction efficiency was achieved and was maintained with low mechanical shear degradation of λ-DNA for 1 h in the turbulent flow. In contrast, an abrupt decrease in the drag reduction efficiency was observed after SPD injection. This can be explained as follows: during the short period when the concentration of SD is sufficiently high, the original coil form of λ-DNA is changed into the globular state, which hardly produces any DR. Furthermore, they indicated that despite the different flow conditions, all λ-DNA molecules had the same half-dimension, indicating that discrete variation of the DNA conformation can significantly alter the flow characteristics.

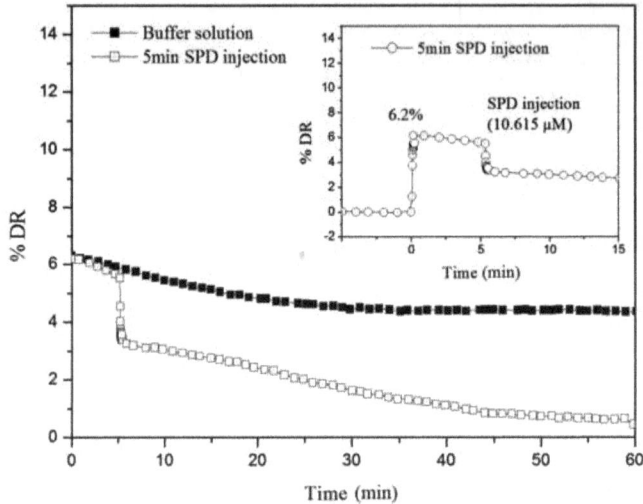

**Figure 21.** The percent DR vs. time for 1.35 wppm λ -DNA in buffer solution at 1157 rpm ($N_{Re} = 5.9 \times 10^5$) and 25 °C with and without (SPD). The inset shows the magnification of initial change of DR by SPD injection, reprinted with permission from [100]. Copyright John Wiley and Sons, 2005.

Recently, Ueberschar et al. [101] compared the fluid friction of single micrometer-sized blank and DNA-grafted PS microspheres under shear flow, and found that a thick DNA brush layer grafted onto PS microspheres reduced the frictional drag significantly, especially in dilute λ-DNA solution. This effect was more pronounced depending on the average molecular weight of the grafted DNA macromolecules, and was predicted to find use in chip devices or biomedical engineering.

## 5. Applications of Polymer-Induced Drag Reduction

With the recognition of polymers as prospective turbulent drag reducers, many possible applications were put forward, including in the fields of crude oil transportation through a pipeline, sewage, floodwater disposal, biomedical areas, oil well, heating circuits, and so on. In this section, we review a few of these applications, and hope that the principles used in these fields can be applied to other areas.

Drag reduction is considered to be a very important application prospect in the crude oil industry. The introduction of drag reducing polymeric additives in the long-distance crude oil pipeline has attracted huge attention. Evans [102] reported improved mechanical shear stability in non-aqueous systems using a tri-*n*-butylstannyl fluoride drag reducer. Belokin et al. [103] demonstrated that certain soluble high-molecular organosilicone polymers, such as poly(dimethyl siltrimethylene) and poly(dimethyl silmethylene), are very effective for reduction of turbulent frictional resistance in kerosene, and there are many ongoing studies [104–106]. In addition, a number of other applications of drag in oilfields have been studied, including enhanced oil recovery (EOR) applications [107,108] and hydraulic fracturing [77,109].

Another major application of polymer-induced DR in sewers has also aroused the interest of researchers. Forester et al. [110] analyzed the effect of a polymeric additive on DR in a pipe flow (where the pipe diameter was 254 mm), and suggested the potential application of drag reduction to sewer systems. Experimental evidence suggested that the discharge capacities of sewers can be increased significantly by appropriate polymer dosing. Detailed studies of this application have been published by Sellin et al. [111] for a number of sewer situations. The cost estimates, pollution, and toxicity problem are also discussed in their study.

Many tests have confirmed that a soluble polymer coating can be used to reduce the friction of ship models. Thurston and Jones [112] reported reduction of friction drag by application of soluble coatings to underwater vessels. Tests were performed in both fresh water and saline water, although the coating location was confined to the stagnation point of the nose. The DR achieved in the model was 33% in fresh water and 30% in seawater. Choi et al. [113] evaluated the effect of compliant coating on the turbulent boundary layer thickness in an aqueous channel flow, and observed significant turbulent DR, induced by the compliant coatings. Their investigation indicated that in order to build up the compliant coating for turbulent skin-frictional DR, the proper combination of material characteristics was necessary.

The use of biocompatible polymeric DR additives in biomedical areas has also been extensively studied [114], and blood-soluble drag reducing polymers have been demonstrated to influence the hydrodynamics in the blood flow of animals when injected intravenously at very low polymer concentration. Many experiments with intravenous drag reduction polymers showed positive hemorheological effects in several animal systems [115–117]. These studies indicate that the use of drag reducing polymers in blood has great potential for improving blood flow, and thus affords the possibility for prevention or treatment of circulatory system diseases.

As is well known, polymer induced drag reduction has a wide range of applications in many fields, which provides high impetus for continued study of the drag reduction phenomenon, and we believe that there are yet unearthed application prospects.

## 6. Conclusions

This article summarizes the recent research on typical synthetic and natural water-soluble polymers and their application as promising drag reducing agents, along with the research achievements in terms of the drag reduction capability. A selective overview of the proposed mechanisms of the polymer-induced turbulent DR and the current state of the experimental and simulation techniques are discussed, focusing on the drag reducing capability and complexity of water-soluble polymers in aqueous flows. The DR efficiency of water-soluble polymers could be dramatically altered by slight changes in experimental factors, including temperature, Reynolds number, pH values, pipe diameter, etc. or in molecular parameters including the molecular weight, polydispersity, intrinsic viscosity, concentration, etc. Therefore, the introduction of relative drag reduction capability for different polymers could be an alternative way covering specific molecular parameters and experimental conditions.

Regarding the importance of polymer structure and chemical composition to the DR, comparison of the DR efficiency of flexible and rigid polymers demonstrates the importance of molecular conformation. Among various polymeric drag reducing agents, synthetic water-soluble polymers can be fabricated with control of the molecular weight, structure, molecular weight distribution, and the characteristics of the ionic groups. Natural water-soluble polymers are inexpensive, biodegradable, fairly shear stable, and can be easily obtained from agricultural resources. However, the biodegradability of natural water-soluble polymers limits their application. It is evident that both synthetic polymers and natural polymers have their own optimum drag reduction conditions, which are closely associated with the rheological, physical, and/or chemical characteristics. Therefore, future work could be extended to improve the drag reduction efficiency of water-soluble polymers by combining the best characteristics of various polymers, e.g., by grafting synthetic polymers onto the backbone of natural polymers. It is believed that these new water-soluble polymers will exhibit more excellent drag reduction capability, thereby expanding the range of applications.

**Acknowledgments:** This research was financially supported by Inha University.

**Conflicts of Interest:** The authors declare no conflict of interest.

## References

1. White, C.M.; Mungal, M.G. Mechanics and prediction of turbulent drag reduction with polymer additives. *Annu. Rev. Fluid Mech.* **2008**, *40*, 235–256. [CrossRef]
2. Ranade, V.V.; Mashelkar, R.A. Turbulent mixing in dilute polymer-solutions. *Chem. Eng. Sci.* **1993**, *48*, 1619–1628. [CrossRef]
3. Abubakar, A.; Al-Wahaibi, T.; Al-Wahaibi, Y.; Al-Hashmi, A.R.; Al-Ajmi, A. Roles of drag reducing polymers in single- and multi-phase flows. *Chem. Eng. Res. Des.* **2014**, *92*, 2153–2181. [CrossRef]
4. Toms, B.A. Some observations on the flow of linear polymer solutions through straight tubes at large Reynolds numbers. *Proc. First Int. Congr. Rheol.* **1948**, *2*, 135–141.
5. Manzhai, V.N.; Nasibullina, Y.R.; Kuchevskaya, A.S.; Filimoshkin, A.G. Physico-chemical concept of drag reduction nature in dilute polymer solutions (the Toms effect). *Chem. Eng. Process.* **2014**, *80*, 38–42. [CrossRef]
6. Brostow, W. Drag reduction in flow: Review of applications, mechanism and prediction. *J. Ind. Eng. Chem.* **2008**, *14*, 409–416. [CrossRef]
7. Greene, H.L.; Mostardi, R.F.; Nokes, R.F. Effects of drag reducing polymers on initiation of atherosclerosis. *Polym. Eng. Sci.* **1980**, *20*, 499–504. [CrossRef]
8. Usui, H.; Li, L.; Suzuki, H. Rheology and pipeline transportation of dense fly ash-water slurry. *Korea-Aust. Rheol. J.* **2001**, *13*, 47–54.
9. Zhang, Y.; Schmidt, J.; Talmon, Y.; Zakin, J.L. Co-solvent effects on drag reduction, rheological properties and micelle microstructures of cationic surfactants. *J. Colloid Interface Sci.* **2005**, *286*, 696–709. [CrossRef] [PubMed]
10. Wei, J.J.; Kawaguchi, Y.; Li, F.C.; Yu, B.; Zakin, J.L.; Hart, D.J.; Zhang, Y. Drag-reducing and heat transfer characteristics of a novel zwitterionic surfactant solution. *Int. J. Heat Mass Tranf.* **2009**, *52*, 3547–3554. [CrossRef]
11. Li, F.C.; Wang, D.Z.; Kawaguchi, Y.; Hishida, K. Simultaneous measurements of velocity and temperature fluctuations in thermal boundary layer in a drag-reducing surfactant solution flow. *Exp. Fluids* **2004**, *36*, 131–140. [CrossRef]
12. Mizunuma, H.; Kobayashi, T.; Tominaga, S. Drag reduction and heat transfer in surfactant solutions with excess counterion. *J. Non-Newton. Fluid* **2010**, *165*, 292–298. [CrossRef]
13. Brunchi, C.E.; Morariu, S.; Bercea, M. Intrinsic viscosity and conformational parameters of xanthan in aqueous solutions: Salt addition effect. *Colloid Surf. B* **2014**, *122*, 512–519. [CrossRef] [PubMed]
14. Qi, Y.; Kawaguchi, Y.; Lin, Z.; Ewing, M.; Christensen, R.N.; Zakin, J.L. Enhanced heat transfer of drag reducing surfactant solutions with fluted tube-in-tube heat exchanger. *Int. J. Heat Mass Tranf.* **2001**, *44*, 1495–1505. [CrossRef]
15. Fu, Z.; Otsuki, T.; Motozawa, M.; Kurosawa, T.; Yu, B.; Kawaguchi, Y. Experimental investigation of polymer diffusion in the drag-reduced turbulent channel flow of inhomogeneous solution. *Int. J. Heat Mass Transf.* **2014**, *77*, 860–873. [CrossRef]
16. Yang, J.W.; Park, H.; Chun, H.H.; Ceccio, S.L.; Perlin, M.; Lee, I. Development and performance at high Reynolds number of a skin-friction reducing marine paint using polymer additives. *Ocean Eng.* **2014**, *84*, 183–193. [CrossRef]
17. Sadicoff, B.L.; Brandão, E.M.; Lucas, E.F. Rheological Behaviour of Poly (Acrylamide-G-Propylene Oxide) Solutions: Effect of Hydrophobic Content, Temperature and Salt Addition. *Int. J. Polym. Mater.* **2000**, *47*, 399–406. [CrossRef]
18. Bhambri, P.; Narain, R.; Fleck, B.A. Thermo-responsive polymers for drag reduction in turbulent Taylor–Couette flow. *J. Appl. Polym. Sci.* **2016**, *133*, 44191. [CrossRef]

19. Patterson, G.K.; Zakin, J.L.; Rodriguez, J.M. Drag Reduction-Polymer Solutions, Soap Solutions and Solid Particle Suspensions in Pipe Flow. *Ind. Eng. Chem.* **1969**, *61*, 22–30. [CrossRef]
20. Wu, X.; Wang, D.; Gao, Y.; Zhao, S.S.; Zheng, W.L. Mechanism analysis on powerful chip and drag reduction of the polymer drilling fluid. *Procedia Eng.* **2014**, *73*, 41–47.
21. Chagas, B.S., Jr.; Machado, D.L.P.; Haag, R.B.; Sousa, C.R.D.; Lucas, E.F. Evaluation of hydrophobically associated polyacrylamide-containing aqueous fluids and their potential use in petroleum recovery. *J. Polym. Sci. Part A Polym. Chem.* **2004**, *91*, 3686–3692. [CrossRef]
22. Eshrati, M.; Al-Hashmi, A.R.; Al-Wahaibi, T.; Al-Wahaibi, Y.; Al-Ajmi, A.; Abubakar, A. Drag reduction using high molecular weight polyacrylamides during multiphase flow of oil and water: A parametric study. *J. Petrol. Sci. Eng.* **2015**, *135*, 403–409. [CrossRef]
23. Yusuf, N.; Al-Wahaibi, T.; Al-Wahaibi, Y.; Al-Ajmi, A.; Al-Hashmi, A.R.; Olawale, A.S.; Mohammed, I.A. Experimental study on the effect of drag reducing polymer on flow patterns and drag reduction in a horizontal oil-water flow. *Int. J. Heat Fluid Flow* **2012**, *37*, 74–80. [CrossRef]
24. Abubakar, A.; Al-Wahaibi, Y.; Al-Hashmi, A.R.; Al-Wahaibi, Y.; Al-Ajmi, A.; Eshrati, M. Empirical correlation for predicting pressure gradients of oil-water flow with drag-reducing polymer. *Exp. Therm. Fluid Sci.* **2016**, *79*, 275–282. [CrossRef]
25. Astarita, G. Possible interpretation of mechanism of drag reduction in viscoelastic liquids. *Ind. Eng. Chem. Fundam.* **1965**, *4*, 354–356. [CrossRef]
26. Gadd, G.E. Reduction of turbulent drag in liquids. *Nature* **1971**, *230*, 29–31. [CrossRef]
27. Shenoy, A.V. A review on drag reduction with special reference to micellar systems. *Colloid Polym. Sci.* **1984**, *262*, 319–337. [CrossRef]
28. Tulin, M.P.B. Hydrodynamic aspects of macromolecular solutions. In Proceedings of the 6th Symposium on Naval Hydrodynamics, Washington, DC, USA, 28 September–4 October 1966; pp. 3–18.
29. Lumley, J.L. Drag reduction in turbulent flow by polymer additives. *J. Polym. Sci. Macromol. Rev.* **1973**, *7*, 263–290. [CrossRef]
30. Min, T.; Yoo, J.Y.; Choi, H.; Joseph, D.D. Drag reduction by polymer additives in a turbulent channel flow. *J. Fluid Mech.* **2003**, *486*, 213–238. [CrossRef]
31. Kostic, M. The ultimate asymptotes and possible causes of friction drag and heat transfer reduction phenomena. *J. Energy Heat Mass Transf.* **1994**, *16*, 1–14.
32. Armstrong, R.; Jhon, M.S. Turbulence induced change in the conformation of polymer molecules. *J. Chem. Phys.* **1983**, *79*, 3143–3147. [CrossRef]
33. Hidema, R.; Suzuki, H.; Hisamatsu, S.; Komoda, Y.; Furukawa, H. Effects of the extensional rate on two-dimensional turbulence of semi-dilute polymer solution flows. *Rheol. Acta* **2013**, *52*, 949–961. [CrossRef]
34. Brostow, W. Drag reduction and mechanical degradation in polymer solutions in flow. *Polymer* **1983**, *24*, 631–638. [CrossRef]
35. Pereira, A.S.; Soares, E.J. Polymer degradation of dilute solutions in turbulent drag reducing flows in a cylindrical double gap rheometer device. *J. Non-Newton. Fluid Mech.* **2012**, *179–180*, 9–22. [CrossRef]
36. Den Toonder, J.M.J.; Hulsem, M.A.; Kuiken, G.D.C.; Nieuwstadt, F.M. Drag reduction by polymer additives in a turbulent pipe flow: numerical and laboratory experiments. *J. Fluid Mech.* **1997**, *337*, 193–231. [CrossRef]
37. White, C.M.; Somandepalli, V.S.R.; Mungal, M.G. The turbulence structure of drag-reduced boundary layer flow. *Exp. Fluids* **2004**, *36*, 62–69. [CrossRef]
38. Zadrazil, I.; Bismarck, A.; Hewitt, G.F.; Markides, C.N. Shear layers in the turbulent pipe flow of drag reducing polymer solutions. *Chem. Eng. Sci.* **2012**, *72*, 142–154. [CrossRef]
39. Cai, W.H.; Li, F.C.; Zhang, H.N.; Li, X.B.; Yu, B.; Wei, J.J.; Kawaguchi, Y.; Hishida, K. Study on the characteristics of turbulent drag-reducing channel flow by particle image velocimetry combining with proper orthogonal decomposition analysis. *Phys. Fluids* **2009**, *21*, 115103. [CrossRef]
40. Warholic, M.D.; Massah, H.; Hanratty, T.J. Influence of drag-reducing polymers on turbulence: effects of Reynolds number, concentration and mixing. *Exp. Fluids* **1999**, *27*, 461–472. [CrossRef]
41. Itoh, M.; Tamano, S.; Yokota, K.; Ninagawa, M. Velocity measurement in turbulent boundary layer of drag-reducing surfactant solution. *Phys. Fluids* **2005**, *17*, 075107. [CrossRef]
42. Wei, T.; Willmarth, W.W. Modifying turbulent structure with drag-reducing polymer additives in turbulent channel flows. *J. Fluid Mech.* **1992**, *245*, 619–641. [CrossRef]

43. Wedgewood, L.E.; Ostrov, D.N.; Bird, B.R. A finitely extensible beadspring chain model for dilute polymer solutions. *J. Non-Newton. Fluid Mech.* **1991**, *40*, 119–139. [CrossRef]
44. Li, F.C.; Cai, W.H.; Zhang, H.N.; Wang, Y. Inuence of polymer additives on turbulent energy cascading in forced homogeneous isotropic turbulence studied by direct numerical simulations. *Chin. Phys. B* **2012**, *21*, 114701. [CrossRef]
45. Eshrati, M.; Al-Hashmi, A.R.; Ranjbaran, M.; Al-Wahaibi, T.; Al-Wahaibi, Y.; Al-Ajmi, A.; Abubakar, A. Prediction of water-soluble polymer drag reduction performance in multiphase flow using fuzzy logic technique. *Int. J. Adv. Chem. Environ. Eng. Biol. Sci.* **2016**, *3*, 1507–1515.
46. Zadeh, L.A. Similarity relations and fuzzy orderings. *Inform. Sci.* **1971**, *3*, 177–200. [CrossRef]
47. Amrouchene, Y.; Kellay, H. Polymer in 2D turbulence: Suppression of large scale fluctuation. *Phys. Rev. Lett.* **2002**, *89*, 088302. [CrossRef] [PubMed]
48. Jun, Y.; Zhang, J.; Wu, X.L. Polymer effect on small and large scale two-dimensional turbulece. *Phys. Rev. Lett.* **2006**, *96*, 024502. [CrossRef] [PubMed]
49. Kraichnan, R.H. Inertial ranges in Two-dimensional turbulence. *Phys. Fluids* **1967**, *10*, 1417–1423. [CrossRef]
50. Hidema, R.; Suzuki, H.; Hisamatsu, S.; Komoda, Y. Characteristic scales of two-dimensional turbulence in polymer solutions. *AIChE J.* **2014**, *60*, 1854–1862. [CrossRef]
51. Virk, P.S.; Merrill, E.W.; Mickley, H.S.; Smith, K.A. The Toms phenomenon: Turbulent pipe flow of dilute polymer solutions. *J. Fluid Mech.* **1967**, *30*, 305–328. [CrossRef]
52. Little, R.C.; Patterson, R.L.; Ting, R.Y. Characterization of the Drag Reducing Properties of Poly(ethylene oxide) and Poly(acrylamide) Solutions in External Flows. *J. Chem. Eng. Data* **1976**, *21*, 281–283. [CrossRef]
53. Little, R.C. Drag reduction in capillary tubes as a function of polymer concentration and molecular weight. *J. Colloid Interface Sci.* **1971**, *37*, 811–818. [CrossRef]
54. Little, R.C. Flow Properties of Polyox Solutions. *Ind. Eng. Chem. Fund.* **1969**, *8*, 557–559. [CrossRef]
55. Choi, H.J.; Jhon, M.S. Polymer induced Turbulent Drag Reduction. *Ind. Eng. Chem. Res.* **1996**, *35*, 2993–2998. [CrossRef]
56. Kim, C.A.; Sung, J.H.; Choi, H.J.; Kim, C.B.; Chun, W.; Jhon, M.S. Drag Reduction and Mechanical Degradation of Poly(ethylene oxide) in Seawater. *J. Chem. Eng. Jpn.* **1999**, *32*, 803–811. [CrossRef]
57. Choi, H.J.; Kim, C.A.; Sung, J.H.; Kim, C.B.; Chun, W.; Jhon, M.S. Universal drag reduction characteristics of saline water-soluble poly(ethylene oxide) in a rotating disk apparatus. *Colloid Polym. Sci.* **2000**, *278*, 701–705. [CrossRef]
58. Lim, S.T.; Hong, C.H.; Choi, H.J.; Lai, P.Y.; Chan, C.K. Polymer turbulent drag reduction near the theta point. *Eur. Phys. Lett.* **2007**, *80*, 58003. [CrossRef]
59. Sung, J.H.; Kim, C.A.; Choi, H.J.; Hur, B.K.; Kim, J.G.; Jhon, M.S. Turbulent drag reduction efficiency and mechanical degradation of poly(acrylamide). *J. Macromol. Sci. Phys.* **2004**, *B43*, 507–518. [CrossRef]
60. Sandoval, G.A.B.; Trevelin, R.; Soares, E.J.; Silveira, L.; Thomaz, F.; Pereira, A.S. Polymer degradation in turbulent drag reducing flows in pipes. *Therm. Eng.* **2015**, *14*, 03–06.
61. Pereira, A.S.; Andrade, R.M.; Soares, E.J. Drag reduction induced by flexible and rigid molecules in a turbulent flow into a rotating cylindrical double gap device: Comparison between Poly(ethylene oxide), Polyacrylamide, and Xanthan Gum. *J. Non-Newton. Fluid* **2013**, *202*, 72–87. [CrossRef]
62. Lim, S.T.; Choi, H.J. λ-DNA induced turbulent drag reduction and its characteristics. *Macromolecules* **2003**, *36*, 5348–5354. [CrossRef]
63. Zhang, X.; Liu, L.; Cheng, L.; Guo, Q.; Zhang, N. Experimental study on heat transfer and pressure drop characteristics of air–water two-phase flow with the effect of polyacrylamide additive in a horizontal circular tube. *Int. J. Heat Mass Tranf.* **2013**, *58*, 427–440. [CrossRef]
64. Terao, K. Poly(acrylic acid) (PAA). In *Encyclopedia of Polymeric Nanomaterials*; Springer: Berlin/Heidelberg, Germany, 2014; pp. 1–6.
65. Montante, G.; Laurenzi, F.; Paglianti, A.; Magelli, F. A study on some effects of a drag-reducing agent on the performance of a stirred vessel. *Chem. Eng. Des. Des.* **2011**, *89*, 2262–2267. [CrossRef]
66. Kim, J.T.; Kim, C.A.; Zhang, K.; Jang, C.H.; Choi, H.J. Effect of polymer–surfactant interaction on its turbulent drag reduction. *Colloid Surf. A* **2011**, *391*, 125–129. [CrossRef]
67. Kim, O.K.; Choi, L.S.; Long, T.; McGrath, K.; Armistead, J.P.; Yoon, T.H. Unusual complexation behavior of poly(acrylic acid) induced by shear. *Macromolecules* **1993**, *26*, 379–384. [CrossRef]

68. Zhang, K.; Choi, H.J.; Jang, C.H. Turbulent drag reduction characteristics of poly(acrylamide-*co*-acrylic acid) in a rotating disk apparatus. *Colloid Polym. Sci.* **2011**, *289*, 821–1827. [CrossRef]
69. Zhang, K.; Lim, G.H.; Choi, H.J. Mechanical degradation of water-soluble acrylamide copolymer under a turbulent flow: Effect of molecular weight and temperature. *J. Ind. Eng. Chem.* **2016**, *33*, 156–161. [CrossRef]
70. Cole, D.P.; Khosravi, E.; Musa, O.M. Efficient water-Soluble drag reducing star polymers with improved mechanical stability. *J. Polym. Sci. Polym. Chem.* **2016**, *54*, 335–344. [CrossRef]
71. Pinschmidt, R.K.; Renz, W.L.; Carroll, W.E.; Yocoub, K.; Drescher, J.; Nordquist, A.F.; Chen, N. N-Vinylformamide—Building Block for Novel Polymer Structures. *J. Macromol. Sci. Pure Appl. Chem.* **1997**, *A34*, 1885–1905. [CrossRef]
72. Gu, L.; Zhu, S.; Hrymak, A.N.; Pelton, R.H. Kinetics and modeling of free radical polymerization of *N*-vinylformamide. *Polymer* **2001**, *42*, 3077–3086. [CrossRef]
73. Marhefka, J.N.; Marascalco, P.J.; Chapman, T.M.; Russell, A.J.; Kameneva, M.V. Poly(*N*-vinylformamide)—A drag-reducing polymer for biomedical applications. *Biomacromolecules* **2006**, *2*, 1597–1603. [CrossRef] [PubMed]
74. Mishra, M.M.; Yadav, M.; Sand, A.; Tripathy, J.; Behari, K. Water soluble graft copolymer (κ-carrageenan-g-*N*-vinyl formamide): Preparation characterization and application. *Carbohydr. Polym.* **2010**, *80*, 235–241. [CrossRef]
75. Mccormick, C.L.; Hester, R.D.; Morgan, S.E.; Safieddine, A.M. Water-Soluble Copolymers. 30. Effects of Molecular Structure on Drag Reduction Efficiency. *Macromolecules* **1990**, *23*, 2124–2131. [CrossRef]
76. Mumick, P.S.; Welch, P.M.; Salazar, L.C.; Mccormick, C.L. Water-Soluble Copolymers. 56. Structure and Solvation Effects of Polyampholytes in Drag Reduction. *Macromolecules* **1994**, *27*, 323–331. [CrossRef]
77. Brun, N.L.; Zadrazil, I.; Norman, L.; Bismarck, A.; Markides, C.N. On the drag reduction effect and shear stability of improved acrylamide copolymers for enhanced hydraulic fracturing. *Chem. Eng. Sci.* **2016**, *146*, 135–143. [CrossRef]
78. Lumley, J.L. Drag reduction by additives. *Ann. Rev. Fluid Mech.* **1969**, *1*, 367–384. [CrossRef]
79. Reis, L.G.; Oliveira, I.P.; Pires, R.V.; Lucas, E.F. Influence of structure and composition of poly(acrylamide-g-propylene oxide) copolymers on drag reduction of aqueous dispersions. *Colloid Surf. A* **2016**, *502*, 121–129. [CrossRef]
80. Camail, M.; Margaillan, A.; Martin, I. Copolymers of *N*-alkyl- and *N*-arylalkylacrylamides with acrylamide: influence of hydrophobic structure on associative properties. Part I: viscometric behaviour in dilute solution and drag reduction performance. *Polym. Int.* **2009**, *58*, 149–154. [CrossRef]
81. Al-Hashmi, A.; Al-Maamari, R.; Al-Shabibi, I.; Mansoor, A.; Al-Sharji, H.; Zaitoun, A. Mechanical stability of high-Molecular-weight polyacrylamides and an (acrylamido tert-butyl sulfonic acid)-acrylamide copolymer used in enhanced oil recovery. *J. Appl. Polym. Sci.* **2014**, *131*, 40921. [CrossRef]
82. Akbar Ali, S.K.; Singh, R.P. An investigation of the flocculation characteristics of polyacrylamide-grafted chitosan. *J. Appl. Polym. Sci.* **2009**, *114*, 2410–2414.
83. Singha, V.; Tiwaria, A.; Tripathia, D.N.; Sanghib, R. Microwave assisted synthesis of Guar-g-polyacrylamide. *Carbohydr. Polym.* **2004**, *58*, 1–6. [CrossRef]
84. Kim, C.A.; Lim, S.T.; Choi, H.J.; Sohn, J.I.; Jhon, M.S. Characterization of drag reducing guar gum in a rotating disk flow. *J. Appl. Polym. Sci.* **2002**, *83*, 2938–3944. [CrossRef]
85. Singh, R.P.; Pal, S.; Krishnamoorthy, S.; Adhikary, P.; Ali, S.A. High-technology materials based on modified polysaccharides. *Pure Appl. Chem.* **2009**, *81*, 525–547. [CrossRef]
86. Deshmukh, S.R.; Chaturvedi, P.N.; Singh, R.P. The turbulent drag reduction by graft copolymers of guargum and polyacrylamide. *J. Appl. Polym. Sci.* **1985**, *30*, 4013–4018. [CrossRef]
87. Behari, K.; Kumar, R.; Tripathi, M.; Pandey, P.K. Graft Copolymerization of Methacrylamide onto guar Gum using a potassium chromate/malonic acid redox pair. *Macromol. Chem. Phys.* **2001**, *202*, 1873–1877. [CrossRef]
88. Bewersdorff, H.W.; Singh, R.P. Rheological and drag reduction characteristics of xanthan gum solutions. *Rheol. Acta* **1988**, *27*, 617–627. [CrossRef]
89. Katzbauer, B. Properties and applications of xanthan gum. *Polym. Degrad. Stabil.* **1998**, *59*, 81–84. [CrossRef]
90. Sohn, J.I.; Kim, C.A.; Choi, H.J.; Jhon, M.S. Drag-reduction effectiveness of xanthan gum in a rotating disk apparatus. *Carbohydr. Polym.* **2001**, *45*, 61–68. [CrossRef]

91.  Hong, C.H.; Choi, H.J.; Zhang, K.; Renou, F.; Grisel, M. Effect of salt on turbulent drag reduction of xanthan gum. *Carbohydr. Polym.* **2015**, *121*, 342–347. [CrossRef] [PubMed]

92.  Andrade, R.M.; Pereira, A.S.; Soares, E.J. Drag reduction in synthetic seawater by flexible and rigid polymer addition into a rotating cylindrical double gap device. *J. Fluid Eng.* **2016**, *138*, 021101. [CrossRef]

93.  Pushpamalar, V.; Langford, S.J.; Ahmad, M.; Lim, Y.Y. Optimization of reaction conditions for preparing carboxymethyl cellulose from sago waste. *Carbohydr. Polym.* **2006**, *64*, 312–318. [CrossRef]

94.  Biswal, D.R.; Singh, R.P. Characterisation of carboxymethyl cellulose and polyacrylamide graft copolymer. *Carbohydr. Polym.* **2004**, *57*, 379–387. [CrossRef]

95.  Abdulbari, H.A.; Shabirin, A.; Abdurrahman, H.N. Bio-polymers for improving liquid flow in pipelines—A review and future work opportunities. *J. Ind. Eng. Chem.* **2014**, *20*, 1157–1170. [CrossRef]

96.  Deshmukh, S.R.; Sudhakar, K.; Singh, R.P. Drag-reduction efficiency, shear stability, and biodegradation resistance of carboxymethyl cellulose-based and starch-based graft copolymers. *J. Appl. Polym. Sci.* **1991**, *43*, 1091–1101. [CrossRef]

97.  Rana, T.M. Illuminating the silence: Understanding the structure and function of small RNAs. *Nat. Rev. Mol. Cell Biol.* **2007**, *8*, 23–36. [CrossRef] [PubMed]

98.  Choi, H.J.; Lim, S.T.; Lai, P.Y.; Chan, C.K. Turbulent drag reduction and degradation of DNA. *Phys. Rev. Lett.* **2002**, *89*, 088302. [CrossRef] [PubMed]

99.  Lim, S.T.; Park, S.J.; Chan, C.K.; Choi, H.J. Turbulent drag reduction characteristics induced by calf-thymus DNA. *Physica A* **2005**, *350*, 84–88. [CrossRef]

100.  Lim, S.T.; Choi, H.J.; Chan, C.K. Effect of turbulent flow on coil-globule transition of lambda-DNA. *Macromol. Rapid. Commun.* **2005**, *26*, 1237–1240. [CrossRef]

101.  Ueberschär, O.; Wagner, C.; Stangner, T.; Kühne, K.; Gutsche, C.; Kremer, F. Drag reduction by DNA-grafting for single microspheres in a dilute lambda-DNA solution. *Polymer* **2011**, *52*, 4021–4032. [CrossRef]

102.  Evans, A.P. A new drag-reducing polymer with improved Shear Stability for nonaqueous systems. *J. Appl. Polym. Sci.* **1974**, *18*, 1919–1925. [CrossRef]

103.  Belokon, V.S.; Bespalova, N.B.; Vdovin, V.M.; Vlasov, S.A.; Kalashnikov, V.N.; Ushakov, N.V. Reduction of the hydrodynamic friction of hydrocarbons by means of small additions of certain organosilicon polymers. *J. Eng. Phys. Thermophys.* **1979**, *36*, 1–3. [CrossRef]

104.  Rodrigues, R.K.; Folsta, M.G.; Martins, A.L.; Sabadini, E. Tailoring of wormlike micelles as hydrodynamic drag reducers for gravel-pack in oil field operations. *J. Petrol. Sci. Eng.* **2016**, *146*, 142–148. [CrossRef]

105.  Mortazavi, S.M.M. Correlation of polymerization conditions with drag reduction efficiency of poly(1-hexene) in oil pipelines. *Iran. Polym. J.* **2016**, *25*, 731–737. [CrossRef]

106.  Wang, Z.H.; Yu, X.Y.; Li, J.X.; Wang, J.G.; Zhang, L. The use of biobased surfactant obtained by enzymatic syntheses for wax deposition inhibition and drag reduction in crude Oil Pipelines. *Catalysts* **2016**, *6*, 61. [CrossRef]

107.  Taylor, K.C.; Nasr-El-Din, H.A. Water-soluble hydrophobically associating polymers for improved oil recovery: A literature review. *J. Petrol. Sci. Eng.* **1998**, *19*, 265–280. [CrossRef]

108.  Wever, D.A.Z.; Picchioni, F.; Broekhuis, A.A. Polymers for enhanced oil recovery: A paradigm for structure–property relationship in aqueous solution. *Prog. Polym. Sci.* **2011**, *36*, 1558–1628. [CrossRef]

109.  Wang, L.; Wang, D.; Shen, Y.; Lai, X.; Guo, X. Study on properties of hydrophobic associating polymer as drag reduction agent for fracturing fluid. *J. Polym. Res.* **2016**, *23*. [CrossRef]

110.  Forester, R.H.; Larson, R.E.; Hayden, J.W.; Wetzel, J.M. Effects of polymer addition on friction in a 10-inch diameter pipe. *J. Hydronaut.* **1969**, *3*, 59–62. [CrossRef]

111.  Sellin, R.H.J. Increasing sewer capacity by polymer dosing. *Proc. Inst. Civ. Eng.* **1977**, *63*, 49–67. [CrossRef]

112.  Thurston, S.; Jones, R.D. Experimental model studies of non-newtonian soluble coatings for drag reduction. *AIAA* **1964**. [CrossRef]

113.  Choi, K.S.; Yang, X.; Clayton, B.R.; Glover, E.J.; Altar, M.; Semenov, B.N.; Kulik, V.M. Turbulent drag reduction using compliant surfaces. *Proc. Math. Phys. Eng. Sci.* **1997**, *453*, 2229–2240. [CrossRef]

114.  Greene, H.L.; Nokes, R.F.; Thomas, L.C. Biomedical implications of drag reducing agents. *Biorheology* **1971**, *7*, 221–223. [PubMed]

115.  Polimeni, P.L.; Coleman, P. Enhancement of aortic blood flow with a linear anionic macropolymer of extraordinary molecular length. *J. Mol. Cell. Cardiol.* **1985**, *17*, 721–724. [CrossRef]

116. Sawchuk, A.P.; Unthank, J.L.; Dalsing, M.C. Drag reducing polymers may decrease atherosclerosis by increasing shear in areas normally exposed to low shear stress. *J. Vasc. Surg.* **1999**, *30*, 761–764. [CrossRef]

117. Kameneva, M.V.; Wu, Z.J.; Uraysh, A.; Repko, B.; Litwak, K.N.; Billiar, T.R.; Fink, M.P.; Simmons, R.L.; Griffith, B.P.; Borovetz, H. Blood soluble drag-reducing polymers prevent lethality from hemorrhagic shock in acute animal experiments. *Biorheology* **2004**, *41*, 53–64. [PubMed]

*processes*

MDPI

*Article*

# Design of Cross-Linked Starch Nanocapsules for Enzyme-Triggered Release of Hydrophilic Compounds

Fernanda R. Steinmacher [1,2], Grit Baier [2], Anna Musyanovych [2,3], Katharina Landfester [2], Pedro H. H. Araújo [1] and Claudia Sayer [1,*]

[1] Chemical Engineering and Food Engineering Department, Federal University of Santa Catarina—UFSC, CP 476, Florianópolis 88040-900, Brazil; fersteinmacher@gmail.com (F.R.S.); pedro.h.araujo@ufsc.br (P.H.H.A.)

[2] Max Planck Institute for Polymer Research, Ackermannweg 10, 55128 Mainz, Germany; grit.baier@online.de (G.B.); Anna.Musyanovych@imm.fraunhofer.de (A.M.); landfest@mpip-mainz.mpg.de (K.L.)

[3] Nanoparticle Technologies Department, Fraunhofer ICT-IMM, Carl-Zeiss-Str. 18-20, 55129 Mainz, Germany

\* Correspondence: claudia.sayer@ufsc.br; Tel.: +55-48-3721-2516

Academic Editor: Alexander Penlidis

Received: 8 March 2017; Accepted: 2 May 2017; Published: 6 May 2017

**Abstract:** Cross-linked starch nanocapsules (NCs) were synthesized by interfacial polymerization carried out using the inverse mini-emulsion technique. 2,4-toluene diisocyanate (TDI) was used as the cross-linker. The influence of TDI concentrations on the polymeric shell, particle size, and encapsulation efficiency of a hydrophilic dye, sulforhodamine 101 (SR 101), was investigated by Fourier transform infrared (FT-IR) spectroscopy, dynamic light scattering (DLS), and fluorescence measurements, respectively. The final NC morphology was confirmed by scanning electron microscopy. The leakage of SR 101 through the shell of NCs was monitored at 37 °C for seven days, and afterwards the NCs were redispersed in water. Depending on cross-linker content, permeable and impermeable NCs shell could be designed. Enzyme-triggered release of SR 101 through impermeable NC shells was investigated using UV spectroscopy with different α-amylase concentrations. Impermeable NCs shell were able to release their cargo upon addition of amylase, being suitable for a drug delivery system of hydrophilic compounds.

**Keywords:** inverse mini-emulsion; interfacial polymerization; aqueous-core nanocapsules; high-efficiency encapsulation; enzyme-triggered release

## 1. Introduction

The development of new strategies for the delivery of hydrophilic drugs is emerging as an important research field [1–6] due to several facts: water-soluble drugs are often easily degradable in the body, poor cellular penetration of macromolecules, toxicity of small molecules, and unsuitable biodistribution [1]. These limitations can be overcome by the use of nanocarriers that offer protection against degradation or oxidation until the drugs reach the targeted tissues.

The importance of polymeric nanoparticles is now widely recognized in order to conceive drug carriers and controlled-release systems due to their versatility and unique features, such as different compositions, morphologies, reduced particle sizes and high surface area, allowing surface modification [7] in order to design the triggered release, for example, pH-responsive [8] and enzyme-responsive nanoparticles [9,10]. Thus, compounds, materials and, especially, the surface of drug carriers should be biocompatible, nontoxic and, sometimes, also biodegradable [11]. Therefore, the use of well-known polysaccharides for the preparation of drug delivery systems has advantages regarding safety, toxicity, and availability [12,13].

Recently, starch-based nanocapsules prepared by the emulsion solvent evaporation method were used for topical application and showed very good skin compatibility, an absence of allergenic potential and, additionally, improved skin permeation of lipophilic bioactive molecules [13]. However, regarding oral administration, starch-based drug delivery systems experience different environments in the body, especially, environmental conditions that favor enzymatic biodegradation. Although amylase is the major component of parotid saliva, most degradation of starch results from pancreatic enzymes in the small intestine, and not from salivary amylases [14–16]. Efforts have been made to inhibit or reduce the enzymatic degradation. In this way, starch-based nanoparticles have been obtained using modified [17] and cross-linked starch [18].

Well-established methods usually used for the encapsulation of lipophilic compounds result in low encapsulation efficiency of hydrophilic molecules, due to the rapid partitioning of the drug to the external aqueous phase [4]. Double-emulsion solvent diffusion techniques and emulsion solvent evaporation are the most common methods used for the encapsulation of hydrophilic compounds in nanoparticles, but usually result in low encapsulation efficiency [4,19]. Higher loading efficiency of hydrophilic compounds was recently reported using a novel organic solvent-free double-emulsion/melt dispersion technique [5].

High encapsulation efficiency of hydrophilic drugs was achieved using the inverse mini-emulsion polymerization technique. The high stability of droplets/particles created by the mini-emulsion technique allows performing the polymerization reaction inside the droplet, as well as at their interface [20]. A relatively recent strategy is the use of interfacial polymerization, which allows the preparation of aqueous-core nanocapsules. Polyurea, polythiourea, or polyurethane shells were successfully obtained by interfacial polyaddition at the droplet interface using starch, dextran, 1,6-hexanediol, and 1,6-diaminohexane, for example [2,3,18,21,22]. The core composed of water optimizes the drug solubility inside the nanoparticle, providing high encapsulation efficiency.

NaCl is the most common salt applied in inverse mini-emulsion polymerization, although hydrophilic metal salts composed of transition metal cations, such as $Fe^{2+}$, $Fe^{3+}$, $Co^{2+}$, $Ni^{2+}$, and $Cu^{2+}$, may present advantages for further application [23]. Copper (Cu) is an element essential for almost all organisms, including bacteria, but Cu overload is toxic in most systems. Studies show Cu accumulates in macrophage phagosomes infected with bacteria, suggesting that Cu is mobilized in mammals to control bacterial growth [24]. In fact, Wolschendorf et al. [25] found that dietary supplementation with Cu resulted in the accumulation of Cu in lung granulomas of *Mycobacterium tuberculosis*-infected guinea pigs, and coincided with a reduction in the bacterial burden.

The effective carrier for drug delivery should be optimized in terms of chemical composition, surface morphology, size, shape, and be able to release its payload in a controlled/predictable manner. Therefore, it is of great importance to have knowledge about how the chemical compositions affect the degradation kinetics of the carriers in order to program the properties of a carrier during the synthesis for each particular application. Hamdi and Ponchel [26] synthesized starch microspheres using epichlorohydrin as a crosslinking agent, and studied enzymatic degradation by α-amylase. It was suggested that degradation profiles were dependent on the initial size distribution of the microspheres. The aim of this study was to prepare starch nanocapsules by interfacial inverse (water-in-oil) mini-emulsion polymerization using 2,4-toluene diisocyanate as a cross-linker and to evaluate the permeability of the capsule's shell upon enzyme (amylase) degradation by adjusting the chemical parameters during the crosslinking reaction. Our focus was also to evaluate whether the permeability of NC shells can be modulated by the encapsulation of a small hydrophilic molecule, sulforhodamine 101 (SR 101), in aqueous-core NCs for different amounts of cross-linker.

## 2. Materials and Methods

### 2.1. Materials

For the synthesis of starch NCs, all chemicals were used without further purification. Hydrophilic potato starch and 2,4-toluene diisocyanate (TDI) were purchased from Fluka (Neu-Ulm, Germany) and Sigma Aldrich (Steinheim, Germany), respectively. Cyclohexane (Sigma Aldrich, Steinheim, Germany) was used as continuous phase. Sodium chloride (NaCl, VWR, BDH Prolabo, Darmstadt, Germany) and copper II sulfate pentahydrate ($CuSO_4 \cdot 5H_2O$, Cromoline, Diadema, Brazil) were used as osmotic costabilizers of the mini-emulsion droplets. Polyglycerol polyricinoleate (PGPR, $Mw$ = 5870 g/mol, DANISCO, Kopenhagen, Denmark) was employed as a hydrophobic surfactant. Sulforhodamine 101 (SR 101, $Mw$ = 606.71 g/mol, BioChemica, Aldrich, Taufkirchen, Germany) was used as a hydrophilic dye. Sodium dodecyl sulfate (SDS, Merck, Darmstadt, Germany) and Tween 80 (Vetec, São Paulo, Brazil) were used as surfactants to redisperse the cross-linked aqueous-core NCs in the aqueous phase. The α-amylase from *Bacillus subtilis* was purchased from Fluka (Neu-Ulm, Germany). The phosphate buffer solution (PBS at pH 7.4 and 0.05 M) was freshly prepared from monobasic and dibasic sodium phosphate (Vetec, São Paulo, Brazil).

### 2.2. Preparation of Starch Nanocapsules

Starch nanocapsules (NCs) were prepared by interfacial polymerization by inverse mini-emulsion at 60 °C for 2 h, based on a procedure proposed in the literature [3,18], except for the fact that the dispersed phase was prepared from gelatinizing 0.1 g of starch under stirring for 30 min at 90 °C in a mixture of 50 mg of sodium chloride and 1.3 g of water. When copper salt was used, NaCl was partially replaced by previously-dried $CuSO_4 \cdot 5H_2O$. Subsequently, the macroemulsion was formed by adding the continuous phase, composed of 7.5 g of cyclohexane and PGPR (varied in the range from 10 to 20 wt% related to the disperse phase), to the disperse phase and by stirring over 1 h at room temperature. The mini-emulsion was created by sonication for 3 min at 70% of amplitude in a pulsed regime (20 s on, 10 s pause) using a Branson Sonifier W-450-Digital (Thermo Fisher Scientific, WA, USA) under ice cooling to prevent the evaporation of the continuous phase. A clear solution of cyclohexane (5 g), PGPR (30 mg), and TDI (varied in the range from 80 to 200 mg), previously prepared, was added dropwise to the mini-emulsion for 1 min. The reaction was performed for 2 h at 60 °C or for 24 h at 25 °C, under magnetic stirring. Figure 1 illustrates the different steps. Dye-loaded cross-linked starch NCs were prepared, in brief, replacing the water that composes the dispersed phase by a 0.02 wt% SR 101 aqueous solution.

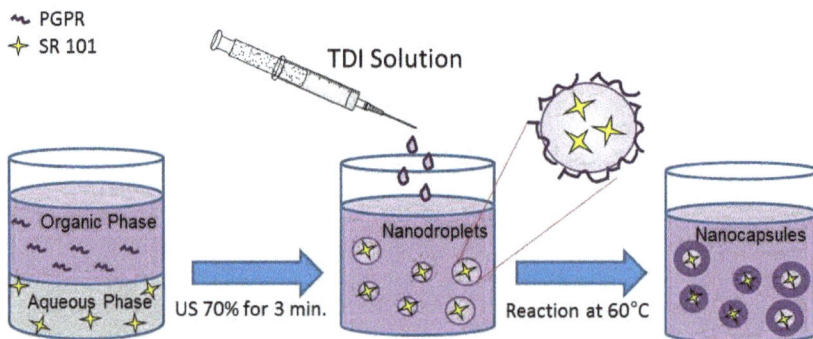

**Figure 1.** Preparation of aqueous-core NCs by interfacial polymerization via inverse mini-emulsion.

In order to redisperse the NCs in water, the following procedure was applied: 1 g of NCs dispersed in cyclohexane were redispersed in 5 g of 0.3 wt% SDS aqueous solution after 30 min in a sonication bath (25 kHz) and magnetically stirred at 1000 rpm over 8 h at room temperature for evaporation of the organic solvent.

*2.3. Characterization*

The intensity average particle size was measured by dynamic light scattering (DLS, NanoSizer Nano S, Malvern, UK) of diluted dispersions. The morphology of the NCs was evaluated by field emission scanning electron microscopy (FESEM) (LEO (Zeiss) 1530 Gemini, Oberkochen, Germany) at an accelerating voltage of 0.5 kV. Generally, the samples were prepared by diluting the NCs in cyclohexane, then one droplet of the sample was placed onto silica wafers and dried under ambient conditions.

Fourier transform infrared (FT-IR) spectroscopy was performed to evaluate the reaction between the cross-linker TDI and hydroxyl groups of starch during the mini-emulsion polymerization. The sample powder was obtained by freeze-drying the particles dispersion for 24 h at $-60\,^{\circ}$C under reduced pressure. The dried sample was pressed with KBr to form a pellet. Spectra were recorded using a IFS 113v spectrometer (Bruker, Billerica, MA, USA). When mentioned, attenuated total reflectance (ATR) mode was used. For that, a film of the sample was obtained drying the final mini-emulsion in a convection oven at $60\,^{\circ}$C and spectra were recorded using a Tensor 27 spectrometer (Bruker, Billerica, MA, USA).

Interfacial tension measurements were carried out by using a series of colloidal dispersions of varying type and amount of hydrophilic salts. The measurements were carried out using a drop shape analysis system (Ramé-hart Model 250 Standard Goniometer/Tensiometer, Ramé-hart, Succasunna, NJ, USA). The equipment was calibrated by measuring the pure air-water surface tension until a value of 72.8 mN m$^{-1}$ was obtained. Single droplets (50 µL) of the dispersed phase were formed at the end of a steel needle (1.84 mm), placed in the oil phase within a cuvette, and images were recorded using a digital camera over a period of time at 21 $^{\circ}$C. The profile of the droplet in each image was detected automatically using the analysis software package and fitted to the Young-Laplace equation to obtain interfacial tension values as a function of time. At least three independent measurements were taken.

The encapsulation efficiency was studied using a fluorescence spectrometer (NanoDrop ND-3300, Thermo Fisher Scientific, WA, USA) after redispersion of NCs in water. SR 101 was chosen as the hydrophilic fluorescent dye due its stable fluorescence (red fluorophore λexc/em: 583/603 nm) during polymerization. Sonication and exposition to different temperatures did not influence the intensity of its fluorescence signal. After the synthesis step, NCs dispersed in cyclohexane were freeze-dried, and the encapsulation efficiency was determined by redispersing 15 µg of the dried sample in 1 g of 0.3 wt% SDS aqueous solution. The NCs were collected by centrifugation for 20 min at 10,000 rpm. The fluorescence signal of the supernatant related to a calibration curve provided the concentration of unloaded dye. The encapsulation efficiency ($EE$ (%)) was calculated as the difference between total concentration of dye in the sample ($W_{total,sample}$) and the concentration of free dye ($W_{free}$) using the following equation:

$$EE\ (\%) = \frac{\left(W_{total,sample} - W_{free}\right)}{W_{total,sample}} \times 100 \tag{1}$$

The encapsulation efficiency of the dye after redispersion, and the leakage by migration of the hydrophilic dye to the continuous aqueous phase, were calculated by relating the fluorescence signal of the supernatant obtained from the sample after centrifugation at 4000 rpm for 30 min. The leakage of SR 101 was monitored for seven days at 37 $^{\circ}$C. Aliquots were taken at certain times and centrifuged at 4000 rpm for 30 min. The fluorescence signal of the supernatant related to a calibration curve provided the concentration of non-entrapped dye ($W_{free}$). The cumulative loss of SR 101 ($1 - EE$ (%)) was calculated using the following equation:

$$1 - EE\ (\%) = \frac{\left(W_{free}\right)}{W_{total,sample}} \times 100 \qquad (2)$$

where $W_{total,sample}$ stands for the total concentration of dye in the sample. All experiments were repeated three times and for each sample the encapsulation efficiency was calculated from two measurements.

The activity of α-amylase was determined based on colorimetric measurements at 37 °C and pH 7.4, 0.05 M PBS. The reaction was started by adding 0.5 mL of α-amylase solution to 10 mL of gelatinized 0.1 g/mL starch solution in pH 7.4 and 0.05 M PBS. The final enzyme concentration was 3.61 µg/mL of α-amylase. The reaction was allowed to proceed at 37 °C and stopped by adding 0.5 mL of reacted starch dispersion to 5 mL of 0.1 M HCl solution, and 0.5 mL of the terminated reaction solution was added to 5 mL of iodine reagent (0.2% iodine and 2% potassium iodine) aqueous solution. The starch-iodine solution turned deep blue in the presence of unconverted starch and the color developed was determined by measuring the absorbance at 620 nm (UV-VIS spectrophotometer, AJX-1900, Micronal, São Paulo, Brazil). The α-amylase activity was calculated by relating the absorbance of the undigested starch solution with the absorbance of the digested starch solution.

For the release assays, the NCs were redispersed into an aqueous solution using the following procedure: 2 g of NCs dispersed in cyclohexane (solid content around 3%) were added to 10 g of Tween 80 solution (1 g of surfactant in 10 g of 0.05 M PBS pH 7.4) after 30 min in a sonication bath (25 kHz) and magnetically stirred at 1000 rpm over 8 h at room temperature for the evaporation of organic solvent. After redispersion, α-amylase previously dissolved in PBS was added to the NCs. The final enzyme concentrations studied were 36, 18, 9, 0.9, and 0 mg/mL. The samples were gently shaken at 37 °C and the release kinetics of SR 101 were taken upon enzymatic degradation of the shell of the NCs. Aliquots of 1 mL were taken periodically and the enzymatic reaction was stopped by adding 0.5 mL of a 0.1 M HCl solution. The NCs were sedimented by centrifugation for 30 min at 4800 rpm (1467× *g*, MiniSpin Eppendorf, Hamburg, Germany) and the supernatant was immediately analyzed by a UV-VIS spectrophotometer (AJX-1900, Micronal) at 584 nm, as shown in Figure 2. Firstly, it was observed that the intensity of the absorbance peak and maximum absorbance wavelength of SR 101 dissolved in HCl solution were the same compared to an aqueous solution. However, after 24 h some degradation of the signal was observed, emphasizing the importance of carrying out the measurements immediately after the dissolution in HCl solution.

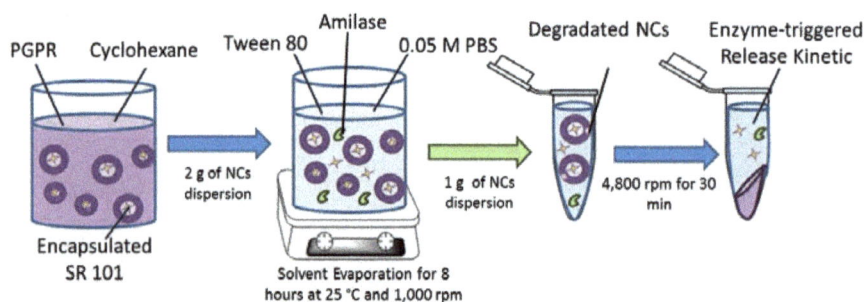

**Figure 2.** Schematic illustration of the procedure for determination of enzyme triggered release of SR 101 from cross-linked starch NCs.

## 3. Results and Discussion

Cross-linked starch NCs were prepared by interfacial polymerization firstly at 60 °C for 2 h using different types and amounts of osmotic costabilizer, as well as different surfactant, TDI, and starch concentrations. Secondly, the effect of a lower reaction temperature was investigated. Finally, shell permeability of NCs was evaluated and impermeable shell cross-linked starch NCs were designed for enzyme-triggered release of hydrophilic compounds. The stability of the mini-emulsions of aqueous nanodroplets was investigated at 60 °C for 2 h. Results indicate that the mini-emulsions are stable during the reaction time at 60 °C. Since the final NCs are composed of a rigid shell, all of the NCs studied and stored at room temperature were stable. Although some NCs precipitated during storage, they could easily be redispersed simply by shaking by hand and no irreversible coagulation was observed.

### 3.1. Effect of Surfactant Concentration

The effect of surfactant concentration on the mean particle size was investigated in NCs obtained using 0.1 g of starch gelatinized in the aqueous phase, composed of 1.3 g of water and 50 mg of sodium chloride. The continuous phase of the mini-emulsion was composed of 7.5 g of cyclohexane and PGPR. The amount of PGPR varied in the range from 10 to 20 wt% related to the dispersed phase. The reaction started by adding a cross-linker solution composed of 5.0 g of cyclohexane, 30 mg of PGPR, and 120 mg of TDI. The amounts of the dispersed phase and cross-linker solution were kept constant. Table 1 summarizes the effect of surfactant (PGPR) concentration on the final particle diameter of NCs dispersed in cyclohexane and redispersed in an aqueous solution stabilized with SDS.

**Table 1.** Average particle size of cross-linked starch NCs prepared by interfacial polymerization in inverse mini-emulsion at 60 °C for 2 h using 120 mg of TDI. The dispersed phase was composed of 0.1 g of starch, 1.3 g of water, and 50 mg of NaCl and the continuous phase was composed of 7.5 g of cyclohexane and PGPR.

| PGPR (%) * | Final Average Particle Size (nm) ** | |
|:---:|:---:|:---:|
| | Dispersed in Cyclohexane | Redispersed in Aqueous Solution |
| 10 | 200 ± 1 | 197 ± 2 |
| 15 | 181 ± 3 | 194 ± 3 |
| 20 | 159 ± 2 | 230 ± 7 |

* wt% related to the disperse phase; ** Mean ± SD, $n \geq 3$.

It can be observed that the mean particle size decreased with the increase of surfactant concentration for NCs dispersed in cyclohexane. This behavior was expected and the influence of surfactant concentration on particle size is well discussed in the literature [2,20]. Higher amounts of surfactant are able to provide colloidal stability to larger surface areas and, thus, smaller particles can be formed. However, when the NCs were redispersed in the aqueous solution, the mean particle size slightly increased for NCs with 15% of PGPR and sharply for that with 20 wt% of PGPR NCs, except for samples obtained with 10 wt% of PGPR, in which the particle size remained unchanged. This different behavior can be attributed to the swelling process during the redispersion step. The different swelling ability of NCs might be explained considering the interfacial polymerization mechanism. Larger particle diameters provide lower total surface area of NCs, $4.4 \times 10^{19}$ nm$^2$ for NCs with 10 wt% PGPR compared to $5.5 \times 10^{19}$ nm$^2$ for 20 wt% PGPR NCs and, thus, present thicker shells. Baier et al. [18] investigated the influence of surfactant concentration on the wall thickness of the cross-linked starch NC shells by SEM analysis. According to the results obtained by the authors, the thickness of the shell decreases with higher amounts of surfactant. Due to the osmotic pressure inside the NCs provided by NaCl, the swelling process is favored, increasing the particle size of the NCs with a thinner shell after redispersion in an aqueous phase, as observed for the samples obtained with 15 wt% and 20 wt% of PGPR.

FT-IR measurements were performed to investigate the shell composition and confirmed the chemical reaction between OH groups from starch and surfactant molecules with NCO groups of TDI. Figure 3a shows the spectra of cross-linked starch NCs, whose shell is composed of urethane and urea groups [18]. The carbonyl vibration at 1734 cm$^{-1}$ and the N-H vibration at 1544 cm$^{-1}$ are strong evidence for the formation of urethane groups. The vibration at 1646 cm$^{-1}$ (the carbonyl of urea groups) indicates that the side reaction of isocyanate with water occurred leading to urea units. The flat signal at 2276 cm$^{-1}$ indicates complete consumption of NCO groups of TDI. Figure 3b shows, in detail, the spectra and the growth of the characteristic peak of the urethane group (1734 cm$^{-1}$) with higher PGPR concentration and smaller particle size (159 nm with 20 wt% of PGPR related to the dispersed phase). This fact might be attributed to the locus of interfacial polymerization; at the beginning of the reaction starch concentration at the interface and, thus, also that of water, was the same in smaller or larger droplets. Nevertheless, smaller droplets presented higher area/volume ratio and higher surface area led to faster interfacial reaction rates between NCO groups of TDI and OH groups from starch due the increased contact between both phases. The mobility of starch macromolecules through the shell to react with TDI at the interface is lower than that of small water molecules and, therefore, thicker shells obtained with larger particles provide more resistance. Consequently, the formation if urethane groups was favored in smaller particles. Figure 4 shows a schematic illustration of interfacial polymerization resulting in cross-linked starch NCs. Since urea bonds are stiffer and form stronger hydrogen bonding, smaller particles with lower percentages of urea bonds swell more when redispersed in water, as observed in Table 1, for NCs with 20 wt% of PGPR.

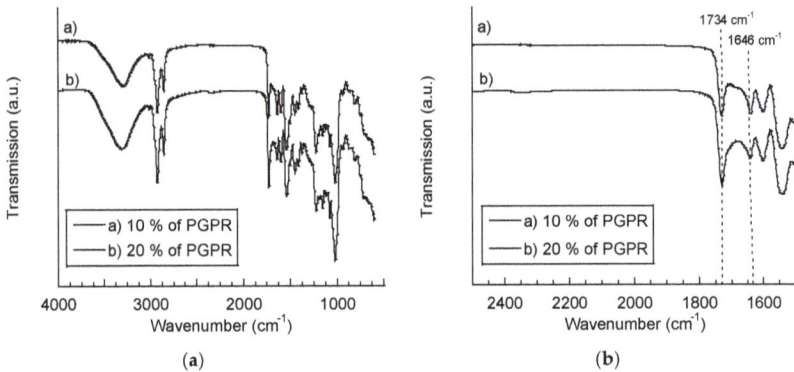

**Figure 3.** The effect of surfactant concentration on shell composition. Results were obtained by FT-IR using ATR mode. (**a**) Spectra between 4000 and 500 cm$^{-1}$; and (**b**) spectra in detail between 2500 and 1500 cm$^{-1}$.

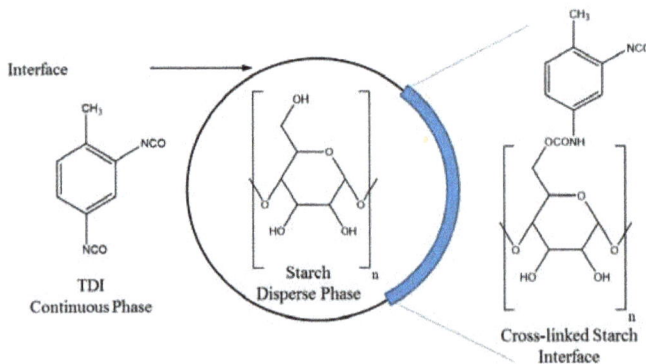

**Figure 4.** Schematic illustration of interfacial polymerization between NCO groups from TDI and OH groups from starch by inverse mini-emulsion.

## 3.2. Effect of Osmotic Agent Type

Hydrophilic salts have been used as osmotic agents (costabilizer) in order to improve the nanodroplet stability. Aqueous nanodroplets stabilized using sodium chloride and copper sulfate as osmotic agents in the dispersed phase are composed of 0.1 g of starch gelatinized in 1.3 g of water and 150 mg of PGPR as the surfactant in the continuous phase were, afterwards, cross-linked with TDI. Sodium chloride was partially replaced by $CuSO_4$, maintaining similar ionic strength. Ten milligrams of NaCl ($1.71 \times 10^{-4}$ mol) were replaced by 7 mg of $CuSO_4$ ($4.82 \times 10^{-5}$ mol).

Table 2 summarizes the formulations and shows the effect of the osmotic agent type on the average particle size of cross-linked starch NCs using different TDI amounts. It can be observed that smaller NCs were obtained using copper salt independently of the TDI concentration after polymerization. The small decrease of particle size containing $CuSO_4$ might be due to the dissociation ability, as well as to the fact that the hydrophilic salt can influence the interfacial properties of the dispersed and continuous phases and interact with the other components of the dispersed phase [23]. However, it was observed that the type and content of the salt had only a minor influence on the interfacial tension values between the phases, as shown in Table 3. In addition, while for those NCs obtained with NaCl, particle sizes remained virtually unchanged after redispersion in aqueous solution, and the particle sizes of NCs obtained with $CuSO_4$, independently of TDI concentration, increased after redispersion in aqueous solution (from about 135 nm in cyclohexane to 180 nm in water). These results indicate that these particle's shells are permeable, favoring swelling of the particle.

**Table 2.** The average particle size of cross-linked starch NCs prepared by interfacial polymerization inverse mini-emulsion at 60 °C for 2 h using 7.5 g of cyclohexane and 150 mg of PGPR as the continuous phase, and the dispersed phase was composed of 0.1 g of starch, 1.3 g of water, and salt.

| Disperse Phase | | TDI Solution | Final Average Particle Size (nm) * | |
|---|---|---|---|---|
| NaCl (mg) | CuSO$_4$ (mg) | TDI (mg) | Dispersed in Cyclohexane | Redispersed in Aqueous Solution |
| 50 | - | 80 | 200 ± 2 | 195 ± 2 |
| 40 | 7 | 80 | 135 ± 1 | 181 ± 2 |
| 50 | - | 120 | 209 ± 1 | 213 ± 4 |
| 40 | 7 | 120 | 139 ± 2 | 181 ± 2 |

* Mean ± SD, $n \geq 3$.

**Table 3.** The influence of type and concentration of salt on interfacial tension values between aqueous droplets and cyclohexane with 10 wt% of PGPR related to the dispersed phase.

| NaCl (mol/g) * | CuSO$_4$ (mol/g) * | $\gamma$ (mN/m) ** |
|---|---|---|
| $6.58 \times 10^{-4}$ | - | 3.04 ± 0.09 |
| $9.21 \times 10^{-4}$ | - | 3.08 ± 0.03 |
| $5.26 \times 10^{-4}$ | $4.82 \times 10^{-5}$ | 3.56 ± 0.02 |

* Molar concentration related to water; ** Mean ± SD, $n \geq 3$.

The effect of the type of salt on the NC shell composition was evaluated using 120 mg of TDI. A comparison between FT-IR spectra of samples is shown in Figure 5. The shell composition of particles obtained using $CuSO_4$ presents a higher fraction of urethane groups, as can be observed by the growth of its characteristic peak at 1734 cm$^{-1}$. Smaller particle sizes of reactions using $CuSO_4$ resulted in a higher interfacial area, favoring the cross-linking reaction between starch and TDI forming urethane groups. Since the side reaction between NCO groups from TDI and OH groups from water were reduced by partially replacing NaCl by $CuSO_4$, as can be observed by the decrease of the characteristic peak of carbonyl of urea groups at 1646 cm$^{-1}$, as mentioned in the previous section, particles swell more after redispersion in water (shown in Table 2).

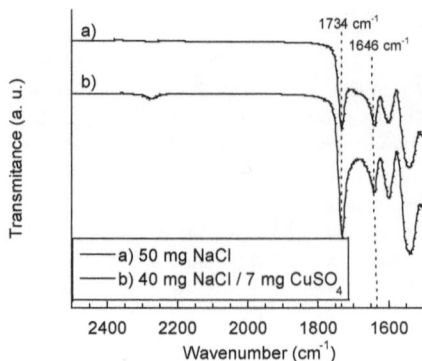

**Figure 5.** The effect of the co-stabilizer on the shell composition of cross-linked starch NCs prepared with 120 mg of TDI and 10 wt% of PGPR related to the dispersed phase. Results wereobtained by FT-IR using ATR mode.

### 3.3. Influence of the Amount of Starch

The effect of the amount of starch on the mean particle size was investigated using 50 mg of sodium chloride, 1.2 g and 1.3 g of water for 0.2 g and 0.1 g of starch, respectively, at different surfactant concentrations in the continuous phase, composed of 7.5 g of cyclohexane. The polymerization reactions were performed using 120 mg of TDI. Firstly, higher amounts of starch lead to larger particle sizes (from 200 nm to 272 nm increasing the starch amount from 0.1 g to 0.2 g, respectively, applying 10 wt% of PGPR related to the dispersed phase), as shown in Figure 6a. This fact might be attributed to the increase of the viscosity of the dispersed phase when a higher amount of gelatinized starch was used making droplet breakage more difficult during the mini-emulsion preparation. As expected, the mean average size decreased with the increase in the PGPR concentration.

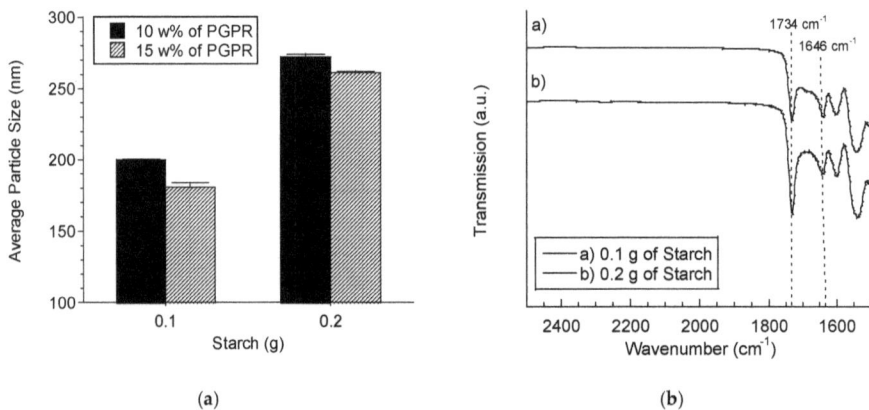

(a)

(b)

**Figure 6.** Effect of starch amount on (**a**) average particle size and (**b**) shell composition of cross-linked starch NCs prepared with 120 mg of TDI and 10 wt% of PGPR related to the dispersed phase. Results obtained by FT-IR using ATR mode (error bars represent standard deviation, $n \geq 3$).

The influence of the starch amount on the NC shell composition was evaluated using 10 wt% of PGPR related to the dispersed phase. A comparison between FT-IR spectra of the samples are shown in Figure 6b. Higher amounts of starch increased the fraction of urethane groups on the shell composition, as can be observed by the growth of its characteristic peak at 1734 cm$^{-1}$. In addition, a higher starch concentration reduces the side reaction between NCO groups from TDI and OH groups from water, as can be observed by the decrease of the characteristic peak of carbonyl of urea groups at 1646 cm$^{-1}$.

### 3.4. Effect of TDI Concentration

Mini-emulsions prepared with 0.1 g of starch, 50 mg of NaCl and 1.3 g of water dispersed into a continuous phase composed of 7.5 g of cyclohexane and 10 wt% of PGPR related to the dispersed phase were polymerized with different amounts of TDI (varying from 80 mg to 200 mg).

Figure 7 shows the influence of the amount of TDI on the average particle size. It can be noticed that the mean diameter of NCs dispersed in cyclohexane increased with higher amounts of TDI. Nevertheless, after the NCs were transferred into the aqueous solution, the particle diameter considerably decreased (for samples prepared with 140 mg and 160 mg of TDI). Figure 8 shows a schematic illustration of a PGPR molecule adsorbed at the aqueous droplet-continuous phase interface of NCs dispersed in cyclohexane. According to Baier et al. [18], the chains of surfactant molecules are free to move in the organic continuous phase, causing the diameter variability. When NCs were redispersed into an aqueous continuous phase, the hydrophobic tail of the surfactant molecules tend to rearrange near to the NC surface. The final average particle diameters in water were around 190 nm, evidencing the absence of inter-nanocapsule crosslinking. It should be emphasized that the fraction of the dispersed phase was around 10% in volume. This content was low enough to avoid coalescence and, thus, inter-nanocapsule crosslinking.

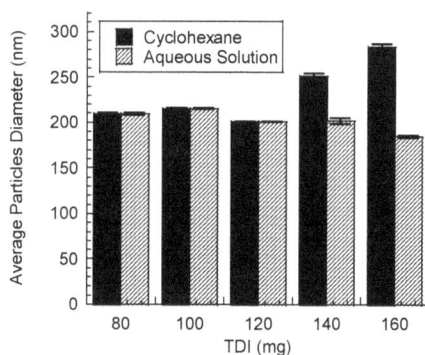

**Figure 7.** The effect of the amount of TDI on the average particle size using 10 wt% of PGPR related to the dispersed phase (error bars represent standard deviation, $n \geq 3$).

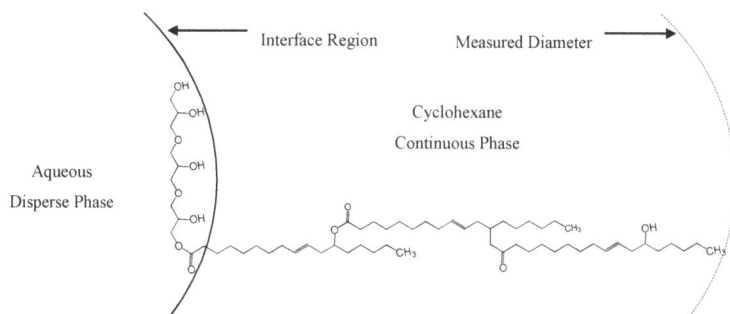

**Figure 8.** Schematic illustration of PGPR molecule adsorbed at the cyclohexane-aqueous droplet interface.

### 3.5. Effect of Reaction Temperature on NC's Shell Composition

Mini-emulsions prepared with 0.1 g of starch, 50 mg of NaCl, and 1.3 g of water dispersed into a continuous phase composed of 7.5 g of cyclohexane and 10 wt% of PGPR related to the dispersed phase were polymerized with 160 mg of TDI solubilized into 30 mg of PGPR in 5 g of cyclohexane solution at different temperatures. Figure 9 shows the influence of the reaction temperature on the capsule's shell composition. It can be observed that decreasing the temperature from 60 °C to 25 °C, a stronger NCO residual peak was observed after 24 h of reaction, due to the incomplete consumption of NCO groups of TDI at 2276 cm$^{-1}$. FT-IR measurements present in Figure 9 were performed using KBr pellets, instead of the ATR mode.

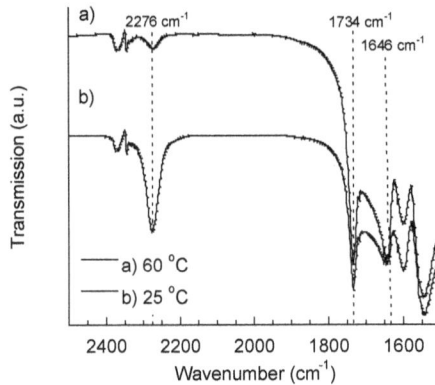

**Figure 9.** The influence of the reaction temperature on the composition of capsule's shells obtained by inverse mini-emulsion polymerization using 0.1 g of starch, 160 mg of TDI, and 10 wt% of PGPR related to the dispersed phase. The results were obtained by FT-IR using pellets pressed with KBr.

In addition to the fact that the reaction rate might be slower at 25 °C, the viscosity inside the aqueous droplet at a lower temperature is also higher than that at 60 °C, enhanced by the previous gelatinization of starch at 90 °C. The higher viscosity is due to the property of starch of forming a viscous gel with water when heated, followed by cooling, consisting of the gelatinized starch embedded in an interconnected network of recrystallized polymer aggregates. Since the interfacial polymerization occurs at the interface region between droplet and continuous phase, and depends on the mobility of the hydrophilic monomer inside the droplet [27], the formation of this viscous gel at 25 °C might reduce the mobility of starch inside the aqueous droplet, resulting in a lower hydroxyl concentration at the interface of the droplet and, consequently, a higher amount of residual isocyanate after the interfacial polymerization after 24 h of reaction time.

### 3.6. Permeability of NC Shells

The characteristics of interfacial polymerization is responsible for the final capsule morphology, composed of an aqueous core and a cross-linked starch shell. Mini-emulsions prepared with 0.1 g of starch, 50 mg of NaCl, and 1.3 g of water dispersed into a continuous phase composed of 7.5 g of cyclohexane and 10 wt% of PGPR related to the dispersed phase were polymerized with 160 mg of TDI. Figure 10 shows a SEM image of nanocapsules with an average size of around 270 nm (measured by DLS). Due to the drying and vacuum effects during the SEM measurements, the aqueous core of the NCs evaporates and the morphology appears similar to deflated balls, confirming the capsule morphology.

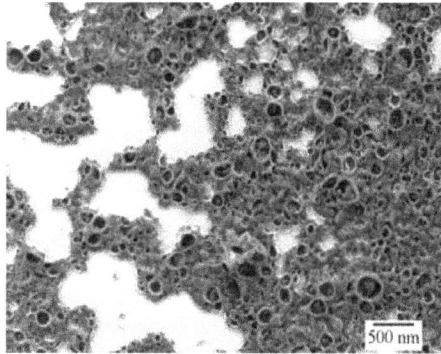

**Figure 10.** SEM image of cross-linked starch NCs (160 mg of TDI).

Encapsulation efficiency of hydrophilic compounds and the permeability of NC shells were assessed for nanocapsules redispersed in aqueous solution. Due to the characteristics of interfacial polymerization using inverse mini-emulsion, high encapsulation efficiencies of the hydrophilic dye SR 101 were found close to 100% for 160 mg and 80 mg of TDI, before redispersion of NCs in water (Figure 11).

The smallest amount of TDI (80 mg) did not provide an entirely sealed shell, and migration of dye to the external aqueous phase occurred leading to a lower encapsulation efficiency of SR 101 (70%) after the NCs were transferred to an aqueous solution. These results revealed that the cross-linking degree affects the ability of the capsule's shell to avoid the premature release of the hydrophilic compound.

The leakage of SR 101 after the NCs were transferred to the aqueous solution was monitored for seven days at 37 °C (Figure 12) and might be attributed to the permeability of the polymeric shell of NCs obtained with lower amounts of TDI (80 mg). Impermeable NCs were prepared using 160 mg of TDI, as shown in Figure 12, since no leakage was detected after the redispersion step of NCs from cyclohexane into the aqueous solution.

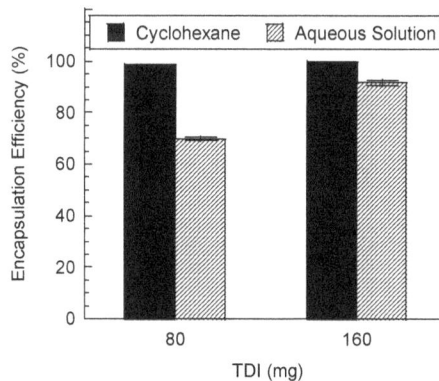

**Figure 11.** Encapsulation efficiency of SR 101 in NCs prepared with different amounts of TDI (error bars represent standard deviation, $n \geq 3$).

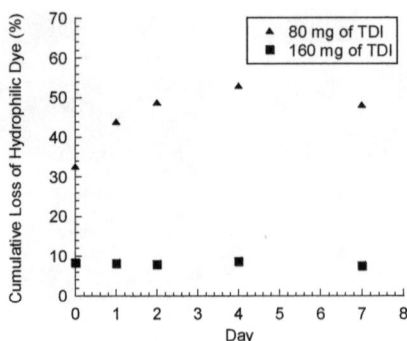

**Figure 12.** Cumulative loss of hydrophilic dye by leakage from the capsules to the aqueous-continuous phase.

### 3.7. Enzyme-Triggered Release of SR 101

The degradation assays of the cross-linked starch NCs were carried out under the same condition as used previously for the determination of the α-amylase activity (described in Section 2.2). At 37 °C and 0.05 M PBS (pH 7.4), enzymes showed an activity of 840 U/mg of enzymes.

The enzyme-triggered release kinetics of SR 101 was evaluated by enzymatic shell degradation of impermeable NCs (160 mg of TDI). The enzymatic degradation occurred using different α-amylase concentrations (36; 18; 9 and 0.9 mg/mL$^{-1}$). The results obtained for the maximum absorbance of SR 101 at 584 nm are shown in Figure 13.

The degradation reaction was allowed to proceed for 48 h. The maximum released SR 101 concentration was achieved after 18 h of the degradation reaction.

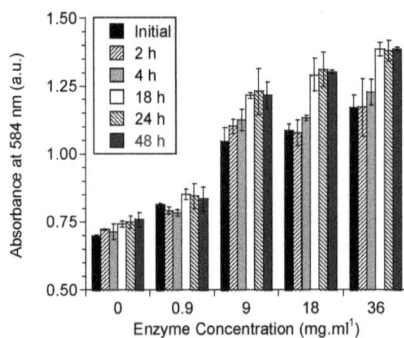

**Figure 13.** Release of SR 101 by enzymatic degradation using different concentrations of α-amylase (0, 0.9, 9, 18, and 36 mg/mL) (error bars represent standard deviation, $n \geq 3$).

The control sample without enzyme treatment presented maximum absorbance at around 0.750 a.u.; small increments might be attributed to damage of the capsule's polymeric shell during the centrifugation step.

The release kinetics were immediately determined after the enzyme solution was added to the NC dispersion. The higher the enzyme concentration is, the greater is the initial absorbance measured (Figure 13). These results indicate that the superficial dye entrapped at the surface of the NCs is rapidly released. After the initial period of degradation reaction (until 18 h), the SR 101 release rate slowed. The lowest enzyme concentration (0.9 mg/mL) did not show a significant release, due to its slow degradation rate.

## 4. Conclusions

Aqueous-core NCs were prepared via the inverse mini-emulsion technique, by interfacial polymerization between gelatinized potato starch and 2,4-toluene diisocyanate (TDI). The NCs were prepared at two different temperatures and the influence of 2,4-TDI on the polymer shell composition, particle size, and encapsulation efficiency was investigated. The ability to redisperse NCs in different continuous phases indicates the high stability of the polymeric shell. In addition, when a higher amount of TDI was used for cross-linking, the leakage of the hydrophilic dye to the aqueous phase after the redispersion in water could be readily minimized. Overall, this work presents a promising possibility for encapsulation of hydrophilic molecules with high encapsulation efficiency.

Biodegradability of NCs was demonstrated by the enzyme-triggered release of SR 101. The release kinetics were investigated using UV-Vis spectroscopy. The starch NCs were degraded by applying different enzyme concentrations. Enzyme concentrations below 0.9 mg/mL led to a slow release rate, due to the low degradation rate of the cross-linked capsule's shell.

**Acknowledgments:** The authors thank the financial support from CAPES—Coordenação de Aperfeiçoamento de Pessoal de Nível Superior e Tecnológico, CNPq—Conselho Nacional de Desenvolvimento Científico e Tecnológico and BMBF—Bundesministerium für Bildung und Forschung.

**Author Contributions:** A.M., K.L., P.H.H.A., and C.S. conceived the experiments; F.R.S. performed the experiments; and G.B. helped to analyze the data.

**Conflicts of Interest:** The authors declare no conflict of interest.

## References

1. Vrignaud, S.; Benoit, J.-P.; Saulnier, P. Strategies for the nanoencapsulation of hydrophilic molecules in polymer-based nanoparticles. *Biomaterials* **2011**, *32*, 8593–8604. [CrossRef] [PubMed]
2. Rosenbauer, E.-M.; Landfester, K.; Musyanovych, A. Surface-active monomer as a stabilizer for polyurea nanocapsules synthesized via interfacial polyaddition in inverse mini-emulsion. *Langmuir* **2009**, *25*, 12084–12091. [CrossRef] [PubMed]
3. Crespy, D.; Stark, M.; Hoffmann-Richter, C.; Ziener, U.; Landfester, K. Polymeric Nanoreactors for Hydrophilic Reagents Synthesized by Interfacial Polycondensation on Mini-emulsion Droplets. *Macromolecules* **2007**, *40*, 3122–3135. [CrossRef]
4. Cohen-Sela, E.; Chorny, M.; Koroukhov, N.; Danenberg, H.D.; Golomb, G. A new double emulsion solvent diffusion technique for encapsulating hydrophilic molecules in PLGA nanoparticles. *J. Control. Release* **2009**, *133*, 90–95. [CrossRef] [PubMed]
5. Becker Peres, L.; Becker Peres, L.; de Araújo, P.H.H.; Sayer, C. Solid lipid nanoparticles for encapsulation of hydrophilic drugs by an organic solvent free double emulsion technique. *Colloids Surf. B Biointerfaces* **2016**, *140*, 317–323. [CrossRef] [PubMed]
6. Arpicco, S.; Battaglia, L.; Brusa, P.; Cavalli, R.; Chirio, D.; Dosio, F.; Gallarate, M.; Milla, P.; Peira, E.; Rocco, F.; et al. Recent studies on the delivery of hydrophilic drugs in nanoparticulate systems. *J. Drug Deliv. Sci. Technol.* **2016**, *32*, 298–312. [CrossRef]
7. Baier, G.; Siebert, J.M.; Landfester, K.; Musyanovych, A. Surface click reactions on polymeric nanocapsules for versatile functionalization. *Macromolecules* **2012**, *45*, 3419–3427. [CrossRef]
8. He, C.W.; Parowatkin, M.; Mailaender, V.; Flechtner-Mors, M.; Ziener, U.; Landfester, K.; Crespy, D. Sequence-Controlled Delivery of Peptides from Hierarchically Structured Nanomaterials. *ACS Appl. Mater. Interfaces* **2017**, *9*, 3885–3894. [CrossRef] [PubMed]
9. Baier, G.; Cavallaro, A.; Vasilev, K.; Mailänder, V.; Musyanovych, A.; Landfester, K. Enzyme responsive hyaluronic acid nanocapsules containing polyhexanide and their exposure to bacteria to prevent infection. *Biomacromolecules* **2013**, *14*, 1103–1112. [CrossRef] [PubMed]
10. Baier, G.; Cavallaro, A.; Friedemann, K.; Müller, B. Enzymatic degradation of poly (L-lactide ) nanoparticles followed by the release of octenidine and their bactericidal effects. *Nanomed. Nanotechnol. Biol. Med.* **2014**, *10*, 131–139. [CrossRef] [PubMed]

11. Landfester, K. Preparation of Polymer and Hybrid Colloids by Mini-emulsion for Biomedical Applications. In *Colloidal Polymers Synthesis and Characterization*; Elaissari, A., Ed.; Marcel Dekker, Inc.: New York, NY, USA, 2003.

12. Liu, Z.; Jiao, Y.; Wang, Y.; Zhou, C.; Zhang, Z. Polysaccharides-based nanoparticles as drug delivery systems. *Adv. Drug Deliv. Rev.* **2008**, *60*, 1650–1662. [CrossRef] [PubMed]

13. Marto, J.; Gouveia, L.F.; Gonçalves, L.M.; Gaspar, D.P.; Pinto, P.; Carvalho, F.A.; Oliveira, E.; Ribeiro, H.M.; Almeida, A.J. A Quality by design (QbD) approach on starch-based nanocapsules: A promising platform for topical drug delivery. *Colloids Surf. B Biointerfaces* **2016**, *143*, 177–185. [CrossRef] [PubMed]

14. Rodrigues, A.; Emeje, M. Recent applications of starch derivatives in nanodrug delivery. *Carbohydr. Polym.* **2012**, *87*, 987–994. [CrossRef]

15. Humphrey, S.P.; Williamson, R.T. A review of saliva: Normal composition, flow, and function. *J. Prosthet. Dent.* **2001**, *85*, 162–169. [CrossRef] [PubMed]

16. Chourasia, M.K.; Jain, S.K. Polysaccharides for colon targeted drug delivery. *Drug Deliv.* **2004**, *11*, 129–148. [CrossRef] [PubMed]

17. Baier, G.; Baumann, D.; Siebert, J.M.; Musyanovych, A.; Mailänder, V.; Landfester, K. Suppressing unspecific cell uptake for targeted delivery using hydroxyethyl starch nanocapsules. *Biomacromolecules* **2012**, *13*, 2704–2715. [CrossRef] [PubMed]

18. Baier, G.; Musyanovych, A.; Dass, M.; Theisinger, S.; Landfester, K. Cross-linked starch capsules containing dsDNA prepared in inverse mini-emulsion as "nanoreactors" for polymerase chain reaction. *Biomacromolecules* **2010**, *11*, 960–968. [CrossRef] [PubMed]

19. Hans, M.L.; Lowman, A.M. Biodegradable nanoparticles for drug delivery and targeting. *Curr. Opin. Solid State Mater. Sci.* **2002**, *6*, 319–327. [CrossRef]

20. Asua, J.M. Mini-emulsion polymerization. *Prog. Polym. Sci.* **2002**, *27*, 1283–1346. [CrossRef]

21. Baier, G.; Friedemann, K.; Leuschner, E.M.; Musyanovych, A.; Landfester, K. PH stability of poly(urethane/urea) capsules synthesized from different hydrophilic monomers via interfacial polyaddition in the inverse mini-emulsion process. *Macromol. Symp.* **2013**, *331–332*, 71–80. [CrossRef]

22. Jagielski, N.; Sharma, S.; Hombach, V.; Maila, V.; Rasche, V.; Landfester, K. Nanocapsules Synthesized by Mini-emulsion Technique for Application as New Contrast Agent Materials. *Macromol. Chem. Phys.* **2007**, *208*, 2229–2241. [CrossRef]

23. Cao, Z.; Wang, Z.; Herrmann, C.; Landfester, K.; Ziener, U. Synthesis of narrowly size-distributed metal salt/poly(HEMA) hybrid particles in inverse mini-emulsion: Versatility and mechanism. *Langmuir* **2010**, *26*, 18008–18015. [CrossRef] [PubMed]

24. Shi, X.; Darwin, K.H. Copper Homeostasis in *Mycobacterium tuberculosis*. *Metallomics* **2015**, *7*, 929–934. [CrossRef] [PubMed]

25. Wolschendorf, F.; Ackart, D.; Shrestha, T.B.; Hascall-Dove, L.; Nolan, S.; Lamichhane, G.; Wang, Y.; Bossmann, S.H.; Basaraba, R.J.; Niederweis, M. Copper resistance is essential for virulence of *Mycobacterium tuberculosis*. *Proc. Natl. Acad. Sci. USA* **2011**, *108*, 1621–1626. [CrossRef] [PubMed]

26. Hamdi, G.; Ponchel, G. Enzymatic Degradation of Epichlorohydrin Crosslinked Starch Microspheres by α-Amylase. *Pharmac. Res.* **1999**, *16*, 867–875.

27. Morgan, P.W. *Interfacial Polymerization*; John Wiley & Sons, Inc.: Hoboken, NJ, USA, 2011; Volume 1929.

**processes**

*Article*

# Comparison of Polymer Networks Synthesized by Conventional Free Radical and RAFT Copolymerization Processes in Supercritical Carbon Dioxide

Patricia Pérez-Salinas [1], Gabriel Jaramillo-Soto [1], Alberto Rosas-Aburto [1], Humberto Vázquez-Torres [2], María Josefa Bernad-Bernad [3], Ángel Licea-Claverie [4] and Eduardo Vivaldo-Lima [1,*]

[1] Facultad de Química, Departamento de Ingeniería Química, Universidad Nacional Autónoma de México, Ciudad de México 04510, Mexico; perez.patricia077@gmail.com (P.P.-S.); jaramillo2000@hotmail.com (G.J.-S.); alberto_rosas_aburto@comunidad.unam.mx (A.R.-A.)
[2] Departamento de Física, Universidad Autónoma Metropolitana-Unidad Iztapalapa, Av. San Rafael Atlixco No. 186, Col. Vicentina, Ciudad de México 09340, Mexico; hvto@xanum.uam.mx
[3] Facultad de Química, Departamento de Farmacia, Universidad Nacional Autónoma de México, Ciudad de México 04510, Mexico; bernadf@comunidad.unam.mx
[4] Instituto Tecnológico de Tijuana, Centro de Graduados e Investigación en Química, A.P. 1166, Tijuana 22000, B.C., Mexico; aliceac@tectijuana.mx
* Correspondence: vivaldo@unam.mx; Tel.: +52-55-5622-5256

Academic Editor: Alexander Penlidis
Received: 3 April 2017; Accepted: 4 May 2017; Published: 9 May 2017

**Abstract:** There is a debate in the literature on whether or not polymer networks synthesized by reversible deactivation radical polymerization (RDRP) processes, such as reversible addition-fragmentation radical transfer (RAFT) copolymerization of vinyl/divinyl monomers, are less heterogeneous than those synthesized by conventional free radical copolymerization (FRP). In this contribution, the syntheses by FRP and RAFT of hydrogels based on 2-hydroxyethylene methacrylate (HEMA) and ethylene glycol dimethacrylate (EGDMA) in supercritical carbon dioxide (scCO$_2$), using Krytox 157 FSL as the dispersing agent, and the properties of the materials produced, are compared. The materials were characterized by differential scanning calorimetry (DSC), swelling index (SI), infrared spectroscopy (FTIR) and scanning electron microscopy (SEM). Studies on ciprofloxacin loading and release rate from hydrogels were also carried out. The combined results show that the hydrogels synthesized by FRP and RAFT are significantly different, with apparently less heterogeneity present in the materials synthesized by RAFT copolymerization. A ratio of experimental ($Mc_{exp}$) to theoretical ($Mc_{theo}$) molecular weight between crosslinks was established as a quantitative tool to assess the degree of heterogeneity of a polymer network.

**Keywords:** supercritical carbon dioxide; RAFT polymerization; hydrogels; polymer network homogeneity; solubility in supercritical fluids

## 1. Introduction

One of the most challenging areas of polymer science and engineering is the synthesis, characterization and development of applications of polymer networks [1] (pp. 145–319). The reason for this is the difficulty in dissolving, processing or manipulating the polymer network after its synthesis. Many authors reported their way to analyze and handle these materials in their respective fields, trying to understand their behavior [1–5].

Polymer science is currently diversified into different fields. Many researchers focus their research on controlling the structure and molecular weight of polymer molecules synthesized using reversible deactivation radical polymerization (RDRP) techniques [6–9]. Other authors, including ourselves, have combined the use of RDRP with the utilization of supercritical fluids, mainly carbon dioxide, as a unique solvent in polymer synthesis [5,10–14].

Some of the advantages of using compressed fluids in organic synthesis, and specifically supercritical carbon dioxide, include their innocuousness, their easiness of removal and recovery and the fact that they are inexpensive and easy to acquire. On the other hand, the disadvantages for their use include the initial high cost of investment in equipment, since reactors and other process equipment should withstand moderate to high pressures and moderate to high temperatures. Although there is information available on the solubility of chemical compounds in supercritical carbon dioxide [15,16], solubility data for some monomers, such as HEMA or EGDMA, and their polymers, are not available in the open literature.

There are few reports on the use of compressed fluids in the synthesis of polymer networks using RDRP controllers. The information available about the properties and performance of those materials is rather limited because the characterization of polymer networks is not straightforward. One specific aspect about the characterization of polymer networks synthesized by copolymerization of vinyl/divinyl monomers that remains unsolved is the determination of their heterogeneity, understood as the regioregularity of polymer chains between crosslinks. Several approaches have been proposed for this purpose, but the combined use of experimental data and theoretical calculations seems to be the most effective way to understand their behavior. One of such approaches is the calculation of the mean molecular weight between crosslinks from swelling index data, using the Flory–Rehner equation [2,17–19].

Working with supercritical fluids also requires the knowledge of the thermodynamic aspects of the reacting mixture, such as the knowledge or construction of a pressure vs. temperature ($P$ vs. $T$) diagram. This diagram is built considering the actual composition of the reacting mixture, thus generating a curve that indicates which zone is related to liquid-vapor equilibrium or to supercritical conditions, where all components are in one phase and liquid and vapor densities are the same.

In this contribution, we analyze the issue of the reduced heterogeneity of polymer networks synthesized by RAFT copolymerization of vinyl/divinyl monomers in supercritical carbon dioxide ($scCO_2$) by combining the information obtained from different characterization techniques: DSC, measurement of the swelling index (SI), FTIR, SEM and loading/controlled release of ciprofloxacin. A pressure-temperature thermodynamic diagram for our specific reacting mixture (monomers, initiator, dispersing agent and solvent), constructed with the ASPEN® software, was used in our analysis of the results. The heterogeneity of the polymer networks is evaluated with the use of a polymer network homogeneity parameter ($H$), defined as the ratio of theoretical ($Mc_{theo}$) to experimental ($Mc_{exp}$) molecular weights between crosslinks.

## 2. Materials and Methods

### 2.1. Reagents

HEMA (Sigma-Aldrich Química, S.L., Toluca, Mexico) and EGDMA (Sigma-Aldrich) were distilled under vacuum. Azobisisobutyronitrile (AIBN) (Akzo Nobel Chemicals S.A. de C.V., Los Reyes La Paz, Mexico) was recrystallized twice from methanol. Carbon dioxide (Praxair, 99.99% purity) was used as received. 4-Cyano-4-(dodecylsulfanylthiocarbonyl) sulfanyl pentanoic acid (RAFT agent) was synthesized following a procedure described previously [11]. Krytox 157 FSL (DuPont), referred to as Krytox in the remainder of this paper, was used as received.

### 2.2. Polymerization System

Polymerizations in $scCO_2$ were conducted in a 38-mL high pressure view cell, equipped with one frontal and two lateral sapphire windows (from Crystal Systems Inc., Salem, MA, USA), which allowed visual observation of the reaction mixture. A 260 dual syringe pump system (from Teledyne ISCO) was used to handle the $CO_2$ and bring it to supercritical conditions. The reactor was charged with monomer, initiator and stabilizer and a magnetic stirrer bar. Then, it was purged with a slow flow of $CO_2$ and pressurized with $CO_2$ until a given pressure, lower than the desired reaction pressure. Next, the reactor was placed into a warm bath and heated to the desired reaction temperature. Once this temperature was reached and controlled, pressure was increased to the desired reaction pressure by slowly loading additional $CO_2$. Reactions were carried out at 65 °C and 172.4 bar. Further information about the reaction system is found elsewhere [11].

Samples were classified into two main groups: those synthesized by FRP and the ones synthesized by RAFT copolymerization. Half of the samples contained stabilizer (Krytox 157 FSL, 5 wt %), and the other half were synthesized without it. All samples were run by duplicate, as shown in Table 1. Therefore, the following pairs are duplicates: G311 and G313, G312 and G314, G315 and G317, as well as G316 and G318.

**Table 1.** Summary of experimental conditions for the FRP or RAFT copolymerization of HEMA/EDGMA in supercritical carbon dioxide (scCO$_2$) (*T* = 65 °C, *P* = 173 bar, *t* = 24 h, 22% *w/v* CO$_2$).

| Sample | HEMA (mmol) | EGDMA (mmol) | AIBN (mmol) | RAFT Agent (mmol) | Krytox (mmol) |
|--------|-------------|--------------|-------------|-------------------|---------------|
| G311 | 25 | 1.25 | 0.1 | 0 | 5 wt%/HEMA |
| G312 | 25 | 1.25 | 0.1 | 0 | 0 |
| G313 | 25 | 1.25 | 0.1 | 0 | 5 wt%/HEMA |
| G314 | 25 | 1.25 | 0.1 | 0 | 0 |
| G315 | 25 | 1.25 | 0.1 | 0.05 | 5 wt%/HEMA |
| G316 | 25 | 1.25 | 0.1 | 0.05 | 0 |
| G317 | 25 | 1.25 | 0.1 | 0.05 | 5 wt%/HEMA |
| G318 | 25 | 1.25 | 0.1 | 0.05 | 0 |

*2.3. Polymer Network Characterization*

Pendant double bond consumption was measured by FTIR using the area for carboxylic groups as an internal reference. Polymer network samples were powdered and mixed with potassium bromide (KBr), compressed and analyzed in an FTIR Perkin Elmer spectrometer.

Glass transition temperature (*Tg*) was measured by modulated differential scanning calorimetry (MDSC). Besides *Tg*, some of the parameters obtained during the characterization experiment (e.g., width and slope of a modulated heat flow versus derivative modulated temperature—Lissajous figure) can in principle correlate with the crosslinking density distribution of the polymer network. A TA Instruments Model 2920 DSC apparatus was employed. For these analyses, 10 mg of sample were placed into the aluminum pan and covered with the corresponding lid. Three cycles were programmed in the DSC. In the first and third cycles, the sample was heated from −40 °C up to 220 °C at a 10 °C/min rate. In the second cycle, the sample was chilled from 220 °C up to −40 °C at a constant cooling rate of 10 °C/ min. The results for *Tg* and energy from the reversible heat flow chart obtained during the third cycle were selected for reporting. These results are in principle more accurate and free of any interference related to molecular arrangements or sample preparation than the ones obtained from the other cycles [3,4].

Swelling index and gel fraction after 48 h of contact time with water were measured for selected samples. For swelling index tests, 15 mg of polymer network were placed into a previously weighted test tube. Both tube and sample were weighted again. Fifty milliliters of solvent were poured into the tube with the sample and remained in contact for specific times. Some samples remained immersed during 48 h. After each time, samples were removed from solvent and centrifuged at 15,000 rpm. Solvent was decanted, and the tubes with swelled sample were weighted. Swelling index was calculated from the difference between the weight of swelled sample with tube and the weight of the tube, divided by the weight of the tube with dried sample minus the weight of the tube. Measurement of gel content follows almost the same procedure. However, instead of decanting the solvent, it was completely removed, and the sample with the tube was dried at 80 °C for 48 h and weighed. Gel content was calculated by subtracting the weight of dried sample in the tube from the weight of the tube and dividing by the weight of the original sample.

In the case of ciprofloxacin loading and desorption tests, 15 mg of polymer network were placed into a flask containing 5 mL of a solution 0.02 M of ciprofloxacin in water. Samples were shaken for 400 min. Small aliquots of the solution were taken periodically. They were analyzed by UV-visible spectroscopy ($\lambda$ = 276.2 nm) to determine antibiotic concentration. The polymer sample was then immersed in 3 mL of distillate water using a Franz cell. Donating-receiving parts were separated using a Millipore® HNWP 0.45-$\mu$m membrane. Samples were taken from the receiving part during 24 h. Aliquots were analyzed in a UV-visible spectrophotometer ($\lambda$ = 276.2 nm). Desorption kinetics charts were built from these data.

Six polymer network samples of 3 ± 0.2 mg each were placed into flasks with different concentrations of ciprofloxacin (5 mL of water in each sample): $1 \times 10^{-5}$, $3 \times 10^{-5}$, $6 \times 10^{-5}$, $8 \times 10^{-5}$, $2 \times 10^{-4}$ and $4 \times 10^{-4}$ M. Each vial was shaken for 24 h. The samples were then filtrated. The remaining solution was analyzed in a UV-visible spectrophotometer at $\lambda$ = 276.2 nm. Isotherm charts were built from these data.

SEM imaging was used to observe the morphologies of the synthesized polymer networks. A JEOL 5900-LV microscope was used. Samples were powdered and placed over a carbon patch for SEM analyses. Image-Pro Plus was used to analyze the particles.

### 2.4. Estimation of $M_c$ from the Flory–Rehner Equation

Average molecular weight between crosslinks ($M_c$) was estimated based on experimental data of swelling index measured for each hydrogel, according to Equation (1). $\rho$ and $ve$ in Equation (1) are polymer network density and the amount of crosslinks per volume unit. The amount of crosslinks per volume unit is calculated using Equation (2), where $V_r$, $V_1$ and $\chi$ are the volume fraction of the polymer in the swelled gel, the molar volume of solvent and the Flory interaction parameter, respectively. $V_r$ and $\chi$ are calculated using Equations (3) and (4), respectively. $SI$ and $d$ in Equation (3) are the swelling index and solvent density, respectively [17–19].

$$Mc = \frac{\rho}{ve} \tag{1}$$

$$ve = \frac{-\left[\ln(1 - Vr) + Vr + \chi\, Vr^2\right]}{\left[V_1 \left(Vr^{\frac{1}{3}} - \frac{Vr}{2}\right)\right]} \tag{2}$$

$$Vr = \left[1 + (SI - 1)\frac{\rho}{d}\right]^{-1} \tag{3}$$

$$\chi = 0.455 - 0.155\, Vr \tag{4}$$

The values of $M_c$ obtained from Equation (1), using experimental values of the swelling index, are denoted as $Mc_{exp}$ in this paper. A theoretical value of $M_c$ can be calculated from the ratio of EGDMA to HEMA concentrations multiplied by the molecular weight of the average repeating unit. At the initial conditions and considering total conversion, $Mc_{theo}$ = 2603 g/mol.

In this paper, we propose a polymer network homogeneity parameter ($H$) defined as the ratio of $Mc_{theo}$ to $Mc_{exp}$, as shown in Equation (5), to provide a quantitative indicator of the degree of homogeneity of the crosslink density distribution of polymer networks. If the experimental value of $M_c$ approaches the theoretical value of $M_c$, $H$ approaches unity. This means that the polymer network is almost like a homogeneous distribution of HEMA molecules referred to EGDMA molecules, in terms of chain length between crosslinks. The range of values of $H$ can be higher or lower than unity. The reason for this is because the chain length between crosslinks is an average value calculated from a bulk test (swelling index), so the behavior exhibited by the bulk of the network will determine the value of $H$. Values below unity mean that hydrogels behave like highly crosslinked networks. On the contrary, values above unity mean that hydrogels behave as having long chains between crosslinks.

$$H = \frac{Mc_{theo}}{Mc_{exp}} \tag{5}$$

*2.5. Thermodynamic Analysis, Estimation of Solubility Parameters of Components and Solubility in Supercritical $CO_2$*

Although the heterogeneity of polymer networks can be in principle reduced by using RDRP in scCO$_2$, other thermodynamic variables, such as the phase where the reaction takes place (liquid, vapor or supercritical) or the solubility of the components participating in the reaction, may also play a role in the structures obtained. For instance, if supercritical conditions are reached inside the reactor, all of the components remain as one phase, until the first molecule of polymer appears. In contrast, if two phases, liquid and vapor, are present from the beginning, the polymerization will proceed in both phases, and molecular weight development will depend on the kinetic behavior in each phase, thus obtaining two polymer network populations, irrespective of the fact that an RAFT agent is present in the formulation. Since most polymers are insoluble in scCO$_2$, it is important to include a dispersing agent soluble in scCO$_2$ in the formulation, such as Krytox.

Therefore, it is important to evaluate the solubility of components, including Krytox, in supercritical carbon dioxide. The solubility parameter of $CO_2$ ($\delta_{CO2(T,P)}$) was taken from the literature [20–22], and its molar volume ($V_{CO2(T,P)}$) was calculated using an equation of state [21,22]. The solubility parameters for HEMA (25 (J/cm$^3$)$^{1/2}$) and Krytox (6.02 (J/cm$^3$)$^{1/2}$), were taken from the literature [20,23]. In the case of the polymer network, $\delta_{solute}$ (26.6 (J/cm$^3$)$^{1/2}$) was estimated using group contribution theory [24]. One criterion to determine if a solute is soluble in a solvent is to calculate its Flory interaction parameter, $\chi$, using Equation (5); if $\chi < 0.84$, then the component should be soluble in $CO_2$ at the given T and P [24].

$$\chi = 0.34 + \left( \frac{V_{CO2\,(T,P)}}{R\,T} \right) \left( \delta_{Solute} - \delta_{CO2\,(T,P)} \right)^2 \tag{6}$$

$V_{CO2(T,P)}$ in Equation (5) is $CO_2$ molar volume (cm$^3$/mol) at the given temperature and pressure; R is the ideal gas constant; T is temperature; $\delta_{solute}$ is solubility parameter of the compound, monomer or polymer, to be dispersed in scCO$_2$ (J/cm$^3$)$^{1/2}$.

A phase diagram (P versus T) for the reactive system (13.43% HEMA, 1.78% EGDMA, 0.07% AIBN and 84.72% $CO_2$, on a weight basis) was created by us using ASPEN Plus (Version 8.8) [25]. Calculations were carried out using the Peng–Robinson equation of state. The processing/reaction path of the reactive mixture was traced on the P-T diagram.

## 3. Results and Discussion

### 3.1. Thermodynamic Behavior of the Reacting Mixture

Figure 1 shows a P vs. T thermodynamic diagram generated with ASPEN Plus for the reacting mixture. The path followed from initial to reacting conditions is shown in the diagram. At the beginning of the reaction, when the reactor was loaded, there was a vapor-liquid mixture at 30 °C and 1 bar (see the black point shown inside the curve, in Figure 1). As pressure increased up to 72 bars, at 30 °C, the mixture approached the vapor-liquid border, but it still remained a vapor-liquid mixture. When the reactor reached the final reacting conditions, 65 °C and 172.4 bars, the mixture was in the supercritical region (outside the curve described in the chart). However, it is observed in Figure 1 that the reactor operated very close to the borderline between vapor-liquid (inside the curve) and supercritical (outside the curve) regions, so that any small change in T or P could shift the equilibrium to the vapor-liquid region, thus having two-phase polymerization even before polymer started phase separating.

It is also observed in Figure 1 that the solubility parameter for $CO_2$ changes significantly along the reaction path. At the beginning, $\delta_{CO2}$ was too small (0.033 (J/cm$^3$)$^{1/2}$) because $CO_2$ was in the gas phase, and its density and molar volume were also small. However, when pressure increased from one to 72 bars (less than two orders of magnitude), $\delta_{CO2}$ increased three orders of magnitude (reaching a value of $\delta_{CO2} = 10.25$ $(J/cm^3)^{1/2}$). This was the most significant increase in $\delta_{CO2}$, since a further increase in pressure by one order of magnitude (from 72 to 172.4 bars) did not significantly changed $\delta_{CO2}$ ($\delta_{CO2} = 10.03$ $(J/cm^3)^{1/2}$) at the final reacting conditions.

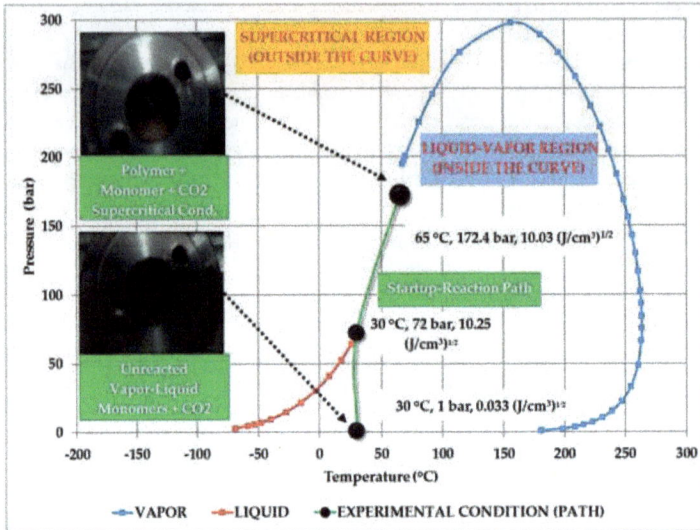

**Figure 1.** Pressure vs. temperature diagram for the system carbon dioxide, HEMA, EGDMA, AIBN, including the path followed by the reactor and the solubility parameter values estimated at each point using an equation of state, generated with ASPEN PLUS.

A plot of calculated $\delta_{CO2}$ versus pressure, using an equation of state [21,22], is shown in Figure 2. The calculations were carried out by us. It is observed that large changes in $\delta_{CO2}$ are obtained when pressure is increased following an isothermal route, whereas a small reduction occurs if an isochoric route is used. Also shown in Figure 2 is the process/reaction path followed by the reacting mixture. The major change in the value of $\delta_{CO2}$ occurs at approximately 72 bar. At 68 bar and 30 °C $\delta_{CO2} = 4.54$ $(J/cm^3)^{1/2}$, but at 72 bar, $\delta_{CO2} = 10.25$ $(J/cm^3)^{1/2}$. The reason is that 30 °C and 72 bar is close to the critical point of $CO_2$, which occurs at 31 °C and 73.8 bar. At the critical point, the densities of gas and liquid $CO_2$ are the same. As density increases, $CO_2$ works as a true solvent. The reaction mixture can be considered dispersed in $CO_2$ beyond this point. However, only when the mixture achieves 65 °C and 172.4 bars, it can be considered as one phase at supercritical conditions.

As explained before, the solubility in $CO_2$ of the components of the reacting mixture is important for the performance of the polymerization and for the heterogeneity of the produced polymer network. The solubilities of the components of the reacting mixture can be estimated from Flory's interaction parameter, $\chi$, calculated using Equation (6). As mentioned earlier, if $\chi > 0.84$, then the solute is soluble in scCO2 [24].

The solubility parameters for monomer and polymer network do not change significantly within the ranges of temperatures and pressures experienced by the reacting mixture. $\delta_{HEMA} = 25$ $(J/cm^3)^{1/2}$, $\delta_{Krytox} = 6.02$ $(J/cm^3)^{1/2}$ and $\delta_{PHEMA-EGDMA} = 27.3$ $(J/cm^3)^{1/2}$. The data used for estimation of $\delta_{P(HEMA-EGDMA)}$ using Fedor's method [24] are summarized in Table 2.

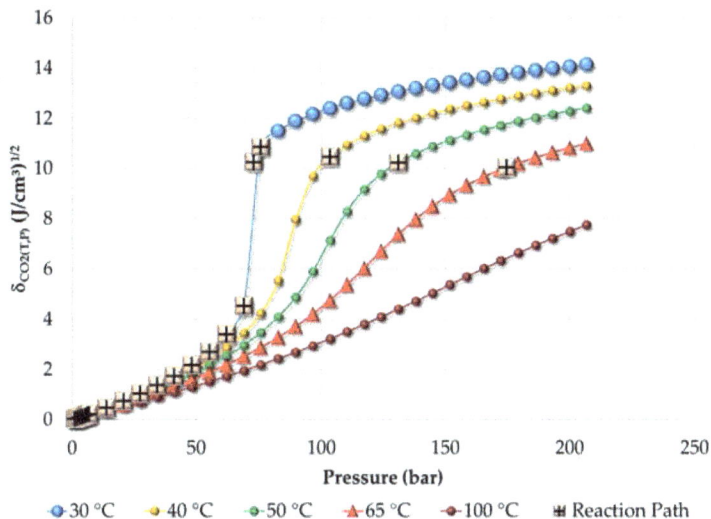

**Figure 2.** Estimation of the solubility parameters of carbon dioxide under subcritical and supercritical conditions, including the reaction path followed by the reactor.

**Table 2.** Summary of group contribution parameters used in Fedor's method [24] for estimation of $\delta_{P(HEMA\text{-}EGDMA)}$ for the polymer network. The composition of HEMA and EGDMA in the polymer network is based on the information from Table 1, namely, HEMA 95.24 mol% and EGDMA 4.76 mol%.

| Molecule | Contribution Groups | Contribution Value for $E_{coh}$ (J/mol) | Contribution Value for $V_{network}$ (cm$^3$/mol) | Frequency |
|---|---|---|---|---|
| | –CH$_3$ | 4710 | 33.5 | 1 |
| | –CH$_2$– | 4940 | 16.1 | 3 |
| | >C< | 1470 | −19.2 | 1 |
| | –CO$_2$– | 18,000 | 18 | 1 |
| | –OH | 29,800 | 10 | 1 |
| | –CH$_3$ | 4710 | 33.5 | 2 |
| | –CH$_2$– | 4940 | 16.1 | 4 |
| | >C< | 1470 | −19.2 | 2 |
| | –CO$_2$– | 18,000 | 18 | 2 |

Figure 3 shows a plot of Flory interaction parameter, $\chi$, versus pressure, for HEMA, poly(HEMA-co-EDGMA) and Krytox 157 FSL. It is observed in Figure 3 that except for Krytox 157 FSL (above 72 bar, which is CO$_2$ critical pressure), all of the components of the reacting mixture are insoluble in CO$_2$ ($\chi > 0.84$). Krytox 157 FSL is assumed to act as a stabilizer, encapsulating the monomers, thus allowing the formation of a homogeneous initial dispersion at supercritical conditions.

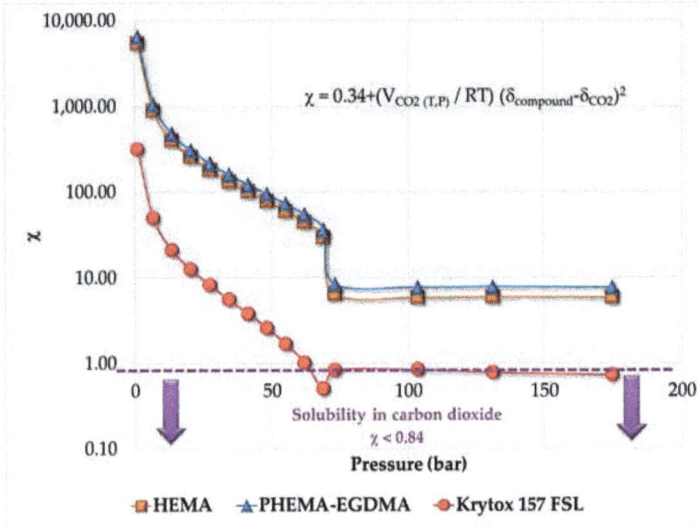

**Figure 3.** Solute-solvent Flory interaction parameter, $\chi$, versus pressure, for solutes HEMA, polymer network and Krytox 157 FSL, in $CO_2$ (solvent). A solute is soluble in the solvent when $\chi < 0.84$.

*3.2. Swelling Index, Gel Content and Homogeneity Parameter*

Figure 4 shows a plot of experimental swelling index versus average molecular weight between crosslinks, *Mc*. The plot was built using Equations (1) to (4) for the samples described in Table 1. Blue square points in Figure 4 represent polymer networks synthesized by RAFT copolymerization, whereas orange circles correspond to polymer networks synthesized by FRP. The red triangle corresponds to the theoretical value of swelling index for an ideal polymer network with an Mc value of 2603 g/mol, calculated from the comonomer composition given in Table 1. As observed in Figure 4, there are two well-defined regions. One is very broad and corresponds to polymer networks synthesized by conventional FRP; the second one is smaller in size and contains hydrogels synthesized by RAFT copolymerization. It is observed that the theoretical value lies within the region of RAFT polymer networks. Figure 4 thus shows a first difference between polymer networks synthesized by FRP and RAFT copolymerization. Hydrogels synthesized by FRP exhibit larger *Mc* values due to higher swelling indexes. The large dispersion in swelling index values seems to indicate a random growth of the polymer network. On the other hand, hydrogels synthesized by RAFT copolymerization exhibit swelling index-*Mc* values close to the theoretical point of (5.36, 2603 g/mol). The lower dispersion of SI versus *Mc* values observed in the case of polymer networks synthesized by RAFT copolymerization seems to describe a more structured and ordered polymer network growth process. Sample G316 seems to be an outlier, since this polymer network was synthesized using an RAFT agent, but it falls within the FRP region. There was no Krytox included in the formulation of sample G316, which may explain the anomalous behavior. However, sample G316 will remain in the analyses of our other characterization studies.

The results of Figure 4 point to the fact that hydrogels synthesized by FRP have broader distributions of *Mc*, compared to those synthesized by RAFT copolymerization. Figure 5 illustrates the concept of heterogeneous *Mc* distribution. Each polymer network contains nodes connected by chain segments of different lengths (molecular weights). The molecular weight between crosslinks (nodes) calculated from experimental swelling index, using Equations (1) to (4), represents an average value. A schematic representation of how the polymer networks synthesized by RAFT and FRP looks as shown in Figure 5.

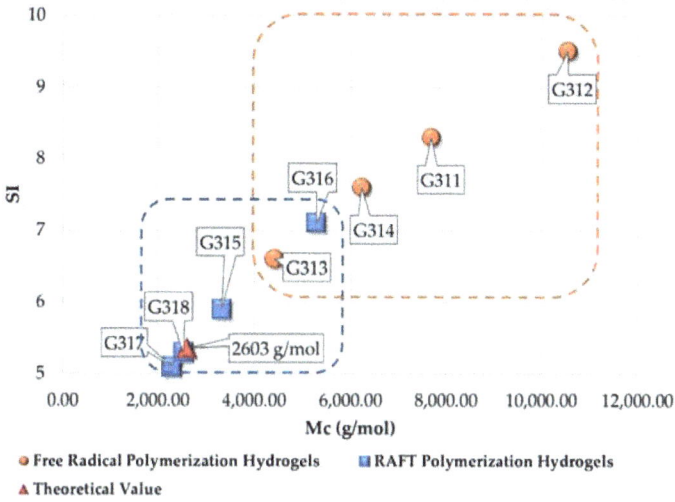

**Figure 4.** Correlation between experimental results of swelling index (SI) and calculated average molecular weight between crosslinks, using Equation (1).

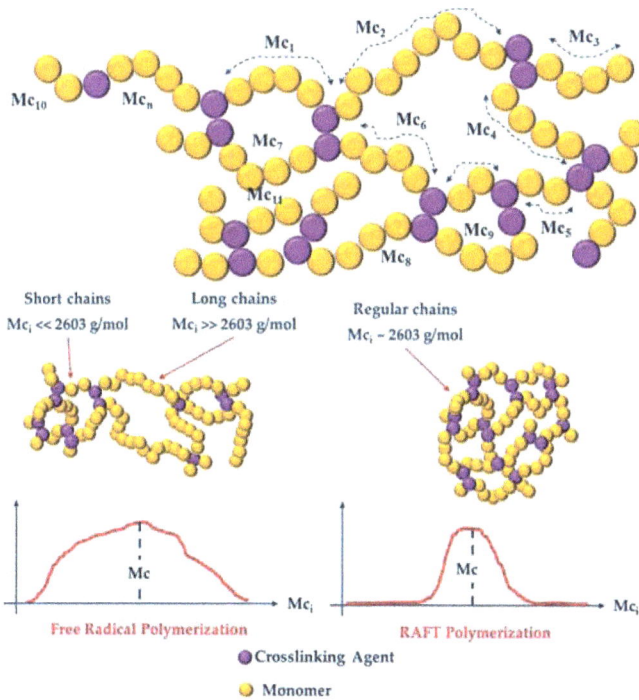

**Figure 5.** Schematic representation of the *Mc* distribution in polymer networks and how they differ for hydrogels synthesized by FRP and RAFT copolymerizations.

Two polymer populations for the case of polymer networks synthesized by FRP are observed in Figure 5. The first population contains short chains with individual $Mc_i$ values smaller than the theoretical value of 2603 g/mol, and the second population contains larger chains with individual $Mc_i$ values higher than 2603 g/mol. In the case of short chains linking crosslink points (nodes), the polymer network is so tight and the pores so small that swelling with water in those regions is negligible. Therefore, the overall swelling measured for these polymer networks synthesized by FRP is attributable to the regions with long chain segments between crosslinks. It is this second population that is the one responsible for increasing the average value of $Mc$ and, thus, causing such large values of SI, as observed in Figure 4.

Except for sample G316, the polymer networks synthesized by RAFT copolymerization lie very close to the theoretical value of SI versus $Mc$, as observed in Figure 4. These results suggest that the polymer chains between crosslinks have almost the same length, as illustrated in Figure 5. As will be evidenced from the characterization results described in the following sections, the polymer networks synthesized by free radical copolymerization of vinyl/divinyl monomers have differentiated properties and perform differently, depending on the presence or absence of a RAFT agent.

As explained earlier, in this contribution, a homogeneity parameter ($H$), defined by Equation (5) is used to assess the homogeneity of polymer networks. The values of SI, $Mc_{exp}$, $Mc_{theo}$ and $H$ for the polymer networks synthesized in this study are summarized in Table 3. It is observed that $H \approx 1.0$ for the samples synthesized by RAFT copolymerization, except sample G316. The presence or absence of Krytox in the reaction does not seem to affect the behavior of the polymer network in terms of SI and $H$.

**Table 3.** Estimation of average molecular weight between crosslinks from the Flory–Rehner equation.

| Polymerization Process | Sample | Swell Index | Mc Experimental (g/mol) | Mc Theoretical (g/mol) | Polymer Network Homogeneity $H \approx 1.0$ | Krytox Content |
|---|---|---|---|---|---|---|
| Free Radical Polymerization | G311 | 8.3 | 7,681 | 2,603; this value was calculated as the HEMA to EGDMA molar ratio | 0.34 | Yes |
| | G312 | 9.5 | 10,541 | | 0.25 | No |
| | G313 | 6.6 | 4,407 | | 0.59 | Yes |
| | G314 | 7.6 | 6,221 | | 0.42 | No |
| | G315 | 5.9 | 3,329 | | 0.78 | Yes |
| RAFT Polymerization | G316 | 7.1 | 5,274 | | 0.49 | No |
| | G317 | 5.1 | 2,289 | | 1.14 | Yes |
| | G318 | 5.3 | 2,530 | | 1.03 | No |

The relationship between gel content and $H$ is shown in Figure 6, where two different populations are clearly observed. The first population corresponds to polymer networks synthesized by FRP and the second one to polymer networks synthesized by RAFT copolymerization. One hundred percent gel content is observed for the polymer networks synthesized by FRP, whereas ~90% gel content was obtained for the materials synthesized by RAFT copolymerization. It seems that in polymer networks synthesized by RAFT copolymerization, the slower polymerization rate causes polymer chains of low to medium molecular weight produced late in the polymerization to encounter mobility restrictions at high viscosities, thus reaching lower limiting conversions, whereas in polymer networks synthesized by FRP monomer disappears quickly, and medium-sized polymer networks rapidly become part of the gel by propagation through pendant double bonds or termination with radicals placed within the polymer network.

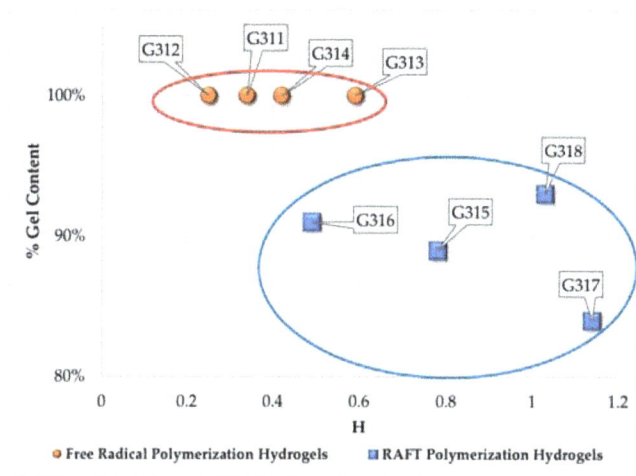

**Figure 6.** Correlation between gel content and the polymer network homogeneity parameter (*H*).

*3.3. Unreacted Pendant Double Bonds from FTIR*

The amount of unreacted pendant double bonds can also be used as an indicator of polymer network heterogeneity. If the polymer network grows in a gradual, ordered way, as would be the case in the presence of an RAFT controller, all pendant double bonds would have the same probability of being reached by a molecule with a free radical reacting unit. If the polymer network is being produced in a non-controlled manner, a significant amount of pendant double bonds could get trapped within the structure of the polymer network, thus becoming inaccessible to polymer chains with active (free radical) segments. Therefore, the amount of unreacted double bonds present in a polymer network is expected to be higher if the synthesis took place by FRP, compared to polymer networks synthesized by RAFT copolymerization. The FTIR spectra of the hydrogels synthesized in this study are shown in Figure 7. Also shown in the lower section of Figure 7 is an enlargement of the region of interest, where the spectra of HEMA and EGDMA are also included, in order to emphasize the region where double bonds are noticeable.

An internal standard was used for quantification purposes. A summary of the assignment of functional groups to the different bands is shown in Table 4. The signal at 1717 cm$^{-1}$ corresponds to carbonyl groups –C=O, which remained unreacted and depend on HEMA and EGDMA initial concentrations only. The area for the –C=O band calculated for each sample was considered as the internal reference ($A_{C=O}$). Also observed in Table 4 are two regions for double bonds. The most promising region to evaluate remaining double bonds was 1635 cm$^{-1}$ in wavelength, so we integrated those peaks ($A_{C=C}$).

The results of integrated areas from FTIR analyses for both regions of interest are summarized in Table 5. The ratio of integrated area for double bonds to integrated area for carbonyl groups is equivalent to the normalized amount of remaining double bonds in the sample $[A_{C=C}/A_{C=O}]_{sample}$. Overall, pendant double bond (PDB) conversion, expressed as percentage, is calculated using Equation (7), where $[A_{C=C}/A_{C=O}]_0$ corresponds to the integrated areas for the HEMA-EDGMA monomer mixture. Conversion results are summarized in Table 5.

$$\% \, PDB \, Conversion = \frac{\left[\frac{A_{C=C}}{A_{C=O}}\right]_0 - \left[\frac{A_{C=C}}{A_{C=O}}\right]_{sample}}{\left[\frac{A_{C=C}}{A_{C=O}}\right]_0} * 100 \tag{7}$$

(a)

(b)

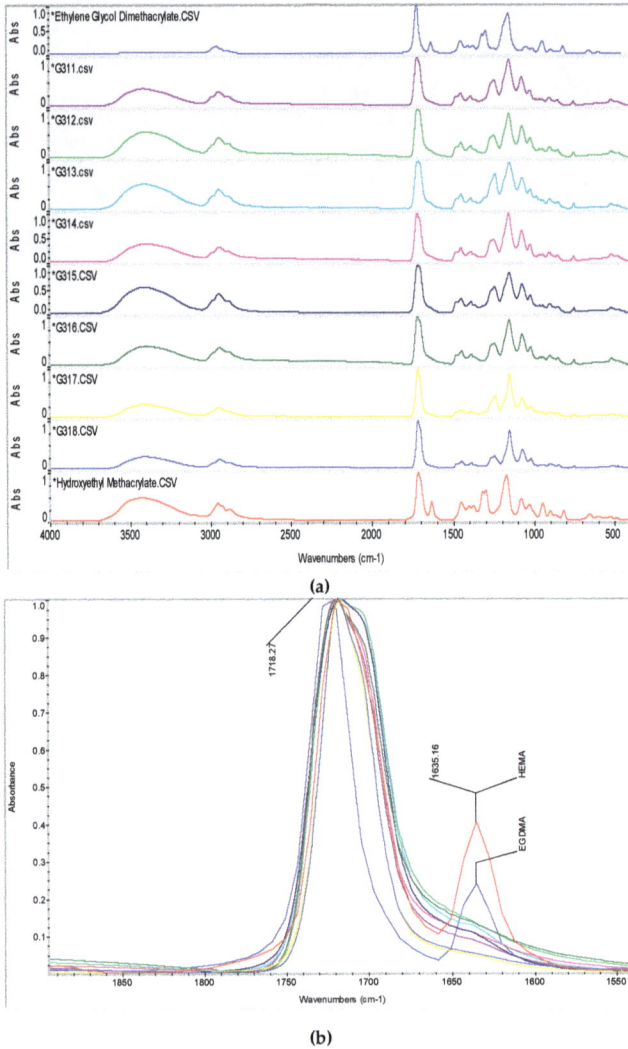

**Figure 7.** FTIR spectra (**a**) and enlarged view of the region of interest (**b**).

**Table 4.** Assignment of functional groups to bands of the FTIR spectra.

| Band (cm$^{-1}$) | Functional Group |
| --- | --- |
| 3,420 | –OH from HEMA |
| 2,990 | –CH from HEMA and EGDMA in polymer |
| 2,950 | –CH$_2$ from HEMA and EGDMA structures |
| 1,717 | –C=O from HEMA and EGDMA structures |
| 1,635 | –C=CH$_2$ remaining from HEMA and EGDMA monomers |
| 1,450 | –CH from HEMA and EGDMA in polymer |
| 1,320 to 1,300 | –C–O– ester from HEMA and EGDMA |
| 1,170 | –C–O– carboxylic derivate |
| 1,080 to 1,030 | –C–O– from the –OH of HEMA |
| 950 | remaining –C=CH$_2$ from HEMA and EGDMA monomers |
| 900 | remaining –C=CH$_2$ from HEMA and EGDMA monomers |

**Table 5.** FTIR quantifications results.

| Sample | Area Measured for Total C=O | Area Measured for Total C=C | Area Ratio between C=C/C=O | % Conversion Total C=C | Mean % Conversion Total C=C |
|--------|------|------|--------|--------|--------|
| G311 | 48.662 | 4.251 | 0.0873 | 64.88% | Hydrogels synthesized by FRP 56.98% |
| G312 | 51.952 | 6.425 | 0.1236 | 50.28% | |
| G313 | 53.434 | 5.890 | 0.1102 | 55.68% | |
| G314 | 47.482 | 4.999 | 0.1052 | 57.67% | |
| G315 | 53.029 | 4.944 | 0.0932 | 62.52% | Hydrogels synthesized by RAFT 66.10% |
| G316 | 49.214 | 6.066 | 0.1232 | 50.44% | |
| G317 | 39.497 | 2.235 | 0.0565 | 77.25% | |
| G318 | 39.128 | 2.511 | 0.0641 | 74.20% | |
| HEMA | 46.718 | 11.620 | 0.2487 | - | - |
| EGDMA | 35.502 | 5.721 | 0.1611 | - | |

As observed in Table 5, the PDB conversions obtained for samples synthesized by FRP are lower than samples synthesized by RAFT copolymerization. That means that the amount of unreacted pendant double bonds is higher in the samples synthesized by FRC, as was expected. Sample G316 using an RAFT agent was once more an outlier, since it falls within the range of values for samples synthesized by FRP.

A plot of percentage double bond conversion measured by FTIR versus *H* is shown in Figure 8. A linear relationship is observed with RAFT hydrogels showing higher double bond conversions (lower residual pendant double bonds) and values of *H* closer to one. It should be pointed out that the correlation shown in Figure 8 comes from two different characterization techniques: *SI*, which is a gravimetric technique, and FTIR spectroscopy. These results strengthen the concept of *H* as an indicator of polymer network heterogeneity.

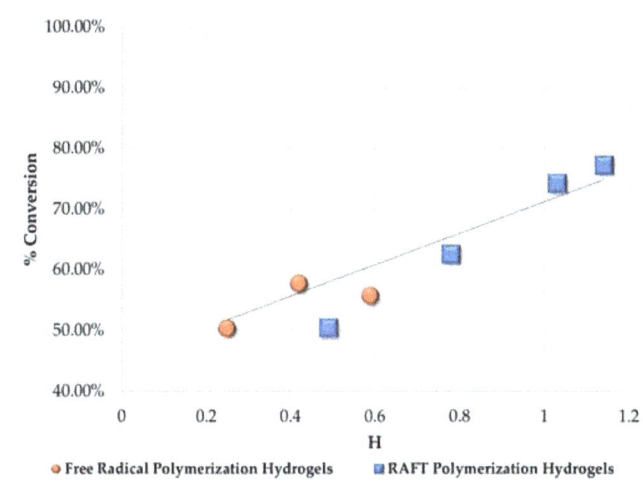

**Figure 8.** Relationship between conversion of pendant double bonds measured by FTIR and *H*.

*3.4. Measurement of Tg by DSC*

The differences in homogeneity (*H* measured from *SI*) for polymer networks synthesized by FRP and RAFT copolymerizations, caused by the reaction path followed (thermodynamic analysis) and pendant double bond conversions achieved (determined from FTIR), led to significant differences in properties and network performance, as evidenced from glass transition temperature (*Tg*) measurements by modulated differential scanning calorimetry (MDSC).

Correlations for *Tg* of polymers and polymer networks are available in the literature [24]. The Nielsen and DiBenedetto equations for calculation of *Tg* for crosslinked polymers are based on the concept of a polymer network having an infinite average molecular weight between crosslinks [24]. In a plot of *Tg* vs. $1/Mc$, $Tg°$ represents the value of *Tg* when $1/Mc \rightarrow 0$. Plots of *Tg* vs. $1/Mc$ for hydrogels synthesized by free FRP and RAFT copolymerizations are shown in Figure 9. Significant differences in *Tg* are observed with values that go from 105 to 133 °C. There is a 6 °C difference between the average values of *Tg* between samples synthesized by FRP and RAFT copolymerizations. The higher average value corresponds to samples synthesized by FRP. It is also observed from the error bars shown in Figure 9 that the standard deviation for samples synthesized by FRP is significantly higher than the corresponding value for polymer networks synthesized by RAFT copolymerization. *Tg* in the case of polymer networks is related to the number of crosslink points. Higher *Mc* values imply having less crosslink points and, therefore, lower *Tg* values. Both groups of samples (FRP and RAFT) follow linear trends with a difference of ~10 °C between the values of $Tg°$, the lower value corresponding to the samples synthesized by FRP. However, the line corresponding to samples synthesized by RAFT copolymerization has a smaller slope. The measured values of *Tg* were used to evaluate the adequacy of the Nielsen and DiBenedetto equations for calculation of *Tg* as a function of *Mc*, as expressed in Equations (8) and (9), respectively, where crosslink density ($x_{polymer\ network}$) is defined by Equation (10) [24]. It should be noted here that $x_{polymer\ network}$ in Equation (10) was defined as the "degree of crosslinking" in van Krevelen [24], but the definition given corresponds to crosslink density. We also changed the denominator of Equation (10), since it is more adequate to refer the calculation to the number of repeating units rather than the number of backbone atoms as defined in van Krevelen [24].

$$Tg_{polymer\ network} = Tg° + \left( \frac{39,000\,\text{K} \cdot \text{mol/g}}{Mc_{polymer\ network}} \right) \tag{8}$$

$$Tg_{polymer\ network} = Tg° + 1.2 \cdot Tg° \cdot \left( \frac{x_{polymer\ network}}{1 - x_{polymer\ network}} \right) \tag{9}$$

$$x_{polymer\ network} = \frac{\text{\# of crosslinks}}{\text{\# of repeating units in chain segment between crosslinks}} \tag{10}$$

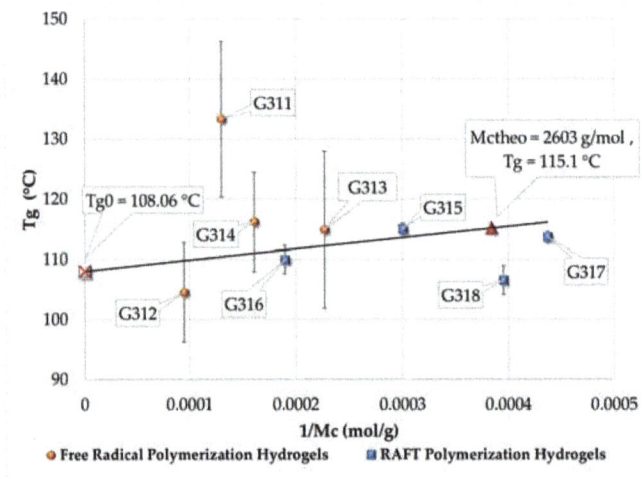

**Figure 9.** Relationship between *Tg* measured by modulated differential scanning calorimetry (MDSC) and $1/Mc$ for hydrogels synthesized by FRP and RAFT copolymerizations.

Figure 10 shows a plot of measured and calculated values of *Tg* versus *H*. The values of *Mc* used in the calculation of *Tg* were the ones obtained experimentally. It is observed that the calculated values of *Tg* using either equation agree well between themselves in the region of *H* < 0.5, but differ significantly when *H* > 0.5. It is also observed in Figure 10 that *Tg* values for polymer networks synthesized by FRP are adequately predicted with the Nielsen equation (10% mean error), whereas DiBenedetto's equation works better for RAFT synthesized polymer networks (3% average error). Except for sample G311, which can be considered as an outlier, the overall agreement between experimental and estimated values of *Tg* using typical correlations is fairly good.

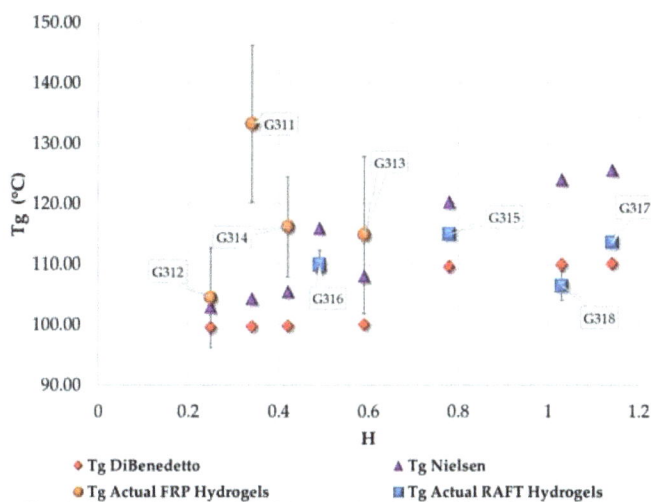

**Figure 10.** Relationship between *Tg* measured by MDSC and *H* and comparison with *Tg* estimates from the Nielsen and DiBenedetto equations.

### 3.5. Analysis of SEM Images

The morphologies of the materials synthesized by FRP and RAFT copolymerizations were observed and analyzed using SEM. SEM micrographs at 90 and 3500 magnifications for FRP and RAFT synthesized hydrogels are shown in Figures 11 and 12, respectively.

SEM images of hydrogels synthesized by FRP are shown in Figure 11. Two main morphologies are observed: solid blocks, as in samples G311 and G313 (Krytox used in the syntheses) and small spheres gathered in bunches, like raspberries, as in samples G314 and G312 (no Krytox used in the syntheses).

As observed in Figure 12, the same two morphologies (blocks and raspberries) were obtained for the particles corresponding to hydrogels synthesized by RAFT copolymerization. Sample G315, synthesized in the presence of Krytox as the dispersant, consisted of solid blocks. Samples G316 and G317 consisted of mostly raspberry spheres, with some blocks.

Particle size distribution data are shown in Figure 13 as mean particle size (MPS) vs. *H*. Clear linear trends, distinct for each population, are observed. Sample G312 for hydrogels synthesized by FRP is the only case not following the linear trend. Minimum and maximum particle sizes, as well as the standard deviation for all of the samples considered in this study are also shown in Figure 13 (see the numbers in boxes).

It is observed in Figure 13 that the trend line for samples synthesized by RAFT copolymerization of HEMA and EDGMA crosses the value of *H* = 1 (theoretical value) at MPS = 25 µm. It would be interesting to carry out additional experiments to corroborate this prediction. This linear trend between MPS and *H* indicates that polymer networks with low *Mc_exp* values (highly crosslinked polymer networks) will result in small (compact) particles. Likewise, polymer networks with high $Mc_{exp}$ values (slightly crosslinked or loose polymer networks) will result in large particles.

**Figure 11.** SEM images for hydrogel samples synthesized by FRP: (**a**) G311 (with Krytox); (**b**) G312 (without Krytox); (**c**) G313 (with Krytox); and (**d**) G314 (without Krytox).

(a)

(b)

(c)

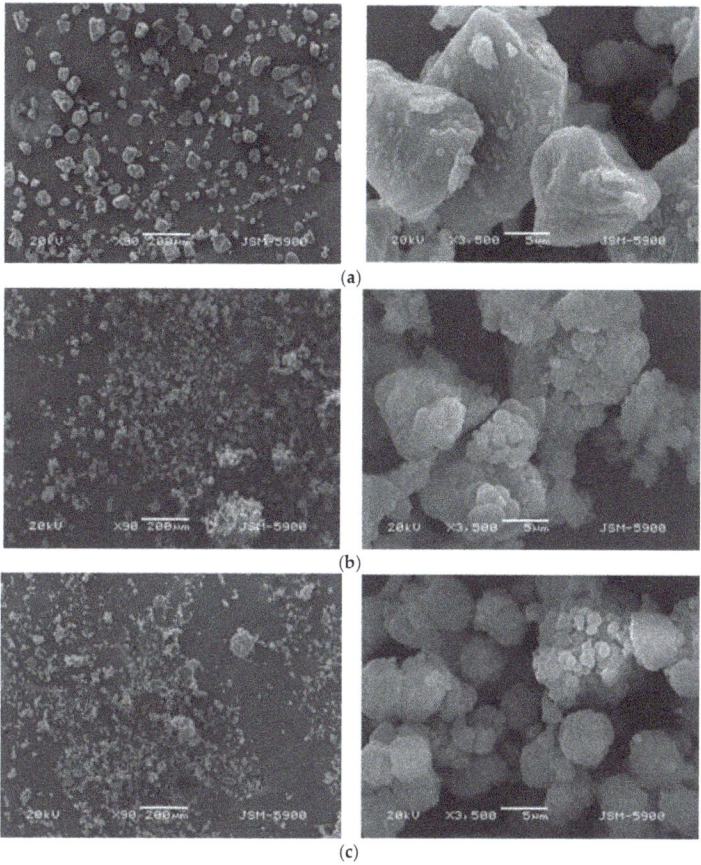

**Figure 12.** SEM images for hydrogel samples synthesized by RAFT copolymerization: (**a**) G315 (with Krytox); (**b**) G316 (without Krytox); and (**c**) G317 (with Krytox).

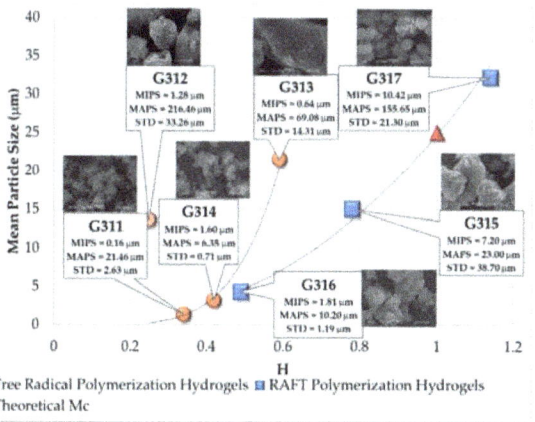

**Figure 13.** Correlation between MPS and *H* for hydrogels synthesized by FRP and RAFT copolymerizations of HEMA and EGDMA in scCO$_2$. Abbreviations: MIPS = minimum particle size; MAPS = maximum particle size; STD = standard deviation.

Regarding morphology, in general terms, all samples tended to form spherical particles, arranged as raspberries. However, in the samples where Krytox was used as the dispersing agent, solid blocks were observed. These differences are influenced by the thermodynamic conditions achieved by the reacting mixture in the presence of Krytox. Further experimentation is needed to clarify the effect of Krytox on the morphology of hydrogel particles synthesized by conventional (FRP) or RAFT copolymerization of HEMA and EDGMA in $scCO_2$.

*3.6. Antibiotic Loading, Adsorption and Release Studies*

The last study carried out with our hydrogels was the loading and release of ciprofloxacin, a fluoroquinolone antibiotic. The objective was to assess if the synthesized materials performed differently in an actual application, depending on the synthetic route. In a previous study using vitamin B12, we found that our FRP and RAFT synthesized materials perform differently [26].

Ciprofloxacin adsorption isotherm plots using hydrogels synthesized by FRP of HEMA and EGDMA in $scCO_2$ are shown in Figure 14. The corresponding profiles using hydrogels synthesized by RAFT copolymerization of HEMA and EGDMA in $scCO_2$ are shown in Figure 15. As observed in Figure 14, type I adsorption isotherms are obtained in the case of hydrogels synthesized by FRP. This means that weak interactions between polymer matrix and the antibiotic are present. On the other hand, as observed in Figure 15, adsorption isotherms types II, V and VI are obtained in the case of hydrogels synthesized by RAFT copolymerization. These types of isotherms are related to pores having restrictions to their cavities, which seems to be the case with blackberry morphologies. This may explain why RAFT synthesized hydrogels have low antibiotic loading levels, compared to hydrogels synthesized by FRP.

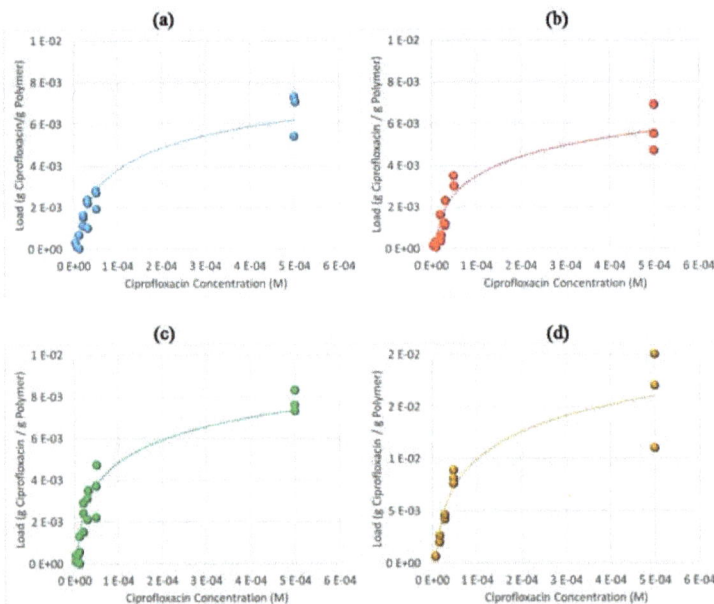

**Figure 14.** Adsorption isotherm charts for hydrogels synthetized by FRP. Samples: (**a**) G311; (**b**) G312; (**c**) G313; and (**d**) G314.

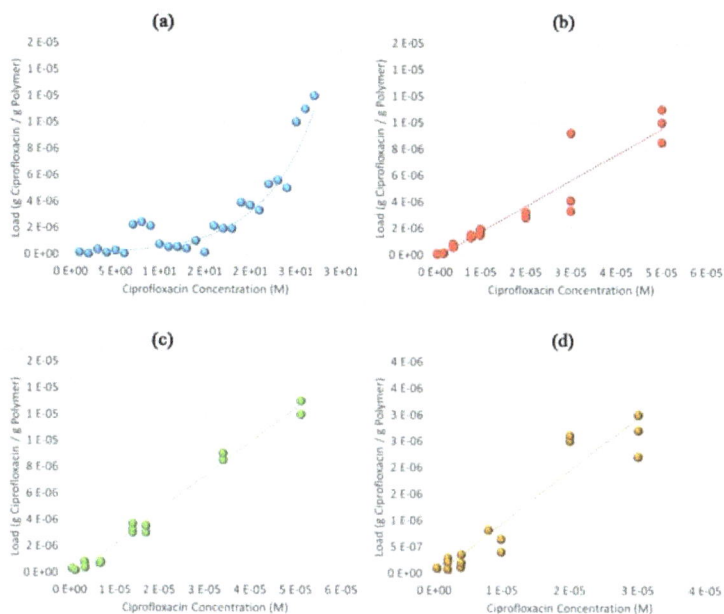

**Figure 15.** Adsorption isotherm charts for hydrogels synthetized by RAFT copolymerization. Samples: (**a**) G315; (**b**) G316; (**c**) G317; and (**d**) G318.

Ciprofloxacin release rates from hydrogels synthesized by FRP and RAFT copolymerizations of HEMA and EGDMA in scCO$_2$ are shown in Figures 16 and 17, respectively. It is observed that the release rate of ciprofloxacin from hydrogels synthesized by FRP is higher than in hydrogels synthesized by RAFT copolymerization. A maximum release rate of 25% at 200 min for sample G312 was obtained in the case of hydrogels synthesized by FRP. The maximum release rate obtained with RAFT hydrogels was 3% at 300 min (sample G315). However, broad dispersion of data is observed in the case of hydrogels synthesized by FRP (maximum release rates from 4 to 25%). If we extrapolate the results shown in Figures 16 and 17, it would take 25,000 min (420 h) to release ~80 to 90% of the total loaded ciprofloxacin from hydrogels synthesized by RAFT copolymerization. This can be considered a true controlled release system.

In order to get a better understanding of the relationship between ciprofloxacin release rate and polymer network homogeneity, a plot of wt% ciprofloxacin released at 400 min vs. *H* is shown in Figure 18. Except for samples with block only morphologies (samples G313 and G315, both synthesized using Krytox), a clear difference in ciprofloxacin release performance between the two types of polymer networks (FRP and RAFT synthesized) is observed (see the trend lines in Figure 18). It seems that samples having block morphologies release higher amounts of ciprofloxacin, compared to the analogous samples with raspberry morphologies. In order to understand this behavior, we have to keep in mind that block morphologies are assumed to be less restrictive for ciprofloxacin escape from the hydrogel matrix. Ciprofloxacin molecules must escape from the block, and no further organized hydrogel matrices are found. Holes between blocks are big enough to consider them as a bulky phase. Raspberry morphologies, on the other hand, represent a major escape challenge, since ciprofloxacin molecules must escape the volume within individual spheres, between neighbor spheres and between raspberry bunches. These spaces are too close among themselves to be considered as a bulky phase. This concept is illustrated in Figure 4 of Pérez-Salinas et al. [26].

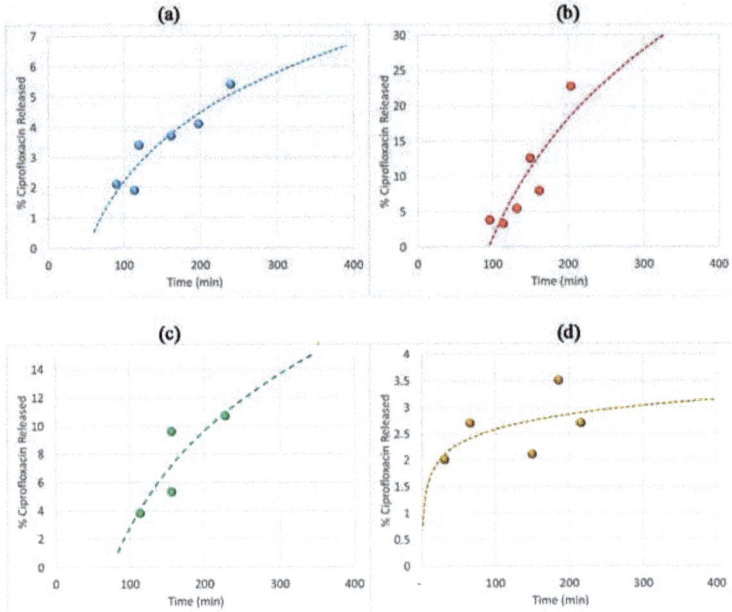

**Figure 16.** Ciprofloxacin release rate from polymer networks synthetized by FRP. Samples: (**a**) G311; (**b**) G312; (**c**) G313; and (**d**) G314.

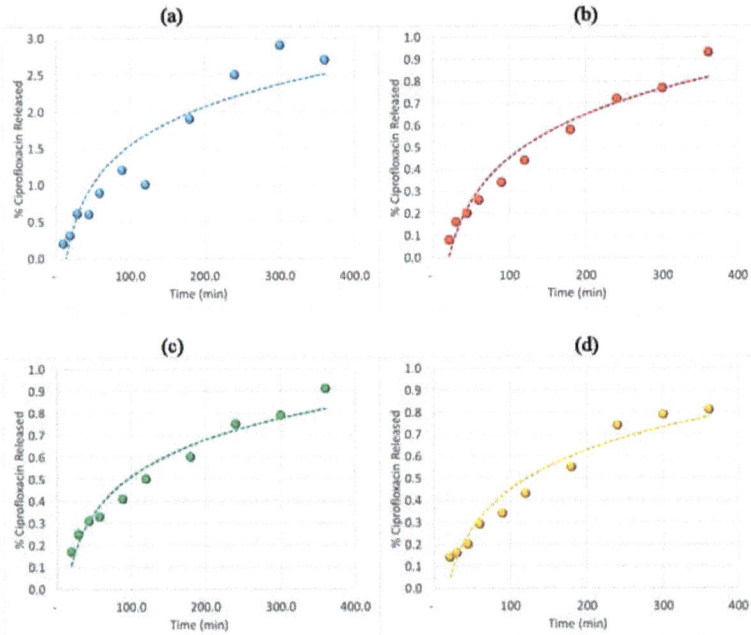

**Figure 17.** Ciprofloxacin release rate from polymer networks synthetized by RAFT copolymerization. Samples: (**a**) G315; (**b**) G316; (**c**) G317; and (**d**) G318.

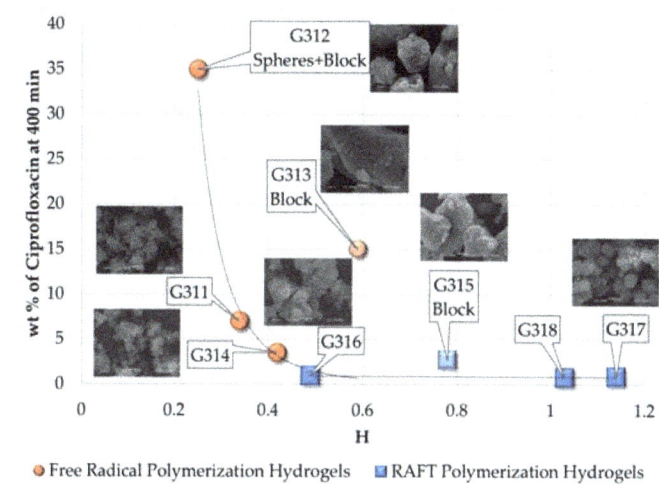

**Figure 18.** Relationship among the amount of ciprofloxacin released at 400 min from hydrogels, morphology and *H*.

## 4. Conclusions

Several considerations must be taken into account in the synthesis of hydrogels in supercritical or near supercritical carbon dioxide. One of such considerations is the thermodynamic behavior of the reacting mixture at different pressures and temperatures. It is important to identify the region of a phase diagram where one is working. It is also important to take into account the solubility in scCO$_2$ of the components present in the reacting mixture, since it can drastically change during the startup of the reaction and even during the reaction itself, depending on temperature and, most importantly, pressure. Characterization of polymer networks is a challenging task. As shown in this contribution, at first glance, most of the data collected from different characterization techniques seemed to be unrelated. However, the use of the polymer network homogeneity parameter (*H*), proposed in this contribution, allowed us to identify and even quantify the differences in the properties and performance between polymer networks synthesized by FRP and RAFT copolymerization of HEMA and EGDMA in scCO$_2$. Since the determination of *H* relies on *SI*, it is very important to get a reliable determination of this property.

One important aspect to take into account when explaining the differences in behavior and performance between polymer networks synthesized by FRP or RAFT copolymerization of vinyl/divinyl monomers in scCO$_2$ is polymerization time. For instance, as observed in Figure 6, an ~100% gel fraction is achieved at 24 h for polymer networks synthesized by FRP, whereas only ~80 to 90% (and large spread of data) has been achieved at the same time when the synthesis proceeds in the presence of an RAFT agent. This means that remaining monomer can last longer in RAFT polymerization, thus swelling the hydrogel in formation, promoting different morphologies and increasing the possibility of significantly changing the thermodynamic behavior of the reacting mixture if a small to moderate variation in pressure or temperature occurs inside the reactor during that time.

The use of Krytox remains unclear in our system since it seemed to promote the formation of hydrogels with solid block morphologies. As pointed out earlier, Krytox is highly soluble in scCO$_2$, but its solubility in HEMA and EGDMA is poor. Further experimentation is needed to fully elucidate the role of Krytox in the synthesis of hydrogels in scCO$_2$.

**Acknowledgments:** Financial support from the following sources is gratefully acknowledged: (a) Consejo Nacional de Ciencia y Tecnología (CONACYT, México), Project CB 239364 and the Ph.D. scholarship granted to P.P.-S.; (b) CONACYT and Essencefleur de México S.A. de C.V., Project FIT 235804; (c) CONACYT and Essencefleur de México S.A. de C.V., Project PEI 220695; (d) DGAPA-UNAM, Project PAPIIT IG100815; (e) Facultad de Química-UNAM, research funds granted to E.V.-L. (PAIP 5000-9078). No funds for covering the costs to publish in open access were received from the above mentioned sources. The help and support from Prof. Enrique Bazúa-Rueda on the use and interpretation of results of ASPEN PLUS for this application is gratefully acknowledged.

**Author Contributions:** Eduardo Vivaldo-Lima conceived of the idea of improving the performance of polymer networks by RAFT synthesis in scCO$_2$, put together and led the research team. Patricia Pérez-Salinas conceived of and carried out the thermodynamic analyses and measurements, including the concept of a polymer network homogeneity parameter; she also led the swelling, gel fraction determination and SEM analyses. Gabriel Jaramillo-Soto synthesized the polymer networks by FRP and RAFT copolymerization of HEMA and EGDMA in scCO$_2$. Humberto Vázquez-Torres, Ángel Licea-Claverie and Ma. Josefa Bernad-Bernad designed with Eduardo Vivaldo-Lima the original research strategy and types of characterization techniques that would be needed. Patricia Pérez-Salinas carried out the FTIR and DSC characterization analyses with guidance from Alberto Rosas-Aburto and Humberto Vázquez-Torres. Ma. Josefa Bernad-Bernad conceived and led the loading, adsorption and release studies. Ángel Licea-Claverie trained and guided Patricia Pérez-Salinas and Alberto Rosas-Aburto in the continuation and analysis of data for these studies. Patricia Pérez-Salinas put together the first complete manuscript draft, which was revised by Eduardo Vivaldo-Lima, Alberto Rosas-Aburto, Humberto Vázquez-Torres, Ángel Licea Claverie and Ma. Josefa Bernad-Bernad.

**Conflicts of Interest:** The authors declare no conflict of interest. The funding sponsors had no role in the design of the study; in the collection, analyses or interpretation of data; in the writing of the manuscript; nor in the decision to publish the results.

# References

1. Bhattacharya, A.; Rawlins, J.W.; Ray, P. *Polymer Grafting and Crosslinking*, 1st ed.; John Wiley & Sons: Hoboken, NJ, USA, 2009; pp. 1–100, 145–319.

2. Menard, K.P. *Dynamic Mechanical Analysis: A Practical Introduction*, 2nd ed.; CRC Press, Taylor & Francis Group: Boca Raton, FL, USA, 2008; pp. 95–122.

3. Seidel, A. *Characterization Analysis of Polymers*, 1st ed.; Wiley-Interscience: Hoboken, NJ, USA, 2008; pp. 649–675.

4. Wunderlich, B. *Thermal Analysis of Polymeric Materials*, 1st ed.; Springer: Berlin/Heidelberg, Germany, 2005; pp. 597–609.

5. Espinosa-Pérez, L.; Hernández-Ortiz, J.C.; López-Domínguez, P.; Jaramillo-Soto, G.; Vivaldo-Lima, E.; Péerez-Salinas, P.; Rosas-Aburto, A.; Licea-Claverie, Á.; Vázquez-Torres, H.; Bernad-Bernad, M.J. Modeling of the Production of Hydrogels from Hydroxyethyl Methacrylate and (Di) Ethylene Glycol Dimethacrylate in the Presence of RAFT Agents. *Macromol. React. Eng.* **2014**, *8*, 564–579. [CrossRef]

6. Scherf, R.; Müller, L.S.; Grosch, D.; Hübner, E.G.; Oppermann, W. Investigation on the homogeneity of PMMA gels synthesized via RAFT polymerization. *Polymer* **2015**, *58*, 36–42. [CrossRef]

7. Moad, G. RAFT (Reversible Addition-Fragmentation Chain Transfer) Crosslinking (Co)polymerization of Multi-Olefinic Monomers to Form Polymer Networks. *Polym. Int.* **2015**, *64*, 15–24. [CrossRef]

8. Matyjaszewski, K. Atom Transfer Radical Polymerization (ATRP): Current Status and Future Perspectives. *Macromolecules* **2016**, *45*, 4015–4039. [CrossRef]

9. Chatgilialoglu, C.; Ferreri, C.; Matyjaszewski, K. Radicals and Dormant Species in Biology and Polymer Chemistry. *ChemPlusChem* **2016**, *81*, 11–29. [CrossRef]

10. Jaramillo-Soto, G.; Vivaldo-Lima, E. RAFT Copolymerization of Styrene/Divinylbenzene in Supercritical Carbon Dioxide. *Aust. J. Chem.* **2012**, *65*, 1177–1185. [CrossRef]

11. García-Morán, P.R.; Jaramillo-Soto, G.; Albores-Velasco, M.E.; Vivaldo-Lima, E. An Experimental Study on the Free-Radical Copolymerization Kinetics with Crosslinking of Styrene and Divinylbenzene in Supercritical Carbon Dioxide. *Macromol. React. Eng.* **2009**, *3*, 58–70. [CrossRef]

12. Kiran, E. Supercritical Fluids and Polymers—The Year in Review—2014. *J. Supercrit. Fluids* **2016**, *110*, 126–153. [CrossRef]

13. Boyère, C.; Jérôme, C.; Debuigne, A. Input of Supercritical Carbon dioxide to Polymer Synthesis: An Overview. *Eur. Polym. J.* **2014**, *61*, 45–63. [CrossRef]

14. Kemmere, M.F. *Supercritical Carbon Dioxide in Polymer Reaction Engineering*, 1st ed.; Kemmere, M.F., Meyer, T., Eds.; Wiley-VCH Verlag GmbH & Co.: Weinheim, Germany, 2005; pp. 15–34.

15. Gupta, R.B.; Shim, J.J. *Solubility in Supercritical Carbon Dioxide*, 1st ed.; CRC Press, Taylor & Francis Group: Boca Raton, FL, USA, 2007; pp. 1–18.
16. Khansary, M.A.; Amiri, F.; Hosseini, A.; Sani, A.H.; Shahbeig, H. Representing Solute Solubility in Supercitical Carbon Dioxide: A Novel Empirical Model. *Chem. Eng. Res. Des.* **2015**, *93*, 355–365. [CrossRef]
17. Brown, R. *Physical Testing of Rubber*, 4th ed.; Springer: New York, NY, USA, 2006; pp. 316–326.
18. Flory, P.J.; Rehner, J. Statistical Mechanics of Crosslinked Polymer Networks. *J. Chem. Phys.* **1943**, *11*, 521–526. [CrossRef]
19. Flory, P.J. Statistical Mechanics of Swelling of Network Structures. *J. Chem. Phys.* **1950**, *18*, 108–111. [CrossRef]
20. Yang, Y. Fluorous Membrane-Based Separations and Reactions. Ph.D. Thesis, University of Pittsburg, Pittsburgh, PA, USA, 22 February 2011.
21. Bush, D. *Equation of State for Windows 95 Software*, version 1.0.14; Georgia Institute of Technology: Atlanta, GA, USA, 2005.
22. Ely, J.F.; Haynes, W.M.; Bain, B.C. Isochoric ($p$, $V_m$, $T$) measurements on $CO_2$ and on ($0.982CO_2 + 0.018N_2$) from 250 to 330 K at pressures to 35 MPa. *J. Chem. Thermodyn.* **1989**, *21*, 879–894. [CrossRef]
23. Mark, J.E. *Physical Properties of Polymers Handbook*, 2nd ed.; Springer: New York, NY, USA, 2007; pp. 289–303.
24. Van Krevelen, D.W.; Tenijenhuis, K. *Properties of Polymers: Their Correlation with Chemical Structure; Their Numerical Estimation and Prediction from Additive Group Contributions*, 4th ed.; Elsevier: Oxford UK, 2009; pp. 129–225.
25. Aspen Technology Inc. *Aspen Plus Software*, version 8.8; Aspen Technology, Inc.: Bedford, MA, USA, 2016.
26. Pérez-Salinas, P.; Rosas-Aburto, A.; Antonio-Hernández, C.H.; Jaramillo-Soto, G.; Vivaldo-Lima, E.; Licea-Claverie, Á.; Castro-Ceseña, A.B.; Vázquez-Torres, H. Controlled Release of Vitamin B-12 Using Hydogels Synthesized by Free Radical and RAFT Copolymerization in scCO₂. *Macromol. Symp.* **2016**, *360*, 69–77. [CrossRef]

| MDPI

*Review*

# Synthesis of Water-Soluble Group 4 Metallocene and Organotin Polyethers and Their Ability to Inhibit Cancer

**Charles E. Carraher Jr. [1,*], Michael R. Roner [2], Jessica Frank [1], Alica Moric-Johnson [2], Lindsey C. Miller [2], Kendra Black [1], Paul Slawek [1], Francesca Mosca [1], Jeffrey D. Einkauf [1] and Floyd Russell [1]**

1 Department of Chemistry and Biochemistry, Florida Atlantic University, Boca Raton, FL 33431, USA; jessfrank1207@gmail.com (J.F.); k.black93@live.com (K.B.); pslawek@fau.edu (P.S.); fmosca@fau.edu (F.M.); jde0703@gmail.com (J.D.E.); frussel3@fau.edu (F.R.)
2 Department of Biology, University of Texas Arlington, Arlington, TX 76010, USA; roner@uta.edu (M.R.R.); amoric@sbcglobal.net (A.M.-J.); lindseym@uta.edu (L.C.M.)
* Correspondence: carraher@fau.edu; Tel.: +1-561-297-2107

Received: 7 August 2017; Accepted: 26 August 2017; Published: 1 September 2017

**Abstract:** Water-soluble metallocene and organotin-containing polyethers were synthesized employing interfacial polycondensation. The reaction involved various chain lengths of poly(ethylene glycol), and produced water-soluble polymers in decent yield. Commercially available reactants were used to allow for easy scale up. The polymers exhibited a decent ability to inhibit a range of cancer cell lines, including two pancreatic cancer cell lines. This approach should allow the synthesis of a wide variety of other water-soluble polymers.

**Keywords:** Group 4 metallocenes; anticancer; poly(ethylene glycol); pancreatic cancer; breast cancer; prostate cancer; interfacial polycondensation; Group 4 metallocene polymers; organotin polyethers

## 1. Polymer Solubility

Polymer solubility is both difficult and low compared with the solubility of smaller molecules, and is limited by the amount of suitable solvent, extent of solubility, and rapidness of solubility. By comparison, the solubility of metal-containing polymers is even more difficult. Over 50 years ago, Bailar, one of the pioneers of inorganic chemistry, described some problems associated with the solubility of metal-containing polymers [1]. Briefly, these are as follows. First, little flexibility is imparted by the metal ion or within its immediate environment; thus, flexibility must arise from the organic moiety. Flexibility increases as the covalent nature of the metal–ligand bond increases. Second, metal ions only stabilize ligands in their immediate vicinity; thus, the chelates should be strong and close to the metal ions. Third, thermal, oxidative, and hydrolytic stability are not directly related; polymers must be designed specifically for the properties desired. Fourth, metal–ligand bonds have sufficient ionic character to permit them to rearrange more readily than typical "organic bonds". Fifth, polymer structure (such as square planar, octahedral, linear, and network) is dictated by the coordination number and stereochemistry of the metal ion or chelating agent. Lastly, employed solvents should not form strong complexes with the metal or chelating agent or they will be incorporated into the polymer structure and/or prevent a reaction from occurring. These problems are present with all metal-containing polymers.

Why is it important to have water solubility for polymers that are considered for use as drugs? Not all drugs must be water soluble to be useful. Many of the front-line drugs, such as ciprofloxacin, are not water soluble. Their solution testing is similar to what we employ in our non-water-soluble polymer testing. The drug is initially dissolved in dimethyl sulfoxide, DMSO, and then water is added and the appropriate testing carried out. However, given the choice, water solubility is advantageous for two important reasons. First, it allows for versatility in the administration of the drug, including water-associated approaches. Second, it eliminates effects that may occur because of the presence of any non-water-solvent systems.

## 2. Poly(Ethylene Glycol) as a Synthetic Template

This paper is part of a focus on water-soluble polymers, and as such the emphasis is to describe approaches that have allowed metal-containing polymers to be water soluble. Recent activities involve the use of poly(ethylene oxide), PEG, to achieve water solubility for metal-containing polymers.

Poly(ethylene glycol) (also called poly(ethylene oxide)), PEG, is considered to be nontoxic and is currently employed in a number of medical-related treatments, including pill coatings and in many commercial laxatives [2,3]. It is intentionally attached to many materials, including drugs, to assist in their water solubility. When attached to certain protein medications, they produce a drug with a longer activity and with a reduced toxicity [4]. For instance, its incorporation into polymers is widely employed to increase the solubility of polymers of biologically important materials [5–7].

PEGs are generally designated some average molecular weight. Since the ethylene glycol unit, $CH_2CH_2O$, has a molecular weight of 44 Da, PEG 200 has about 4.5 ethylene glycol units, that is 200/44. It should be noted that PEG is typically a mixture with the average being whatever the cited molecular weight is. Thus, PEG 200 is a combination of ethylene oxide repeat units, the most prevalent being four ethylene oxide units, molecular weight 176 Da, and five ethylene oxide units, molecular weight 220. The average is 396/2, or approximately 200, which is what this particular product is sold as.

## 3. Synthesis of Metal-Containing Polymers

We have synthesized a variety of metal-containing polymers for different purposes. Our most recent purposes for their synthesis include their ability to be doped, producing near conductors [8–11], and their ability to inhibit unwanted pathogens and infectious agents, including bacteria, viruses, cancers, molds, and yeasts. Some of these efforts have been reviewed for platinum [12–15], organotin [15–19], Group 5 [20,21], Group 15 [22–24], uranium [25], ruthenium [26], and vanadocene-containing [27] polymers.

The majority of polymers made by us are dimethyl sulfoxide, DMSO, soluble at concentrations sufficient to allow for molecular weight analysis via light scattering photometry and cancer cell line and virus inhibition analysis.

For ease of treatment, water-soluble drugs offer greater ease and avoid possible side effects due to the presence of DMSO. Thus, we sought to employ PEG to achieve this. As noted before, PEG is known to assist biologically important molecules to become water soluble. PEG is relatively inexpensive and is considered nontoxic. Further, there are a number of PEG chains available with differing end groups and chain lengths such that tailoring the PEG is possible. In the future, we plan to employ this variability to design systems that allow a focus on Lewis bases that offer good ability to inhibit unwanted pathogens and infectious agents. We are at the initial juncture of this study. Because most of the literature studies employ simple dihydroxyl-capped PEG chains, these are the PEG chains currently being employed [4–7].

The general focus in this review is the formation of water-soluble polymers from products that are not traditionally water soluble. Included in this review is the use of these polymers to inhibit cancer cell growth.

The various water-soluble polymers described in this review are all synthesized employing commercially available materials, allowing for a ready scale up to ton quantities. While the ability to scale up is present, it is normally not straightforward [1]. The polymers are structurally characterized employing typical tools used to define their structure and are familiar to chemists. The main exception is the use of matrix assisted-laser desorption/ionization mass spectrometry, MALDI MS. MALDI MS is widely employed in biochemistry because the most important biological polymers act as if they are water soluble and the MALDI MS system is used for their analysis. However, most polymers are not soluble in volatile liquids that allow an intimate mixture of the matrix material with the product. Thus, we developed a system that allowed non-soluble materials to be analyzed. In particular, organometallic structures, such as those employed by us, have poor stability when struck by the laser radiation employed in MALDI MS, with fragmentation typically occurring at the heteroatom sites in the polymer. This creates a fragmentation of the polymer chain, and it is these fragments that are employed to identify the basic repeat unit. This approach has been reviewed [28–30]. Even with this problem, we have been able to identify fragments to 30,000 Da, though we routinely look at ion fragment clusters from 500 to 5000 Da.

## 4. Organotin Polyether Synthesis

The synthesis of water-soluble polymers was initially achieved employing organotin polyethers derived from reaction of the organotin dihalides with PEG as described in Figure 1 [31–38]. The system is described in some detail because it is the same system employed in the metallocene polyether system. The polymers were synthesized employing the interfacial polycondensation system, where the organotin dihalide was dissolved in an organic liquid, generally heptane, and the PEG was dissolved in water with sodium hydroxide added to neutralize the HCl formed from the reaction. The two phases were rapidly stirred, at about 18,000 rpm, at room temperature. Product was formed in about 15 s as a precipitate. While the product was water soluble, it was only soluble to about 0.25 g for 50 mL of water, so it was not totally soluble at the higher concentration of the reaction system of about one gram for 50 mL of water. The white solid was recovered and washed with heptane and several mL of water. The effectiveness of this simple cleanup procedure gives product without bands associated with heptane in their nuclear magnetic resonance spectra, NMR, and infrared spectra, IR. While molecular weight increases as the PEG chain length increases, molecular weight based on the number of repeat units decreases. This decrease in chain length may involve a decreased ability for the growing chains to readily locate PEG ends to react with. Reactions involving the interfacial system are generally rapid and complicated so that the time required for the reaction to occur is relatively short [1]. Also, the percentage yield generally decreases as the PEG length increases, presumably because the longer PEG chains have a greater water solubility and are more easily lost in the aqueous portion of the reaction system. The reaction aqueous phase and wash was tested and contains both unreacted PEG and organotin polyether. Table 1 contains the results for two of the PEG systems.

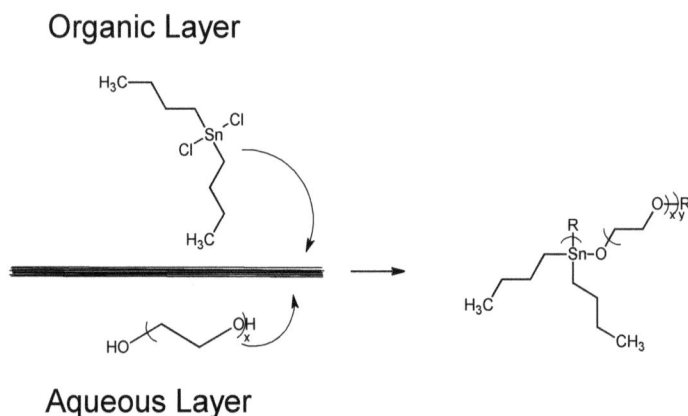

**Figure 1.** Description of the interfacial polycondensation system employed to synthesize water-soluble organotin polyethers, where x represents the PEG chain length and y represents the polymer's average degree of polymerization (PEG: Poly(ethylene glycol)).

**Table 1.** Sample results for the synthesis of water-soluble polyethers from reaction of dibutyltin dichloride with PEG. The specific reaction conditions are dibutyltin dichloride (3.00 mmole) in 30 mL heptane added to rapidly (18,000 rpm; no-load) stirred aqueous solutions (30 mL) containing diol (3.00 mmole) and sodium hydroxide (6.0 mmole) with stirring for 15 s (DMSO: dimethyl sulfoxide; DP: degree of polymerization).

| PEG | %-Yield | Mol. Wt. DMSO | DP DMSO | Mol. Wt. Water | DP Water |
|---|---|---|---|---|---|
| PEG-400 | 61 | $7.6 \times 10^4$ | 120 | $7.4 \times 10^4$ | 120 |
| PEG-10,000 | 6 | $2.5 \times 10^5$ | 24 | $2.2 \times 10^5$ | 22 |

Molecular weight was studied as a function of time for the product dissolved in DMSO and in water. In all cases, the molecular weight remained essentially unchanged for five weeks, with a molecular weight half-life greater than 60 weeks (that is the time for the molecular weight to become half of its original value). Thus, the organotin polyethers exhibit good solution stability in both DMSO and water, well within the time limits needed to accomplish the needed biological and physical measurements.

The polymers were characterized using typical analysis systems, including proton nuclear magnetic resonance spectrometry, various infrared spectroscopy systems, and MALDI mass spectrometry. IR shows the presence of bands characteristic of the formation of Sn-O consistent with the linkage between the PEG and organotin moiety having been formed. Further, IR shows the absence of the Sn-Cl band consistent with its elimination as the PEG reacts with the organotin halide.

Unlike most organic polymers, which do not possess elements that have atoms suitable to evaluate their presence through an investigation of their isotopic abundances, many of the products we work with have metals that do have such isotopes present. Tin has ten isotopes with seven isotopes present in amounts great than 5%, allowing the tin's presence to be determined using these isotopes. MALDI MS is used to make such determinations routinely. Because tin has isotopes, different ion fragments are created that have the same structural formula but vary by the particular tin isotope present. This creates what are referred to as spectral "fingerprints" characteristic of the natural abundance of these isotopes. Table 2 contains matches for ion fragment clusters that contain one and two ion atoms. The matches agree with what is expected, consistent with the presence of tin within these ion fragment clusters. Our focus is on ion fragments with masses of 500 Da and greater. Each of the ion fragment clusters above 500 (all ions are given in daltons, Da, or m/e = 1) are actually clusters of ions that are produced because of the presence of tin atom(s) within each cluster.

Our purpose for synthesizing the polymers was to evaluate their biological properties, and to do so in DMSO and compare the results with the polymers dissolved in only water. Thus, DMSO was initially employed to dissolve the compounds and then water was added to give the desired concentration of the tested compound for one set of studies. An analogous set of studies originally dissolved the tested compounds in water, to which additional water was added to give the desired compound concentrations. This second set is referred to as the "water only" systems. The general biological procedure is described elsewhere [18,19].

Cancer is the leading cause of death globally. The cell lines employed in the current study are given in Table 3. They represent a broad range of important cancers. Two human pancreatic cancer cell lines and two human breast cancer cell lines are included.

We recently found that anticancer activity is brought about by the intact polymer and not through polymer degradation [31]. This is consistent with studies that show that polymers are stable in DMSO with half-chain lives, the time for the chain length to halve, generally in excess of 30 weeks [16–19]. In other studies, we found that the polymer drugs are cytotoxic and cell death is by necrosis [16–19]. Most organometallic compounds associate with polar solvents, such as DMSO, and the biological results may be influenced by the presence of the DMSO [36–38]. For similar organometallic polymers, the results on the influence of DMSO on the tumor were found to be minimal [32–34]. For the current study, the results in water and DMSO are similar, consistent with this [33,34].

**Table 2.** Isotopic abundance matches for two ion fragment clusters containing a single tin atom, top part, and two ion fragment clusters with two tin atoms, bottom part. Bu represents a butyl moiety.

| Known for Sn | | $Bu_2Sn,2O$ | | $(OCH_2CH_2)_4OSnBu_2$ | | $(CH_2CH_2O)_5SnBr_2,Na$ | |
|---|---|---|---|---|---|---|---|
| m/e | Rel. Abu. | m/e | Rel. Abu. Found | m/e | Rel. Abu. Found | m/e | Rel. Abu. Found |
| 116 | 45 | 260 | 40 | 418 | 43 | 477 | 30 |
| 117 | 24 | 261 | 21 | 419 | 29 | 478 | 25 |
| 118 | 75 | 262 | 71 | 420 | 75 | 479 | 79 |
| 119 | 26 | 263 | 29 | 421 | 30 | 480 | 28 |
| 120 | 100 | 264 | 100 | 422 | 100 | 481 | 100 |
| 122 | 14 | 266 | 19 | 424 | 19 | 483 | 20 |
| 124 | 17 | 268 | 20 | 426 | 16 | 485 | 20 |

| Known for 2Sn | | $Bu_2SnO(CH_2CH_2O)_4SnBu_2$ | | $OBu_2SnO(CH_2CH_2O)_4Bu_2SnO$ | |
|---|---|---|---|---|---|
| m/e | Rel. Abu | m/e | Rel. Abu. Found | m/e | Rel. Abu. Found |
| 232 | 12 | 627 | 9 | 664 | 18 |
| 233 | 13 | 628 | 9 | 665 | 14 |
| 234 | 46 | 629 | 46 | 666 | 42 |
| 235 | 36 | 630 | 40 | 667 | 32 |
| 236 | 94 | 631 | 88 | 668 | 94 |
| 237 | 51 | 632 | 51 | 669 | 59 |
| 238 | 100 | 633 | 100 | 670 | 100 |
| 239 | 35 | 634 | 32 | 671 | 39 |
| 240 | 81 | 635 | 82 | 672 | 81 |
| 242 | 32 | 637 | 20 | 674 | 26 |
| 244 | 22 | 639 | 10 | 676 | 21 |

**Table 3.** Cell line characteristics and identification.

| Strain Number | NCI Designation | Species | Tumor Origin | Histological Type |
|---|---|---|---|---|
| 3465 | PC-3 | Human | Prostate | Carcinoma |
| 7233 | MDA MB-231 | Human | Pleural effusion breast | Adenocarcinoma |
| 1507 | HT-29 | Human | Recto-sigmoid colon | Adenocarcinoma |
| 7259 | MCF-7 | Human | Pleural effusion-breast | Adenocarcinoma |
| ATCC CCL-75 | WI-38 | Human | Normal embryonic lung | Fibroblast |
| ATCC CRL-1658 | NIH/3T3 | Mouse | Embryo-continuous cell line of highly contact-inhibited cells | Fibroblast |
| ATCC CCL-1 | L929 | Mouse | Transformed | Fibroblast |
| ATCC CRL-8303 | 143 | Human | Fibroblast | Bone |
| ATCC CCC-81 | Vero | Monkey | Transformed | Africa Green Monkey kidney epithelial |
| ATCC CCL-75.1 | WI-38 VA13 2RA | Human | Transformed | WI-38 Embryo lung fibroblast |
| ATCC CRL-8303 | 143 | Human | Fibroblast | Bone osteosarcoma |
| ATCC CCL-81 | Vero | Monkey | Transformed | African green monkey kidney epithelial |
| ATCC CCL-75.1 | WI-38 VA13 2RA | Human | Transformed | WI-38 embryo lung fibroblast |
| | AsPC-1 | Human | Pancreatic cells | Adenocarcinoma |
| | PANC-1 | Human | Epithelioid Pancreatic cells | Carcinoma |

Different measures are employed in the evaluation of compounds to control cancer growth. The two most widely employed measures are used in the present studies. The first measure employs as a measure effective concentration, EC. EC is the concentration dose needed to reduce the growth of a particular cell line. The concentration of a drug, antibody, or toxicant that induces a response halfway between the baseline and maximum after a specified exposure time is referred to as the 50% response concentration and is given the symbol $EC_{50}$. The second measure of the potential use of compounds in the treatment of cancer is the concentration of drug necessary to inhibit the standard cells compared to the concentration of drug necessary to inhibit the growth of the test cell line. The term chemotherapeutic index, CI, is employed for these measurements. The $CI_{50}$ is the ratio of the $EC_{50}$ for WI-38 cells divided by the $EC_{50}$ for the particular test cell.

A recent focus is pancreatic cancer. Pancreatic cancer afflicts close to 32,000 individuals each year in the United States, with almost all dead within a year. It is the fourth-most leading cause of cancer death. Treatment for pancreatic cancer is rarely successful, as this disease typically metastasizes prior to detection. The pancreatic cancer cell lines we tested were AsPC-1, which is a human adenocarcinoma pancreatic cell line that represents about 80% of the human pancreatic cancers, and PANC-1, which is an epithelioid carcinoma pancreatic cell line which represents about 10% of the human pancreatic cancers. Both are human cell lines, and the pair is widely employed in testing for the inhibition of pancreatic cancer. The dibutyltin PEG polyethers exhibit excellent ability to inhibit both pancreatic cancer cell lines. The values in water and DMSO are similar, so only an average is given for the two products. For the PEG 400 dibutyltin polyether and the AsPC-1 Cell line, the $EC_{50}$ value is 0.006 microgram/mL, in the nanogram/mL range, and the $CI_{50}$ is 47. Values of 2 and greater are considered significant. For the PEG 10,000 dibutyltin product, the $EC_{50}$ value is 0.06 microgram/mL and the $CI_{50}$ value is 16. For the PAN-1 pancreatic cancer for the PEG 400 dibutyltin polymer cell, the $EC_{50}$ value is 0.005 μg/mL and the $CI_{50}$ value is 56. Thus, the two values for the two different cancer cell lines are similar for the two different polymers, consistent with the possibility that the polymers will be active against other pancreatic cancers.

In summary, the organotin PEG polymers exhibit good inhibition towards pancreatic cancer cells when initially dissolved in DMSO or water. The aqueous solubility allows most forms of administration to be employed.

## 5. Group 4 Metallocene Polyethers

Group 4 metallocene-containing small molecules inhibit cancer cell growth. The mechanism by which they accomplish this differs from that found for cisplatin. This allows molecules from both groups to be employed as members of a "cocktail" intersecting cancer growth at differing junctures [39–44].

The first non-platinum metal to undergo clinical trial was titanocene dichloride [45]. While the activity of cisplatin involves interaction with deoxyribonucleic acid, DNA, the activity for titanocene dichloride is related to its ability to react with transferring [45,46].

A lack of solubility is a major problem limiting efforts to employ Group 4 metallocenes as anticancer agents [39–45]. The present effort involving PEG offers an avenue allowing metallocene-containing small and large molecules to be soluble.

We recently described the synthesis and initial cell line results for Group 4 metallocene polymers with structures analogous to the organotin polyethers [47–49] (Figure 2).

As in the case of the organotin products, these materials show good stability in solution for a month, with essentially no loss in molecular weight when dissolved in water or in DMSO.

Table 4 contains the yield, product molecular weights in water and DMSO, and average degree of polymerization, DP-number of repeat units for the product. The first column contains the particular metallocene and the molecular weight for the particular PEG followed by the number of PEG units in that particular PEG. The molecular weight and average chain length of the polymer decrease as the employed PEG length increases. This is the same as what occurred for the organotin products, and is again probably a consequence of the growing chains having increased difficulty in locating the end groups as the PEG chain length increases. The yield trends should be viewed with caution, because we have not yet perfected a good procedure for the recovery of the product. The infrared spectroscopy and proton NMR results are consistent with the proposed repeat unit. IR shows the formation of bands associated with the formation of the M-O linkage.

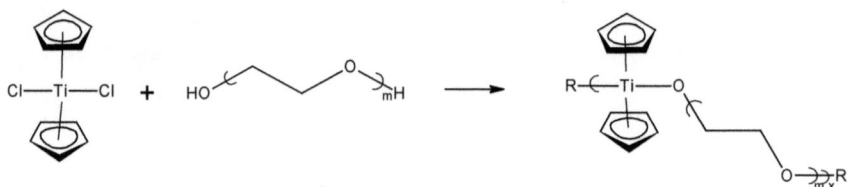

**Figure 2.** Synthesis of titanocene polyethers from reaction of titanocene dichloride and PEG, where R is a simple chain extension.

**Table 4.** Product yield and molecular weight as a function of PEG length and metallocene. Cp: cyclopetadiene group; MW: molecular weight.

| Sample | Percentage Yield | MW ($H_2O$) | MW (DMSO) | DP |
|---|---|---|---|---|
| $Cp_2Ti$ 200/4.5 | 68 | | $5.8 \times 10^6$ | 15,000 |
| $Cp_2Ti$ 400/9 | 42 | | $1.0 \times 10^6$ | 850 |
| $Cp_2Ti$ 1000/27 | 51 | $1.8 \times 10^5$ | $1.9 \times 10^5$ | 150 |
| $Cp_2Ti$ 1500/34 | 52 | $3.6 \times 10^4$ | $3.7 \times 10^4$ | 21 |
| $Cp_2Ti$ 2000/45 | 49 | $3.4 \times 10^4$ | $3.5 \times 10^4$ | 17 |
| $Cp_2Ti$ 3400/77 | 46 | $3.7 \times 10^4$ | $3.7 \times 10^4$ | 11 |
| $Cp_2Zr$ 200/4.5 | 24 | | $2.0 \times 10^6$ | 4700 |
| $Cp_2Zr$ 400/9 | 15 | | $3.8 \times 10^5$ | 600 |
| $Cp_2Zr$ 1000/27 | 30 | $9.0 \times 10^4$ | $9.1 \times 10^4$ | 75 |
| $Cp_2Zr$ 4600/100 | 15 | $6.1 \times 10^4$ | $6.2 \times 10^4$ | 13 |
| $Cp_2Zr$ 8000/180 | 34 | $4.1 \times 10^4$ | $4.2 \times 10^4$ | 5 |
| $Cp_2Hf$ 200/4.5 | 32 | | $6.3 \times 10^6$ | 12,000 |
| $Cp_2Hf$ 400/9 | 32 | $5.3 \times 10^5$ | $5.5 \times 10^5$ | 770 |
| $Cp_2Hf$ 1000/27 | 24 | $1.7 \times 10^5$ | $1.9 \times 10^5$ | 150 |
| $Cp_2Hf$ 4600/100 | | $9.3 \times 10^4$ | $9.5 \times 10^4$ | 14 |
| $Cp_2Hf$ 8000/180 | 38 | $6.6 \times 10^4$ | $6.8 \times 10^4$ | 8 |

As previously noted, because the metals have different natural abundance isotopes it is possible to do an isotope analysis using MALDI MS to compare the known isotopic relative abundances with the observed ion fragments. This analysis is routine and is consistent with the presence of the metals within the various ion fragments. Table 5 contains sample results for the isotopic abundances of ion fragments from the titanocene polymers. The results are in agreement with what is expected and consistent with the presence of the particular metal within the ion fragment clusters.

One of our primarily objectives is the synthesis of water-soluble drugs that inhibit the growth of pancreatic cancer. Table 6 contains results for the three water-soluble metallocene polymers. Cancer cell analysis requires a lower solubility for analysis compared with light scattering photometry. Thus, the reader may note that some cell line results are present in Table 6, but molecular weight values are not given in Table 4 because of the difference in solubility required to obtain molecular weight compared to cell line data.

**Table 5.** Isotopic abundance matches for two ion fragment clusters containing a single titanium atom (top), and two ion fragment clusters containing two titanium atoms per cluster (bottom).

| Known for Ti | | $O(CH_2CH_2O)_4Cp_2Ti(OCH_2CH_2)_4$ | | $O(CH_2CH_2O)_4Cp_2Ti(OCH_2CH_2)_4$ | |
|---|---|---|---|---|---|
| m/e | Rel. Abu. | m/e | Rel. Abu. Found | m/e | Rel. Abu. Found |
| 46 | 11 | 583 | 10 | 539 | 11 |
| 47 | 11 | 584 | 11 | 540 | 11 |
| 48 | 100 | 585 | 100 | 541 | 100 |
| 49 | 8 | 586 | 8 | 542 | 7 |
| 50 | 7 | 587 | 6 | 543 | 7 |
| **Known for 2 Ti** | | $(CH_2CH_2O)_3Cp_2Ti(OCH_2CH_2)_4Cp_2TiO$ | | $(CH_2CH_2O)_3Cp_2TiO(CH_2CH_2O)_4Cp_2Ti(OCH_2CH_2)_4$ | |
| m/e | Rel. Abu. | m/e | Rel. Abu. Found | m/e | Rel. Abu. Found |
| 94 | 22 | 688 | 22 | 875 | 23 |
| 95 | 21 | 689 | 20 | 876 | 21 |
| 96 | 100 | 690 | 100 | 877 | 100 |
| 97 | 16 | 691 | 18 | 878 | 20 |
| 98 | 15 | 692 | 16 | 879 | 18 |

Several observations are apparent. First, none of the metallocene dichlorides and PEGs offer inhibition to the limits tested. Second, the zirconocene polyethers offer the best inhibition of the pancreatic cancer cells based on having the lowest $EC_{50}$ values compared to the hafnocene and titanocene polyethers. Unlike the titanocene and hafnocene polyethers, there is a difference between the ability to inhibit the two cancer cell lines, with a greater, generally a ten-fold lower, ability to inhibit the PANC-1 cell line for the lower PEG chain lengths, but there are similar $EC_{50}$ values for the longer PEG products. The titanocene and hafnocene products exhibit inhibition rates that are similar to each another. For the present polymers, it is the zirconocene products that should have undergone clinical testing. Third, since neither the PEG or the metallocene dichloride exhibit an ability to inhibit pancreatic cell growth, it is the combination of PEG and metallocene that accounts for the ability to inhibit the pancreatic cancers.

In conclusion, water-soluble metallocene polymers have been synthesized. They exhibit decent inhibition of a battery of cancer cells according to the results for the pancreatic cancer cell lines presented.

**Table 6.** EC$_{50}$ results (micrograms/mL) for the water-soluble metallocene polyethers as a function of metallocene, metallocene/PEG polymer, and pancreatic cancer cell line. Reaction conditions were as follows: an aqueous solution containing PEG (0.00100 mole) and sodium hydroxide (0.00200) dissolved in 50 mL of water is added to a rapidly stirred (about 18,000 rpm) chloroform (50 mL) solution containing the metallocene dichloride (0.00100 mole). Stirring continued for 15 s at room temperature (about 28 °C)(EC: effective concentration).

| Titanocene Results | | | | Zirconocene Results | | | | Hafnocene Results | | | |
|---|---|---|---|---|---|---|---|---|---|---|---|
| Compound | WI-38 | PANC-1 | AsPC-1 | Compound | WI-38 | PANC-1 | AsPC-1 | Compound | WI-38 | PANC-1 | AsPC-1 |
| Cp$_2$TiCl$_2$ | >32 | >32 | >32 | Cp$_2$ZrCl$_2$ | >32 | >32 | >32 | Cp$_2$Cp$_2$HfCl$_2$ | >32 | >32 | >32 |
| Cp$_2$Ti/PEG 200 | 1.2 | 0.77 | 0.70 | Cp$_2$Zr/PEG 200 | 0.0019 | 0.17 | 0.011 | Cp$_2$Hf/PEG 200 | 0.45 | 0.52 | 0.59 |
| Cp$_2$Ti/PEG 400 | 0.95 | 0.61 | 0.64 | Cp$_2$Zr/PEG 400 | 0.0022 | 0.091 | 0.12 | Cp$_2$Hf/PEG 400 | 0.45 | 0.52 | 0.59 |
| Cp$_2$Ti/PEG 800 | 1.3 | 0.59 | 0.59 | | | | | | | | |
| Cp$_2$Ti/PEG 1000 | 1.0 | 0.53 | 0.52 | Cp$_2$Zr/PEG 1000 | 0.0023 | 0.091 | 0.17 | Cp$_2$Hf/PEG 1000 | 0.45 | 0.52 | 0.59 |
| Cp$_2$Ti/PEG 1500 | 0.96 | 0.51 | 0.52 | | | | | | | | |
| Cp$_2$Ti/PEG 2000 | 1.1 | 0.60 | 0.57 | | | | | | | | |
| Cp$_2$Ti/PEG 3400 | 1.1 | 0.56 | 0.51 | | | | | | | | |
| | | | | Cp$_2$Zr/4600 | 0.0025 | 0.11 | 0.13 | Cp$_2$Hf/PEG 4600 | 0.45 | 0.52 | 0.59 |
| CpTi/PEG 8000 | 1.1 | 0.62 | 0.53 | Cp$_2$Zr/8000 | 0.0029 | 0.13 | 0.18 | Cp$_2$Hf/PEG 8000 | 0.45 | 0.52 | 0.59 |

## 6. Future and Summary

PEG is widely employed industrially [50]. It is used in a number of laxatives, including MirLax, GlycoLax, and Movicol. It is also employed as a coating for many pills to allow the medication to pass through the harsh acidic stomach environment area unharmed. PEGs are employed as the soft segment in polyurethanes, and in other hard-soft copolymers as the soft portion. Some producers use it in toothpaste products as a dispersant, and it is employed as a lubricant in eye drops. Dr. Pepper adds PEG as an antifoaming agent. PEG has been used to help preserve wood-replacing water, giving the wood increased dimensional stability. Certain gene therapy vectors, such as viruses, can be coated with PEG to protect them from inactivation by immune systems and to de-target them from organs where they could build up and cause a toxic effect. Recently, they have been part of body armor combinations and to impart water solubility to certain electrically conductive polymers.

The use of PEG to enhance the aqueous solubility of materials includes macromolecules. It is well-established, but many venues remain. There remain a number of condensation polymers where aqueous solubility might be useful. These include commercial polymers, such as polyesters and nylons. There are also many speciality materials where water solubility would enhance their areas of potential use.

In our case, we have a number of polymers that exhibit a good ability to inhibit a range of cancer cell lines, viruses, and bacterial agents.

One product we plan to work towards making water soluble is a group of recently synthesized polyesters derived from the reaction of Group 4 metallocene dichlorides with the salts of camphoric acid (Figure 3) [51]. The zirconocene and hafnocene products exhibit good inhibition of a battery of cancer cell lines, including the pancreatic cancer cell lines described in Table 3, to the nanogram/mL range with $CI_{50}$ values in the thousands. Our approach is to initially evaluate PEGs of varying length as co-monomers with camphoric acid to form products and then test the products for water solubility and ability to inhibit cancer cell lines (Figure 4).

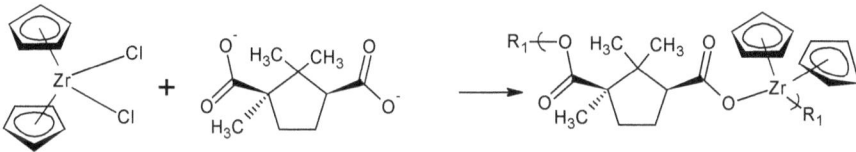

**Figure 3.** Reaction scheme for the reaction between D-camphoric acid and zirconocene dichloride, where $R_1$ represents a simple chain extension.

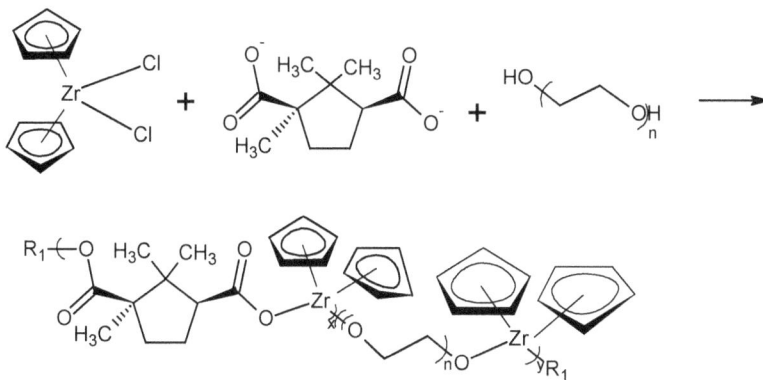

**Figure 4.** Reaction scheme for inclusion of PEG into the $Cp_2Zr$/camphoric acid product.

Some of these products exhibit good thermal stability to 1500 °C so that a simple CH analysis is not satisfactory for structural analysis [30]. Since neither of the Lewis bases have an element that is unique to it, we have begun using X-ray fluorescence spectroscopy to determine the percentage of metal in our products. From this information, the amount of both Lewis bases in the product can be determined.

The synthesis and study of water-soluble metal-containing polymers has begun. The results are promising, and allow for the synthesis of water-soluble products that exhibit good inhibition of a host of cancer cell lines. This is only the beginning of a long journey.

**Conflicts of Interest:** The authors declare no conflict of interest.

## References

1. Carraher, C. *Introduction to Polymer Chemistry*, 4th ed.; Taylor and Francis/CRC Press: New York, NY, USA, 2017.
2. DiPalma, J.; Cleveland, M.; Mark, V.B.; McGowan, J.; Herrera, J. A Randomized, Multicenter Comparison of Polyethylene Glycol Laxative and Tegaserod in Treatment of Patients with Chronic Constipation. *Am. J. Gastroenterol.* **2007**, *9*, 1964–1971. [CrossRef] [PubMed]
3. Sheftel, V.O. *Indirect Food Additives and Polymers: Migration and Toxicology*; CRC: Boca Raton, FL, USA, 2000.
4. Delgado, C.; Francis, G.E.; Fisher, D. Solvent-sensitive nanospheres prepared by self-organization of polymerizing hydrophilic graft chain copolymers. *Drug Carr. Syst.* **1992**, *9*, 249–304.
5. Gerasimov, A.; Ziganshin, M.; Gorbatchuk, V.; Usmanova, L. Increasng the solubility of dipyridamole using polyethylene glycols. *Int. J. Pharm. Sci.* **2014**, *6*, 244–247.
6. Lee, M.; Kim, S.W. Polyethylene glycol-conjugated copolymers for plasmid DNA delivery. *Pharm. Res.* **2005**, *22*, 1–10. [CrossRef] [PubMed]
7. Ansari, M. Investigations of polyethylene glycol mediated ternary molecular inclusion complexes silmarin with beta cyclodextrin. *J. Appl. Pharm. Sci.* **2015**, *5*, 26–31. [CrossRef]
8. Carraher, C.; Battin, A.; Roner, M.R. Effect of bulk doping on the electrical conductivity of selected metallocene polyamines. *J. Inorg. Organomet. Polym. Mater.* **2013**, *23*, 61–73. [CrossRef]
9. Carraher, C.; Battin, A.; Roner, M.R. Effect of Electrical Conductivity Through the Bulk Doping of the Product of Titanocene Dichloride and 2-Nitro-1,4-phenylenediamine. *J. Funct. Biomater.* **2011**, *2*, 18–30. [CrossRef] [PubMed]
10. Battin, A.; Carraher, C.; Roner, M.R. Effect of bulk doping on the electrical conductivity of selected metallocene polyamines. *J. Inorg. Organomet. Polym. Mater.* **2012**, *22*, 1–13.
11. Battin, A.; Carraher, C. Effect of doping by exposure of iodine on the electrical conductivity of the polymer from titanocene dichloride and 2-nitro-p-phenylenediamine. *J. Polym. Mater.* **2008**, *25*, 23–33.
12. Siegmann-Louda, D.; Carraher, C. Polymeric Platinum-Containing Drugs in the Treatment of Cancer. In *Biomedical Applications*; John Wiley & Sons: Hoboken, NJ, USA, 2004.
13. Roner, M.R.; Carraher, C. Cisplatinum Derivatives as Antivirial Agents. In *Inorganic and Organometallic Macromolecules*; Springer: New York, NY, USA, 2008.
14. Carraher, C.; Francis, A. Water-Soluble Cisplatin-Like Chelation Drugs from Chitosan. *J. Polym. Mater.* **2011**, *28*, 189–203.
15. Roner, M.R.; Carraher, C.; Shahi, K.; Barot, G. Antiviral activity of metal-containing polymers organotin and cisplatin-like polymers. *Materials* **2011**, *4*, 991–1012. [CrossRef]
16. Carraher, C.; Siegman-Louda, D. Organotin Macromolecules as Anticancer Drugs. In *Macromolecules Containing Metal and Metal-Like Elements*; John Wiley & Sons: Hoboken, NJ, USA, 2004.
17. Carraher, C. *Organotin Polymers in Macromolecules Containing Metal and Metal-Like Elements*; John Wiley & Sons: Hoboken, NJ, USA, 2004.
18. Roner, M.R.; Carraher, C. Organotin Polyethers as Biomaterials. *Materials* **2009**, *2*, 1558–1598.
19. Carraher, C.; Roner, M.R. Organotin polymers as anticancer and antiviral agents. *J. Organomet. Chem.* **2014**, *751*, 67–82. [CrossRef]
20. Carraher, C. Zirconocene and hafnocene-containing macromolecules. In *Macromolecules Containing Metal and Metal-Like Elements*; John Wiley & Sons: Hoboken, NJ, USA, 2006.

21. Carraher, C. Condensation metallocene polymers. *J. Inorg. Organomet. Polym. Mater.* **2005**, *15*, 121–145. [CrossRef]
22. Carraher, C. Organoantimony-containing polymers. *J. Polym. Mater.* **2008**, *25*, 35–50.
23. Carraher, C. Antimony-containing polymers. In *Inorganic and Organometallic Macromolecules*; Springer: New York, NY, USA, 2008.
24. Carraher, C.; Roner, M.R.; Thbibodeau, R.; Moric-Johnson, A. Synthesis, structural characterization, and preliminary cancer cell study results for poly(amine esters) derived form triphenyl-group VA organometallics and norfloxacin. *Inorg. Chem. Acta* **2014**, *423*, 123–131. [CrossRef]
25. Carraher, C. Uranium-containing polymers. In *Macromolecules Containing Metal and Metal-Like Elements*; John Wiley & Sons: Hoboken, NJ, USA, 2005.
26. Carraher, C.; Murphy, A.T. Ruthenium-containing polymers for solar energy conversion. In *Macromolecules Containing Metal and Metal-Like Elements*; John Wiley & Sons: Hoboken, NJ, USA, 2005.
27. Sabir, T.; Carraher, C. Vanadocene-containing polymers. In *Inorganic and Organometallic Macromolecules*; Springer: New York, NY, USA, 2008.
28. Carraher, C.; Sabir, T.S.; Carraher, C.L. *Inorganic and Organometallic Macromolecules*; Springer: New York, NY, USA, 2008.
29. Carraher, C.; Sabir, T.; Carraher, C.L. Fragmentation matrix assisted-laser desorption/ionization mass spectrometry-basics. *J. Polym. Mater.* **2006**, *23*, 143–151.
30. Carraher, C.; Roner, M.R.; Carraher, C.L.; Crichton, R.; Black, K. Use of mass spectrometry in the characterization of polymers emphasizing metal-containing condensation polymers. *J. Macromol. Sci.* **2015**, *52*, 867–886. [CrossRef]
31. Carraher, C.; Barot, G.; Vetter, S.W.; Nayak, G.; Roner, M.R. Degradation of the organotin polyether derived from dibutyltin dichloride and hydroxyl-capped poly(ethylene glycol) in trypsin and evaluation of trypsin activity employing light scattering photometry and gel electrophoresis. *J. Chin. Adv. Mater. Soc.* **2013**, *1*, 1–6. [CrossRef]
32. Carraher, C.; Barot, G.; Shahi, K.; Roner, M.R. Synthesis, structural characterization, and ability to inhibit cancer cell growth of a series of organotin poly(ethylene glycols). *J. Inorg. Organomet. Polym. Mater.* **2007**, *17*, 595–603.
33. Carraher, C.; Roner, M.R.; Barot, G.; Shahi, K. Comparative anticancer activity of water-soluble organotin poly(ethylene glycol) polyethers. *J. Polym. Mater.* **2014**, *31*, 123–133.
34. Carraher, C.; Barot, G.; Shahi, K.; Roner, M.R. Influence of DMSO on the inhibition of various cancer cells by water soluble organotin polyethers. *J. Chin. Adv. Mater. Soc.* **2013**, *1*, 294–304. [CrossRef]
35. Carraher, C.; Roner, M.R.; Moric-Johnson, A.; Miller, L.; Barot, G.; Sookdeo, N. Ability of Simple Organotin Polyethers to Inhibit Pancreatic Cancer. *J. Macromol. Sci.* **2015**, *53*, 63–67. [CrossRef]
36. Ohtaki, H. Structural studies on solvationi and complexation of metal ions in nonaqueous solutions. *Pure Appl. Chem.* **1987**, *59*, 1143–1150. [CrossRef]
37. Gjevig Jenson, K.; Onfelt, A.; Wallin, M.; Lidumas, V.; Andersen, O. Effects of organotin compounds on mitosis, spindel structure, toxicity, and in vitro microtubule assemble. *Mutagenessis* **1991**, *6*, 409–416. [CrossRef]
38. Corriu, R.; Dabosi, G.; Martineau, M. The nature of the interactioni of nucleophiles such as HMPT, DMSO, DMF and Ph3PO with triorganohalo-silanes, -germanes, and -stannanes and organophosphorus compounds. Mechanism of nucleophile induced racmization and substitution at metal. *J. Organomet. Chem.* **1980**, *186*, 25–37. [CrossRef]
39. Benitez, J.; Guggeri, L.; Tomaz, I.; Pessoa, J.C.; Moreno, V.; Lorenzo, J.; Aviles, F.X.; Garat, B.; Gambino, D. A novel vanadyl complex with a polypyridyl DNA intercalator as ligand: A potential anti-protozoa and anti-tumor agent. *J. Inorg. Biochem.* **2009**, *103*, 1386–1394. [CrossRef] [PubMed]
40. Strohfeldt, K.; Tacke, M. Bioorganometallic fulvene-derived titanocene anti-cancer drugs. *Chem. Soc. Rev.* **2008**, *37*, 1174–1187. [CrossRef] [PubMed]
41. Beckhove, P.; Oberschmidt, O.; Hanauske, A.; Pampillon, C.; Schirrmacher, V.; Sweeney, N.J.; Strohfeldt, K.; Tacke, M. Antitumor activity of titanocene Y against freshly explanted human breast tumor cells and in xenografted MCF-7 tumors in mice. *Anticancer Drugs* **2007**, *18*, 311–315. [CrossRef] [PubMed]
42. Harding, M.M.; Mokdsi, G. Antitumour metallocenes: Structure-activity studies and interactions with biomolecules. *Curr. Med. Chem.* **2000**, *7*, 1289–1303. [CrossRef] [PubMed]

43. Olszewski, U.; Claffey, J.; Hogan, M.; Tacke, M.; Zeillinger, R.; Bednarski, P.; Hamilton, G. Anticancer activity and mode of action of titanocene C. *Investig. New Drugs* **2011**, *29*, 607–614. [CrossRef] [PubMed]
44. Olszewski, U.; Hamilton, G. Mechanisms of cytotoxicity of anticancer titanocenes. *Anticancer Agents Med. Chem.* **2010**, *10*, 302–311. [CrossRef] [PubMed]
45. Roat-Malone, R.M. *Bioinorganic Chemistry*, 2nd ed.; Wiley: New York, NY, USA, 2007; pp. 19–20.
46. Waern, J.B.; Harris, H.H.; Lai, B.; Cai, Z.; Harding, M.M.; Dillon, C.T. Intracellular mapping of the distribution of metals derived from the antitumor metallocenes. *J. Biol. Inorg. Chem.* **2005**, *10*, 443–452. [CrossRef] [PubMed]
47. Carraher, C.; Roner, M.R.; Reckleben, L.; Black, K.; Frank, J.; Crichton, R.; Russell, F.; Moric-Johnson, A.; Miller, L. Synthesis, structural characterization and preliminary cancer cell line results for polymers derived from reaction of titanocene dichloride and various poly(ethylene glycols). *J. Macromol. Sci.* **2016**, *53*, 394–402. [CrossRef]
48. Carraher, C.; Roner, M.R.; Black, K.; Frank, J.; Moric-Johnson, A.; Miller, L.; Russell, F. Synthesis, structural characterization and initial anticancer activity of water soluble polyethers from hafnocene dichloride and poly(ethylene Glycols). *J. Chin. Adv. Mater. Soc.* in press. [CrossRef]
49. Carraher, C.; Roner, M.R.; Frank, J.; Black, K.; Moric-Johnson, A.; Miller, L.; Russell, F. Synthesis and initial anticancer activity of water and dimethyl sulfoxide soluble polyethers from zironocene dichloride and poly(ethylene Glycols). *J. Macromol. Sci. A* in press.
50. Carraher, C. *Introduction of Polymer Chemistry*; CRC Press/Taylor and Francis: New York, NY, USA, 2017.
51. Carraher, C.; Roner, M.R.; Campbell, A.; Moric-Johnson, A.; Miller, L.; Slawek, P.; Mosca, F. Group IVB metallocene polyesters containing camphoric acid and preliminary cancer cell activity. *Int. J. Polym. Mater. Polym. Biomater.* in press.

![processes logo] *processes*

MDPI

*Article*

# Radical Copolymerization Kinetics of Bio-Renewable Butyrolactone Monomer in Aqueous Solution

**Sharmaine B. Luk and Robin A. Hutchinson * ◉**

Department of Chemical Engineering, Queen's University, 19 Division St, Kingston, ON K7L3N6, Canada; sharmaine.luk@queensu.ca
* Correspondence: robin.hutchinson@queensu.ca; Tel.: +1-613-533-3097

Received: 6 September 2017; Accepted: 27 September 2017; Published: 1 October 2017

**Abstract:** The radical copolymerization kinetics of acrylamide (AM) and the water-soluble monomer sodium 4-hydroxy-4-methyl-2-methylene butanoate (SHMeMB), formed by saponification of the bio-sourced monomer $\gamma$-methyl-$\alpha$-methylene-$\gamma$-butyrolactone (MeMBL), are investigated to explain the previously reported slow rates of reaction during synthesis of superabsorbent hydrogels. Limiting conversions were observed to decrease with increased temperature during SHMeMB homopolymerization, suggesting that polymerization rate is limited by depropagation. Comonomer composition drift also increased with temperature, with more AM incorporated into the copolymer due to SHMeMB depropagation. Using previous estimates for the SHMeMB propagation rate coefficient, the conversion profiles were used to estimate rate coefficients for depropagation and termination ($k_t$). The estimate for $k_{t,\text{SHMeMB}}$ was found to be of the same order of magnitude as that recently reported for sodium methacrylate, with the averaged copolymerization termination rate coefficient dominated by the presence of SHMeMB in the system. In addition, it was found that depropagation still controlled the SHMeMB polymerization rate at elevated temperatures in the presence of added salt.

**Keywords:** bio-renewable; depropagation; ionic strength; parameter estimation

---

## 1. Introduction

Water-soluble polymers are used extensively in personal products for hair care [1] and detergents [2], with crosslinked materials utilized as absorbent hydrogels in diapers or feminine products [3]. Other applications include drug delivery [4], flocculation for water recovery in oil sand tailings [5], and metal ion recovery [6]. The bio-derived monomers $\alpha$-methylene-$\gamma$-butyrolactone (MBL) and $\gamma$-methyl-$\alpha$-methylene-$\gamma$-butyrolactone (MeMBL) have previously been homopolymerized [7,8] and copolymerized with styrene (ST) and methyl methacrylate (MMA) [9,10], but in bulk or organic solution. Recently, Kollár et al. demonstrated that MBL could be saponified with sodium hydroxide (NaOH) to make a water-soluble monomer, sodium 4-hydroxy-2-methylene butanoate (SHMB), that was copolymerized with acrylamide (AM) and crosslinker to make superabsorbent hydrogels that exhibited superior water absorbency compared to conventional sodium acrylate:AM materials [11]. In an extension of that work, MeMBL was saponified to sodium 4-hydroxy-4-methyl-2-methylene butanoate (SHMeMB) (see Scheme 1), which was copolymerized with AM and crosslinker to make similar superabsorbent materials [12].

The SHMeMB:AM hydrogels showed higher water absorbency than SHMB:AM hydrogels, a finding attributed to changes in crosslink density caused by reactivity differences between the two systems [12]. Conversion profiles obtained from in situ NMR showed very slow homopolymerization rates of both SHMB and SHMeMB, and a lowered copolymerization rate for SHMeMB:AM compared to SHMB:AM. The lower reactivity of SHMeMB:AM was partially attributed to differences in the system reactivity ratios, estimated as $r_{\text{SHMeMB}} = 0.12$–$0.17$ and $r_{\text{AM}} = 0.95$–$1.10$ for SHMeMB:AM [12] compared to $r_{SHMB} = 0.35 \pm 0.15$ and $r_{AM} = 1.42 \pm 0.40$ for SHMB:AM [11]. Using pulsed-laser polymerization coupled with aqueous-phase size exclusion chromatography (PLP-SEC), the rate of radical chain growth of SHMB:AM copolymers was found to be twice that of SHMeMB:AM under identical conditions. It was not possible, however, to find suitable experimental conditions to directly study SHMeMB homopropagation by the PLP technique. Thus, the copolymerization results were extrapolated to provide the first estimates of the homopropagation rate coefficients ($k_p$), with very low values of 165 and 25 $L \cdot mol^{-1} \cdot s^{-1}$, estimated at 60 °C in aqueous solution for SHMB and SHMeMB, respectively [12].

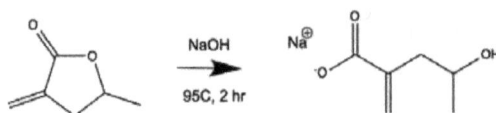

**Scheme 1.** Saponification of γ-methyl-α-methylene-γ-butyrolactone (MeMBL) using NaOH in water for 2 h at 95 °C to form sodium 4-hydroxy-4-methyl-2-methylene butanoate (SHMeMB).

As both SHMB and SHMeMB are fully-ionized water-soluble monomers, their polymerization kinetics in aqueous solution are not well-understood. However, previous studies using PLP-SEC, the IUPAC recommended method for determining $k_p$ [13], have examined the radical polymerization behavior of other water-soluble monomers, including non-ionized to fully ionized acrylic and methacrylic acids [14–17] and acrylamide [18]. It is now known that the polymerization kinetics of these monomers in water differ significantly from those of the same monomers in organic solvents, with the $k_p$ values of non-ionized acrylic acid and methacrylic acid significantly higher in water than in methanol and dimethyl sulfoxide (DMSO) [19]. In addition, the kinetics are greatly affected by monomer concentration in aqueous solution, as studied for acrylic acid [16], methacrylic acid [17], and acrylamide [20]. The dependence of $k_p$ on monomer concentration was attributed to the hydrogen-bonding effects between water, monomer, and radical species. Although $k_p$ is dependent on monomer concentration, it was found that the reactivity ratios of AM and non-ionized acrylic acid copolymerization were constant with concentration [21], although the values are dependent on monomer concentration and ionic strength for copolymerization of AM with fully-ionized AA [22,23]. It should be noted that hydrogen-bonding effects are not only present in aqueous solution, but also influence the reactivity ratios of butyl methacrylate (BMA) and 2-hydroxyethyl acrylate (HEA) copolymerization in organic solution, as the relative reactivity of the two monomers is dependent on solvent choice [24].

Another kinetic mechanism important to this study is depropagation. For most radical polymerizations, the monomer addition to a growing macroradical (propagation) can be considered as an irreversible reaction. However, depropagation, the process by which a single monomer unit is released from the growing radical chain, occurs if there is steric hindrance near the radical site. Under these conditions, the propagation and depropagation mechanisms become a reversible reaction pair, with the relative rates (and hence overall rate of polymerization) a function of temperature and monomer concentration. Some monomers that are known to depropagate are BMA [25], itaconates [26], methyl ethacrylate [27], and α-methyl styrene [28], all studied in organic solution. In the presence of appreciable rates of depropagation, the polymerization does not reach full monomer conversion and the reaction can also influence copolymer composition as well as rate, as seen for the radical copolymerization of methyl ethacrylate and styrene (MEA/ST) [29] and BMA/ST at elevated temperature [25]. It is interesting to note that α-methylene-δ-valerolactone (MVL), a monomer of similar structure to MBL, has been shown to undergo depropagation, with a ceiling temperature of 83 °C [30]. Depropagation of MVL was attributed to its non-planar structure that hinders the radical center, while MBL does not depropagate as it is planar in structure.

In this publication, a series of studies were done to further elucidate the polymerization kinetics of SHMeMB homopolymerization and SHMeMB:AM copolymerization. Polymerizations were conducted at elevated temperatures (60 to 90 °C) to explore the importance of depropagation, using in situ NMR spectroscopy to track monomer conversions. In addition, homopolymerizations were done in the presence of added salt to observe whether the changes in the polymerization rate are similar to those reported for fully ionized acrylic acid (sodium acrylate, NaAA) [31,32] and methacrylic acid (sodium methacrylate, NaMAA) [33]. The homopolymerization conversion profiles were used to estimate the rate coefficients for termination ($k_t$) and depropagation ($k_{dep}$), using the parameter estimation capabilities in the PREDICI® software package [34]. Ultimately, the estimated parameters from SHMeMB homopolymerization were implemented in a kinetic model developed to represent SHMeMB:AM copolymerization behavior.

## 2. Materials and Methods

### 2.1. Materials

The following chemicals were purchased from Sigma-Aldrich, Canada and used as received: acrylamide (AM, >98%), $N,N'$-methylenebis(acrylamide) (BIS, 99%), 2,2'-azobis(2-methyl-propionamidine) dihydrochloride (V-50 initiator, 97%), and sodium hydroxide (NaOH, >97%). 2,2'-Azobis[2-methyl-N-(2-hydroxyethyl)propionamide] (V-86 initiator) was purchased from Wako Pure Chemicals Ltd., USA. Deuterated water ($D_2O$, 99.8% D) and hydrochloric acid (HCl, 36.5% $w/w$) were purchased from Fisher Scientific, Canada and the γ-methyl-α-methylene-γ-butyrolactone (MeMBL, >97%) was provided by DuPont Central Research Laboratories.

### 2.2. Ring-Opening Saponification of MeMBL

The saponification of MeMBL followed the same procedure as in previous studies [11,12]. For 1 g of MeMBL, 10 mol % excess of NaOH was measured and dissolved in 1 g of $D_2O$ in a small vial with a stir-bar. The saponification reaction took place in an oil bath at 95 °C for 2 h, after which the solution was cooled to room temperature and 1 M HCl was added until a pH of 7 was reached. The SHMeMB mixture was then diluted with $D_2O$ to a final monomer concentration of 40 wt % (including mass of sodium ions). This stock solution was mixed with other components to achieve desired concentrations for the in situ NMR studies. The structure of MeMBL was confirmed by NMR, and ring structures were confirmed to be completely opened to make SHMeMB [12].

### 2.3. Preparation for In Situ NMR

The in situ NMR method was used to measure the overall monomer conversion profiles, as well as the variation of monomer and polymer composition with conversion, following procedures described by Preusser and Hutchinson [21]. Near-isothermal conditions for the experiments can be assumed based upon the slow rate of polymerization compared to other systems analyzed in the same setup [21]. Copolymerizations were conducted at 3:7 and 4:6 SHMeMB:AM initial molar ratios and 15 wt % monomer concentration in $D_2O$, with initiator content specified as a weight percent of the total mixture (monomers + $D_2O$). Homopolymerizations were done at 15 and 30 wt % monomer concentration in $D_2O$, and salt was added to the monomer mixture relative to the SHMeMB molar amounts. While the preparation of the SHMeMB monomer added slightly to the ionic strength of the solution (addition of 10 mol % excess NaOH followed by neutralization with HCl), this contribution was not considered when reporting salt concentrations in the discussion section.

Overall conversion $X(t)$ was calculated from the decrease in monomer peak integrations relative to the HOD reference peak (residual solvent peak from $D_2O$ at 4.8 ppm), and individual conversions of SHMeMB and AM were used to calculate SHMeMB monomer ($f_{SHMeMB}$) and polymer mole fractions ($F_{SHMeMB}$), as detailed in our previous study [12].

### 2.4. Kinetic Parameters for PREDICI Parameter Estimation

The parameter estimations for SHMeMB homopolymerizations and SHMeMB:AM copolymerizations were done using PREDICI [34], based on the reaction mechanisms listed in Table 1. For SHMeMB:AM copolymerizations, $k_t$ represents an averaged value for all three termination mechanisms; as will be discussed, the values were estimated from individual experimental monomer conversion profiles to provide a perspective on how the averaged rate coefficient varies with monomer composition. As the termination rate is dominated by the large fraction of SHMeMB radicals in the system, the reaction is assumed to occur solely by disproportionation due to their hindered structure; this assumption does not impact the $k_t$ values estimated from the conversion profiles. Depropagation was also considered in the model, based on the assumption that the reaction occurs only if both the penultimate and terminal monomer units of the growing radical chain are SHMeMB, as captured by the probability factor $p_{11}$ [25].

**Table 1.** Reaction mechanisms for the copolymerization of SHMeMB and acrylamide (AM).

| Reaction Mechanisms | |
|---|---|
| Initiator decomposition | $I \xrightarrow{k_d} 2fR_0^*$ |
| Initiation | $R_0^* + SHMeMB \xrightarrow{k_{p1,1}} SHMeMB_1^*$ |
| | $R_0^* + AM \xrightarrow{k_{p2,2}} AM_1^*$ |
| Propagation | $SHMeMB_n^* + SHMeMB \xrightarrow{k_{p1,1}} SHMeMB_{n+1}^*$ |
| | $SHMeMB_n^* + AM \xrightarrow{k_{p1,2}} AM_{n+1}^*$ |
| | $AM_n^* + SHMeMB \xrightarrow{k_{p2,1}} SHMeMB_{n+1}^*$ |
| | $AM_n^* + AM \xrightarrow{k_{p2,2}} AM_{n+1}^*$ |
| Termination (by disproportionation) | $SHMeMB_n^* + SHMeMB_m^* \xrightarrow{k_t} P_n + P_m$ |
| | $AM_n^* + SHMeMB_m^* \xrightarrow{k_t} P_n + P_m$ |
| | $AM_n^* + AM_m^* \xrightarrow{k_t} P_n + P_m$ |
| Depropagation | $SHMeMB_n^* \xrightarrow{p_{11}k_{dep}} SHMeMB_{n-1}^* + SHMeMB$ |

The known rate coefficients (initiator decomposition, homopropagation, and reactivity ratios) are shown in Table 2. The initiator efficiency ($f$) of V-50 was assumed to be 0.8, and for V-86 it was assumed to be 0.38, as determined in a previous study [35]. The propagation rate expression for AM homopolymerization was determined [18] and modelled [20] previously as a function of both monomer concentration and temperature, yielding a $k_{p,AM}$ value of 86,000 L·mol$^{-1}$·s$^{-1}$ for 15 wt % AM in aqueous solution at 50 °C. Although AM concentration changes with SHMeMB:AM comonomer composition (keeping total monomer content at 15 wt %), the value of $k_{p,AM}$ in the model was kept constant. This assumption is reasonable, as $k_p^{cop}$ is dominated by the low value of $k_{p,SHMeMB}$ and not sensitive to small changes in $k_{p,AM}$. As shown in Table S1 in the supporting information, $k_{p,AM}$ values were calculated at a different total AM wt % (while maintaining 15 wt % monomer concentration) and the estimated values for $k_t$ remained the same. The PLP-SEC estimate for $k_{p,SHMeMB}$, obtained at 60 °C and 15 wt % monomer in the previous study [12], was used in this work.

**Table 2.** Rate expressions for known kinetic coefficients used in model of SHMeMB:AM copolymerization.

| Reaction | Rate Expression | References |
|---|---|---|
| Decomposition of V-50 | $k_d = 9.385 \times 10^{14} \exp(-14{,}890/T(K))$<br>$f = 0.8$ | [36] |
| Decomposition of V-86 | $k_d = 1.24 \times 10^{13} \exp(-14{,}800/T(K))$<br>$f = 0.38$ | [35,37] |
| Propagation of AM | $k_p^0 = 9.5 \times 10^7 \exp(-2189/T(K))$<br>$k_p = k_p^0 \exp(-0.01\,c_{AM}(0.0016T + 1.015))$ | [18] |
| Propagation of SHMeMB | $k_p = 25$ L·mol$^{-1}$·s$^{-1}$ | [12] |
| Reactivity ratios at 50 °C | $r_{AM} = k_{p2,2}/k_{p2,1} = 0.95 \pm 0.01$<br>$r_{SHMeMB} = k_{p1,1}/k_{p1,2} = 0.17 \pm 0.01$ | [12] |

## 3. Results and Discussion

### 3.1. Copolymerization of SHMeMB:AM at Different Temperatures

In the previous study, in situ batch copolymerizations of SHMeMB and AM at 50 °C over a wide range of compositions (initial mole fraction of SHMeMB, $f_{SHMeMB}$, between 0.1 and 0.8) were used to estimate the reactivity ratios of the system summarized in Table 2. In this work, the study was extended to higher temperatures. Conversion profiles measured for experiments with V-50 initiator at an initial monomer content of 15 wt % and SHMeMB:AM molar ratios of 3:7 (50–80 °C) and 4:6 (50–70 °C) are shown in Figure 1. It is evident that the initial rate of polymerization increases with temperature, as expected due to the accelerated radical production rate as well as the increased $k_p$ values. However, monomer conversion plateaus at less than 100% at the higher temperatures. This limiting conversion does not result from initiator depletion, as 27% of the V-50 remains after 3 h at 70 °C, based on literature values for V-50 decomposition kinetics. The conversion plateau occurs at a lower conversion as the initial fraction of SHMeMB is increased, as seen by comparing the profiles for 3:7 and 4:6 SHMeMB:AM after 3 h. Thus, the presence of SHMeMB not only affects the initial rate of polymerization, but also causes the copolymerization rate to significantly slow down as higher conversions are reached at the higher temperatures.

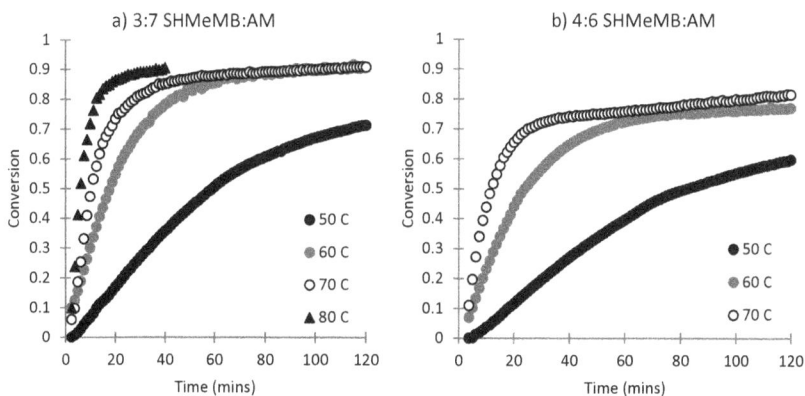

**Figure 1.** Overall monomer conversion profiles for copolymerizations at (**a**) 3:7 and (**b**) 4:6 initial SHMeMB:AM molar ratios at varying temperatures with 15 wt % monomer and 0.5 wt % V-50.

Individual monomer concentration profiles of SHMeMB and AM at 50, 60, and 70 °C are presented as Figure S1 in the supporting information. The plots show that the rate of SHMeMB consumption becomes very slow at the lowered SHMeMB concentrations reached later in the reactions, while the consumption of AM continues. This behavior becomes more evident at higher temperatures and increased SHMeMB content, under which conditions the absolute concentration of AM decreases to values below that of SHMeMB, despite its higher initial value. The limiting SHMeMB conversions suggest that depropagation of SHMeMB monomer may be occurring, leading to a significantly decreased rate of polymerization as AM is consumed.

To further explore this possibility, the comonomer composition drifts with conversion at different temperatures are plotted in Figure 2; the curves are normalized by the initial fraction of SHMeMB in the mixture to provide a better comparison by eliminating slight variations in the initial compositions. If SHMeMB depropagation is important, the value of $f_{SHMeMB}$ would increase more significantly with conversion as temperature is increased due to decreased incorporation of SHMeMB under conditions that favour depropagation, as seen in studies of MEA/ST [29] and BMA/ST [25]. As shown in Figure 2, this behavior is indeed observed for the SHMeMB:AM system as temperature increased from 50 to

80 °C. At higher conversions where monomer concentrations are low, the influence of depropagation on SHMeMB consumption becomes more prominent. The reaction temperature was further increased to 90 °C using a different initiator, V-86, as it has a slower rate of decomposition. At 90 °C, there was further deviation of the drift in $f_{SHMeMB}$ with conversion compared to 50 °C. Reactions with V-50 and V-86 were conducted at the same temperatures to verify that composition drift was consistent using both initiators (see Figure S2).

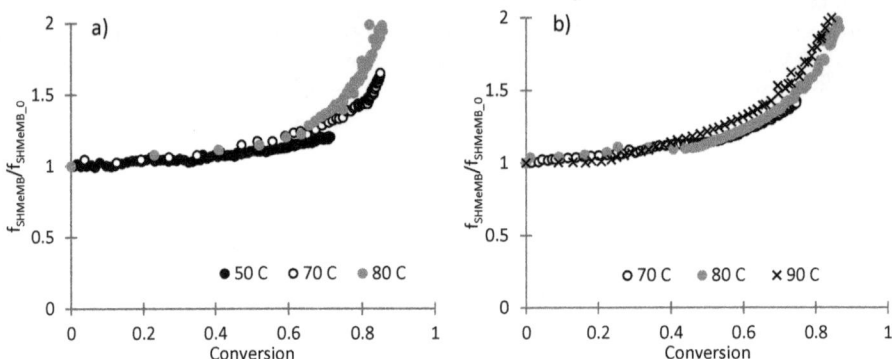

**Figure 2.** Monomer composition drift with conversion for copolymerization with an initial 3:7 SHMeMB:AM molar ratio, 15 wt % monomer and (**a**) 0.5 wt % V-50 initiator, or (**b**) 1.67 wt % V-86 initiator at varying temperatures (90 °C experiment was conducted with 0.5 wt % V-86). Monomer composition was normalized by initial monomer composition to eliminate the influence of slight variations in the comonomer mixture composition.

As a first step to understanding this behavior, monomer composition drifts measured with 3:7 molar ratio of SHMeMB:AM copolymerizations at 70 and 80 °C were fitted to provide reactivity ratio estimates, with results shown in Figure 3. Two methods were used to fit the experimental data: the first uses the previously determined $r_{AM} = 0.951$ so that only one parameter ($r_{SHMeMB}$) was estimated, as depropagation should only influence the addition rate of a SHMeMB monomer to a SHMeMB radical, and thus the effective value of $r_{SHMeMB}$. For the second fitting, both parameters, $r_{AM}$ and $r_{SHMeMB}$, were estimated simultaneously.

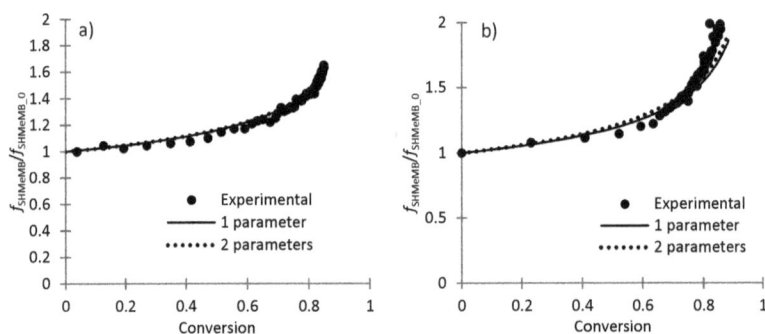

**Figure 3.** Monomer composition drift for copolymerizations of 3:7 molar ratio of SHMeMB:AM with 15 wt % monomer and 0.5 wt % V-50 at (**a**) 70 and (**b**) 80 °C. The solid line represents parameter estimation with $r_{AM}$ fixed at 0.951 (best-fit value at 50 °C), and the dotted line represents parameter estimation to determine both $r_{AM}$ and $r_{SHMeMB}$, with values summarized in Table 3.

As shown by Figure 3, both methods gave good representations to the experimental data, but with drastically different estimates for $r_{SHMeMB}$, as summarized in Table 3. The terminal model reactivity ratio of monomer one ($r_1$) is defined by Equation (1), where $k_{p,11}$ is the homopropagation rate coefficient for addition of monomer one to a monomer one radical and $k_{p,12}$ is the cross-propagation rate coefficient for the addition of monomer two to a monomer one radical.

$$r_1 = \frac{k_{p,11}}{k_{p,12}} \tag{1}$$

When the value of $r_{AM}$ was fixed at 0.951, the parameter estimation forces $r_{SHMeMB}$ to approach zero, indicating that AM monomer addition is greatly favoured over SHMeMB addition to a SHMeMB terminal radical. Estimating $r_{SHMeMB}$ and $r_{AM}$ simultaneously at 70 °C gave an $r_{AM}$ value that was close to the value determined at 50 °C, and lowered the $r_{SHMeMB}$ value to 0.120 (from 0.169 at 50 °C). At 80 °C, $r_{SHMeMB}$ decreased to an even lower value of 0.046 due to more prominent depropagation effects at higher temperatures. While these values seem plausible, the uncertainty in the estimates are large. Nonetheless, they are consistent with the expectations of depropagation.

**Table 3.** Reactivity ratios estimates from copolymerization of 3:7 molar ratio of SHMeMB:AM with 15 wt % monomer and 0.5 wt % V-50 at 70 and 80 °C. The "one parameter" method estimates $r_{SHMeMB}$ with $r_{AM}$ fixed at 0.951, and the "two parameter" method estimates both $r_{AM}$ and $r_{SHMeMB}$.

|  | 70 °C | | 80 °C | |
| --- | --- | --- | --- | --- |
|  | 1 Parameter | 2 Parameters | 1 Parameter | 2 Parameters |
| $r_{SHMeMB}$ | $0.005 \pm 0.008$ | $0.12 \pm 0.22$ | $7 \times 10^{-6} \pm 7 \times 10^{-3}$ | $0.05 \pm 0.17$ |
| $r_{AM}$ | - | $1.04 \pm 0.17$ | - | $1.06 \pm 0.21$ |

It is important to note that the parameter estimation fits the reactivity ratios based on the terminal model (i.e., no depropagation), assuming that the value of $k_{p,SHMeMB}$ remains constant with conversion. In the case of depropagation, the $k_{p,11}$ value should be considered as an effective value, $k_p^{eff}$, dependent on $k_p$, $k_{dep}$ and monomer concentration $[M]$ as shown in Equation (2) [26], such that $r_{SHMeMB}$ would change with conversion.

$$k_p^{eff} = k_p - \frac{k_{dep}}{[M]} \tag{2}$$

Thus, the conversion profiles for SHMeMB homopolymerizations are first used to estimate $k_{dep}$ before returning to analysis of the copolymerization system.

### 3.2. Homopolymerization Kinetics of SHMeMB

To investigate depropagation kinetics further, the in situ NMR technique was used to study homopolymerization of SHMeMB at increased temperature and initiator content (75 °C and 1 wt % V-86), with reaction times (14 h) considerably extended compared to copolymerizations. The conversion profiles, shown in Figure 4, were the same for both initial monomer concentrations (15 and 30 wt %), consistent with reports that monomer concentration did not have a large effect on $k_p$ for other fully ionized monomers such as NaMAA [17]. However, no difference in the final conversions is seen between the two experiments, indicating that monomer concentrations were not yet approaching the equilibrium values at which depropagation would cause a difference in limiting conversion.

The possible effects of depropagation on the homopolymerization of SHMeMB were investigated at higher temperatures. In Figure 4, the initial rate of polymerization is seen to be faster at 90 °C than at 75 °C as expected, but the rate eventually slows down such that the final conversion reached is lower than at 75 °C. The decrease in polymerization rate at 90 °C is not due to the lack of initiator, as there is still 25% remaining after 16 h [37]. Thus, the conversion profiles support the hypothesis that depropagation is affecting SHMeMB polymerization, consistent with the observation of increased AM incorporation into the SHMeMB:AM copolymer observed at elevated temperatures.

It was previously demonstrated that the polymerization rate of fully ionized acrylic acid was influenced by the addition of a salt, such that at a molar ratio of 1:5.7 [AA$^-$]:[NaCl] the polymerization rate of fully ionized AA (NaAA) was comparable to that of non-ionized AA [31]. It was proposed that the screening of charges by the added salt reduced the repulsion between the ionized monomers and ionized radical sites, therefore enhancing the polymerization rate of fully-ionized AA. Thus, NaCl was added to SHMeMB homopolymerizations to examine for a similar effect, as shown in Figure 5. The polymerization rate at 75 °C was found to decrease with added salt, with the rate of polymerization perhaps slightly lower at the 1:1 ratio of SHMeMB]:[NaCl] compared to the 1:0.5 ratio.

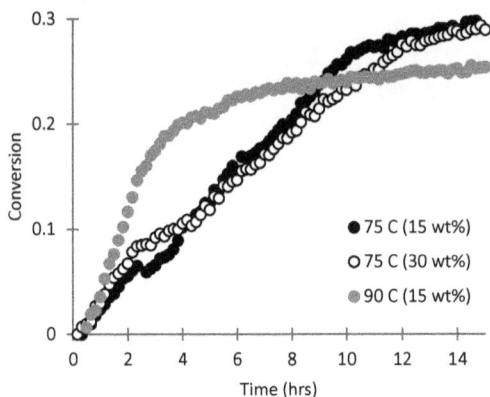

**Figure 4.** Monomer conversion profiles obtained by homopolymerization of SHMeMB at 15 and 30 wt % at 75 °C and 1 wt % V-86, and at 15 wt % and 90 °C.

Although the rate of monomer conversion of NaAA was increased by increasing ionic strength [31], a similar increase was not observed for NaMAA [17]. However, a separate study showed that the $k_p$ of NaMAA did increase with ionic strength [32]. Therefore, it can be concluded that $k_p$ and $k_t$ for ionized monomers are both affected (increased) by the presence of salt, but to different extents, according to the monomer. Individual estimates are not available for SHMeMB, but the conversion profiles in Figure 5a indicate that $k_t$ is enhanced in the presence of NaCl to a greater extent than $k_p$, hence decreasing the overall rate of polymerization at 75 °C. As shown in Figure 5b, the addition of NaCl to the polymerization at 90 °C, however, has no effect on the conversion profile. Depropagation is more important at this elevated temperature, complicating the situation; however, the net effect of the added salt on the rate of conversion is minor.

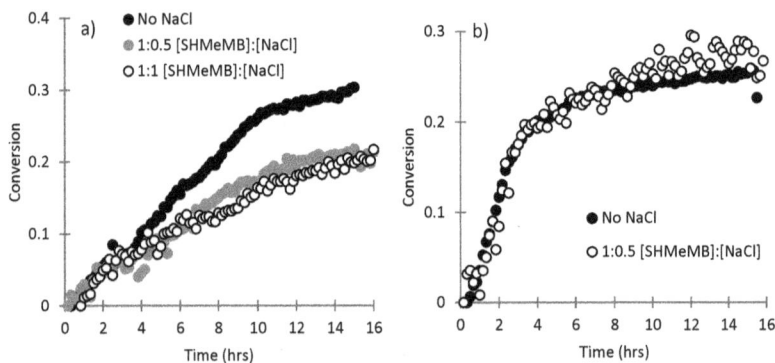

**Figure 5.** Monomer conversion profiles obtained from homopolymerizations of SHMeMB at 15 wt % with (**a**) added NaCl salt at 75 °C and 1 wt % V-86, and (**b**) at 90 °C with added NaCl at 1:0.5 [SHMeMB]:[NaCl] molar ratio.

### 3.3. Parameter Estimation for SHMeMB Homopolymerizations

The SHMeMB homopolymerization conversion profiles presented in Section 3.2 are used in this section to estimate both the termination ($k_t$) and depropagation ($k_{dep}$) coefficients using the data-fitting tools in PREDICI based on the mechanisms shown in Table 1 (initiation, propagation, depropagation and termination). As conversion profiles are a function of the ratios of rate coefficients ($k_{dep}/k_p$ and $k_p/k_t^{0.5}$), the strategy employed was to use the previously-estimated propagation coefficient ($k_p$) of SHMeMB from the PLP-SEC study [12], and to estimate $k_{dep}$ simultaneously with $k_t$. For simplicity, the $k_p$ value of 25 L·mol$^{-1}$·s$^{-1}$ estimated at 60 °C was used, as the activation energy for propagation is not known. Thus, the estimates for $k_t$ and $k_{dep}$ at 75 and 90 °C are lower than the true values, but could be corrected once the temperature dependency of $k_p$ is determined.

The initial fitting of the SHMeMB homopolymerizations curves, shown in Figure 6, was conducted assuming no depropagation occurs ($k_{dep} = 0$). The model fits the experimental data reasonably well at 75 °C until the point at which the rate of polymerization seemed to decrease, around 10 h into the reaction. However, it is evident that the model with no depropagation was not sufficient in fitting the experiment conversion profile at 90 °C, predicting a continued increase in conversion not observed experimentally. The best fit value of $k_t$ is $1.3 \times 10^6$ L·mol$^{-1}$·s$^{-1}$ at both 75 and 90 °C, although the true values would be lower (due to the assumption that depropagation does not occur). Furthermore, $k_t$ does not seem to be a large function of temperature, as the same value was able to fit the initial polymerization rate for both 75 and 90 °C. These estimates of $k_t$ are higher than recently reported values for NaAA of ~$10^5$ L·mol$^{-1}$·s$^{-1}$ [38].

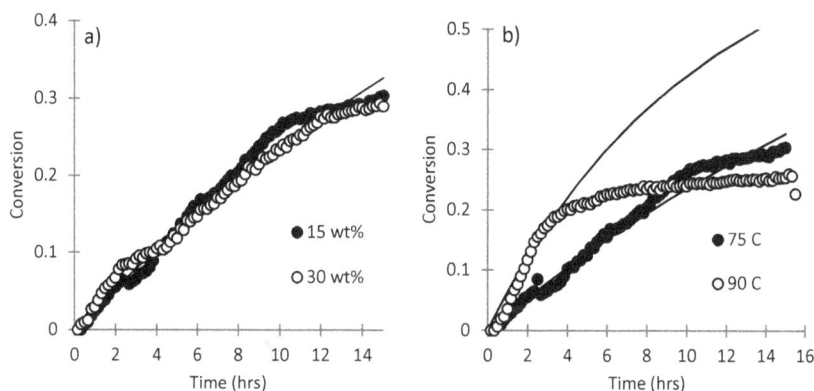

**Figure 6.** Fit of the homopolymerization SHMeMB model assuming no depropagation to monomer conversion profiles obtained at (**a**) 75 °C with 1 wt % V-86 at different monomer concentrations, and (**b**) at different temperatures with 1 wt % V-86 and 15 wt % monomer. Solid lines represent model output, with experimental results indicated by data points.

The experimental conversion profile of SHMeMB homopolymerization at 75 °C was converted using the integrated conversion equation for a batch isothermal reaction, Equation (3), to generate a $k_p/k_t^{0.5}$ vs. conversion plot. In Equation (3), X represents conversion, $k_d$ is the decomposition rate coefficient of V-86 initiator, t is reaction time, $[I]_0$ is initial initiator concentration, and f is initiator efficiency.

$$\frac{k_p}{k_t^{0.5}} = \frac{ln(1-X)}{exp(0.5k_d t) - 1} \sqrt{\frac{k_d}{8[I]_0 f}} \tag{3}$$

Note that Equation (3) is derived assuming that both $k_p$ and $k_t$ are constant with conversion, and that depropagation does not occur in the system. In general, the $k_p/k_t^{0.5}$ ratio calculated from the experimental data at 75 °C, as shown in Figure 7, is fairly constant until it reaches a conversion of 25%, at which point it starts to decrease, likely due to the influence of depropagation. (The $k_p/k_t^{0.5}$ values at <10% conversion were omitted due to scatter of experimental data in the initial stages of the reaction.) Assuming that depropagation was negligible between 0 and 25% conversion and a $k_p$ value of 25 L·mol$^{-1}$·s$^{-1}$, the average value of $k_p/k_t^{0.5}$ of 0.022 (L·mol$^{-1}$·s$^{-1}$)$^{0.5}$ in this region is used to estimate a value of $k_t$ of 1.30 × 10$^6$ L·mol$^{-1}$·s$^{-1}$, in agreement with the value fit to generate the conversion profiles in Figure 6. Also in agreement with Figure 6, the $k_p/k_t^{0.5}$ ratio begins to decrease at a lower conversion at 90 °C, leading to the overestimation of the reaction rate without the consideration of depropagation.

The $k_t$ value of 1.30 × 10$^6$ L·mol$^{-1}$·s$^{-1}$ was estimated using the $k_p/k_t^{0.5}$ equation assuming depropagation was negligible in the early stages of the reaction, but the true extent of depropagation is still unknown at 75 and 90 °C. Using parameter estimation on PREDICI, both $k_t$ and $k_{dep}$ values were simultaneously estimated to be 1.4 ± 1.8 × 10$^5$ L·mol$^{-1}$·s$^{-1}$ and 21 ± 6 s$^{-1}$, respectively, at 75 °C. The estimated value $k_t$ is an order of magnitude lower than estimated using the $k_p/k_t^{0.5}$ plot, but is in reasonable agreement with reported estimates for other ionized monomers [33]. However, the 95% confidence interval encompasses zero, due to the difficulty in estimating both $k_t$ and $k_{dep}$ from a single conversion profile. Thus, the strategy taken was to fix $k_t$ at the value of 1.4 × 10$^5$ L·mol$^{-1}$·s$^{-1}$, and estimate only the $k_{dep}$ values from the conversion profiles obtained with 15 wt % SHMeMB at both 75 and 90 °C. The resulting fits to the conversion profiles are shown in Figure 8, with best fit values for $k_{dep}$ of 20.9 ± 0.6 and 26.8 ± 0.4 s$^{-1}$ at 75 and 90 °C, respectively. The best fit profiles are compared to the curves generated assuming no depropagation (at 75 °C extended to longer time), to further illustrate that the higher $k_t$ value estimated from the $k_p/k_t^{0.5}$ plot did not adequately represent the shape of the curve. Although the estimation at 90 °C did not fit as well to the experimental data, the higher estimated value of $k_{dep}$ indicates that, as expected, depropagation is enhanced at elevated temperatures. Furthermore, the fitting indicates an SHMeMB ceiling temperature (at a standard state of 1 mol·L$^{-1}$) of about 90 °C, at which point $k_p$ and $k_{dep}$ estimates are almost identical. It is interesting to note that this value is close to the ceiling temperature of 83 °C reported for MVL [30], although the latter was polymerized in a closed ring form in organic solvent.

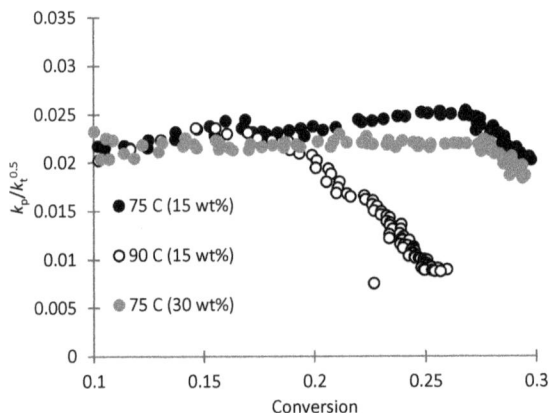

**Figure 7.** $k_p/k_t^{0.5}$ vs. conversion profile of SHMeMB homopolymerization at 75 °C with 15 and 30 wt % monomer concentration and at 90 °C with 15 wt % monomer and 1 wt % V-86 for 15 h.

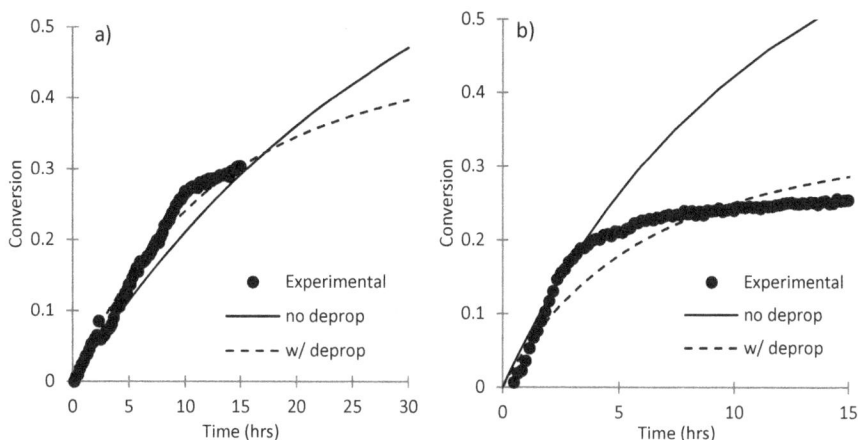

**Figure 8.** Conversion profile of SHMeMB homopolymerization with 15 wt % monomer and 1 wt % V-86 at (**a**) 75 °C and (**b**) at 90 °C. Lines indicate best-fit simulation results with (- - - -) and without (——) depropagation.

Termination and depropagation rate coefficients were also estimated at the higher monomer concentration conditions of 30 wt % SHMeMB at 75 °C with 1 wt % V-86. From previous PLP-SEC studies [39], $k_p^{cop}$ of 10 mol % SHMeMB in SHMeMB:AM mixtures at 10 and 20 wt % were within experimental error. Therefore, a $k_p$ value of 25 L·mol$^{-1}$·s$^{-1}$ was also assumed for SHMeMB homopolymerization at 30 wt %. The value of $k_{dep}$ was also kept constant, and the conversion profile used to estimate a $k_t$ value of $6.15 \pm 0.02 \times 10^5$ L·mol$^{-1}$·s$^{-1}$, with the fitted curve compared to the experimental conversion profile in Figure 9. Even though the observed conversion profiles at 15 and 30 wt % were nearly identical, the estimated values for $k_t$ increased significantly, a finding consistent with a previous study that showed that the $k_t$ of fully ionized NaMAA increased with monomer concentration [40].

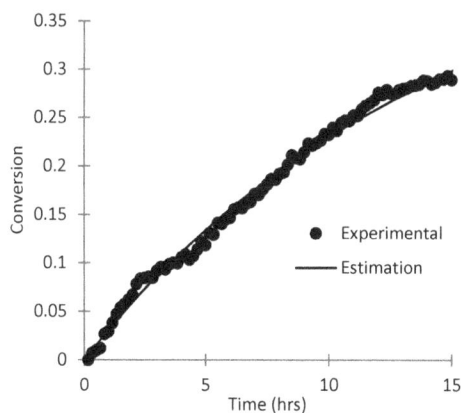

**Figure 9.** Monomer conversion profiles for the homopolymerization of SHMeMB at 75 °C with 1 wt % V-86 and 30 wt % monomer. The solid line represents the estimated conversion profile using parameter estimation.

Finally, parameter estimations were also done for the homopolymerizations of SHMeMB with added NaCl at 75 °C. While depropagation is generally attributed to the steric hindrance near the α-carbon, it is possible that electrostatic repulsion from the ionized carboxylic groups can also affect the mechanism. Nonetheless, it was assumed that $k_{dep}$ remains the same with added salt (21 s$^{-1}$ as determined previously), as the first order depropagation mechanism is less likely to be influenced by the reaction environment than the bimolecular termination and propagation reactions and, as previously stated, both the $k_p$ and $k_t$ values for polymerizations of fully ionized monomers with added salt have been observed to increase with ionic strength [32]. The estimated conversion profiles are compared to the experimental results in Figure 10. As expected, the estimated values for $k_p$ and $k_t$ summarized in Table 4 have high uncertainty due to the difficulty of estimating two rate coefficients from the same conversion profile, with 95% confidence intervals encompassing zero. However, the best fit values are roughly the same for both 1:0.5 and 1:1 molar ratios of [SHMeMB]:[NaCl]. While the estimated values for $k_p$ did not increase greatly with the addition of salt (from 25 to ~30 L·mol$^{-1}$·s$^{-1}$), the estimates for the $k_t$ values increased an order of magnitude to $1 \times 10^6$ L·mol$^{-1}$·s$^{-1}$, significantly larger than the value of $1 \times 10^5$ L·mol$^{-1}$·s$^{-1}$ estimated without salt. Although estimated with high uncertainty, it is interesting to note that this increase in $k_t$ is consistent with the increase estimated for the 30 wt % SHMeMB homopolymerization, suggesting that charge screening provided from either the higher SHMeMB monomer concentration or added salt lowers the electrostatic barrier to radical–radical termination.

**Table 4.** Estimated values for $k_p$ and $k_t$ for homopolymerizations of SHMeMB with added salt at 75 °C with 1 wt % V-86 and 15 wt % monomer assuming a $k_{dep}$ value of 21 s$^{-1}$. Results are shown for reactions done with 1:0.5 and 1:1 molar ratios of [SHMeMB]:[NaCl].

| | 1:0.5 [SHMeMB]:[NaCl] | | 1:1 [SHMeMB]:[NaCl] | |
|---|---|---|---|---|
| | | 95% Confidence | | 95% Confidence |
| $k_p$ (L·mol$^{-1}$·s$^{-1}$) | 30.3 | ±50.9 | 29.2 | ±37.0 |
| $k_t$ (L·mol$^{-1}$·s$^{-1}$) | $9.98 \times 10^5$ | ±$7.96 \times 10^6$ | $1.01 \times 10^6$ | ±$6.32 \times 10^6$ |

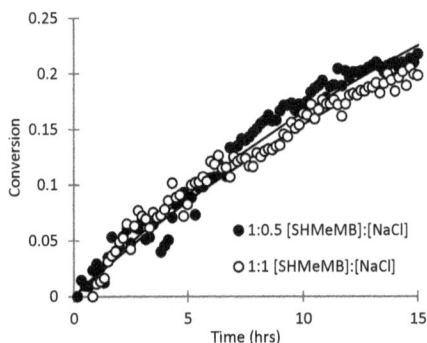

**Figure 10.** Monomer conversion profiles from homopolymerization of SHMeMB with different concentrations of added NaCl salt at 75 °C with 1 wt % V-86 and 15 wt % monomer. The solid lines represent the simulated conversion profiles using parameter estimation.

To summarize, despite considerable uncertainty in the parameter estimations, the analysis of the SHMeMB homopolymerization conversion profiles suggests that the system is characterized by similar $k_t$ values to other ionized monomers, but has very low propagation rates and significant depropagation. A large increase in $k_t$ was required to fit the conversion profiles measured with added salt and with increased monomer concentration, consistent with trends observed in NaMAA [40].

### 3.4. Parameter Estimation for SHMeMB:AM Copolymerizations with Depropagation

The knowledge gained regarding the kinetic behavior of SHMeMB is here applied to the interpretation of the experimental SHMeMB:AM copolymerizations. Details of the PREDICI model, which assumes terminal chain-growth kinetics and SHMeMB depropagation, and uses a single $k_t$ value to represent termination in the two-monomer system, is presented in Section 2.4. Using the coefficients at 50 °C summarized in Table 2, and a SHMeMB $k_{dep}$ value of 21 s$^{-1}$, $k_t$, values of the copolymerization system were estimated at 50 °C for the different molar ratios of SHMeMB and AM studied experimentally. The estimated $k_t$ values of SHMeMB:AM copolymers are plotted as a function of $f_{SHMeMB}$ in Figure 11, with $k_{t,AM}$ and $k_{t,SHMeMB}$ included for reference. The termination rate coefficient for AM ($k_{t,AM}$) was reported as a function of monomer concentration and temperature [20,41]. At 50 °C, $k_{t,AM}$ values at 10 and 20 wt % were calculated to be 5.2 × 10$^8$ and 4.2 × 10$^8$ L·mol$^{-1}$·s$^{-1}$, respectively. Taking the average of the two values gave an estimate of $k_{t,AM}$ of 4.7 × 10$^8$ L·mol$^{-1}$·s$^{-1}$ for 15 wt % monomer, several orders of magnitude higher than the $k_t$ of SHMeMB ($k_{t,SHMeMB}$) estimated as 1.4 × 10$^5$ L·mol$^{-1}$·s$^{-1}$ at 15 wt % and 75 °C in Section 3.4. The fitting of $k_t$ to the SHMeMB:AM copolymerizations at 50 °C with 0.5 wt % V-50 and 15 wt % monomer provided a very good representation of the experimental conversion profiles, as seen in Figure S3.

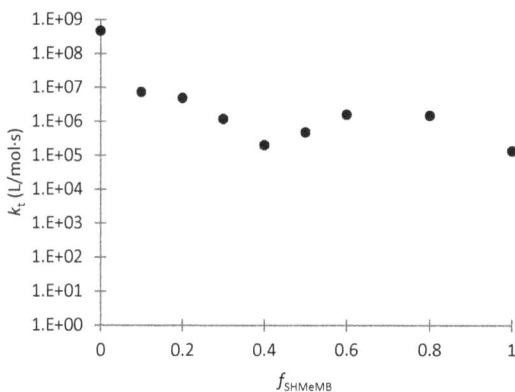

**Figure 11.** $k_t$ of SHMeMB:AM copolymers estimated for copolymerizations done at 50 °C with 0.5 wt % V-50 and 15 wt % monomer.

Figure 11 shows an immediate drop of two orders of magnitude upon addition of SHMeMB as a comonomer to AM, with values further decreasing to ~10$^5$ L·mol$^{-1}$·s$^{-1}$ as $f_{SHMeMB}$ increases to 0.4, at which point the estimates level out. The value of 10$^5$ L·mol$^{-1}$·s$^{-1}$ is similar to the estimated value for $k_{t,SHMeMB}$ in the previous section, and also the value of 3.6 × 10$^5$ L·mol$^{-1}$·s$^{-1}$ reported for homotermination of NaAA at 20 wt % and 50 °C in aqueous solution [38]. Comparable $k_t$ values for SHMeMB and NaAA indicate that the slow termination of two radicals in these systems is dominated by the electrostatic repulsion of the charged species near the radical site, rather than the steric hindrance that leads to the slow propagation and occurrence of depropagation of SHMeMB.

The $k_t$ value estimated for the copolymerization is an averaged value that describes all termination events in the SHMeMB:AM copolymerization, assuming terminal model kinetics. This averaging is described by Equation (4), where $f_{r,SHMeMB}$ is the fraction of SHMeMB terminal radicals, $f_{r,AM}$ is the fraction of AM terminal radicals, and $k_{t,SA}$ is the rate coefficient describing the cross-termination of SHMeMB and AM radicals [28].

$$k_t = k_{t,SHMeMB} f_{r,SHMeMB}^2 + 2k_{t,SA}\, f_{r,SHMeMB} f_{r,AM} + k_{t,AM} f_{r,AM}^2 \qquad (4)$$

According to Equation (4), $k_t$ must be dominated by the value of $k_{t,SHMeMB}$, as its value decreases by two orders of magnitude from the value of $k_{t,AM}$ when only 10 mol % of SHMeMB is added to the mixture. This large drop indicates that even at a low initial SHMeMB monomer fraction, the fraction of SHMeMB radicals ($f_{r,SHMeMB}$) is very high. Indeed, given the reactivity ratios of the system and the high homopropagation rate of AM (see Table 2), it can be calculated that at $f_{SHMeMB} = 0.1$, $f_{r,SHMeMB}$ is greater than 99%, such that SHMeMB-SHMeMB termination is the dominant termination event.

## 4. Conclusions

The difference in reactivity between SHMeMB and SHMB [12] motivated these further kinetic studies to explore the effects of depropagation and added salt on the polymerization rate. A plateau in conversion was observed for SHMeMB:AM copolymerizations conducted at 3:7 and 4:6 molar ratios and elevated temperatures, with monomer composition drift occurring at a faster rate as temperature increased and AM incorporated faster as SHMeMB units became more prone to depropagation. Homopolymerizations of SHMeMB at 75 and 90 °C provided further evidence of depropagation, as conversion reached a lowered equilibrium value at the higher temperature. Upon the addition of salt, the SHMeMB homopolymerization rate decreased at 75 °C, but not at 90 °C, due to the dominating effect of depropagation over the screening of charges provided by counterions.

The experimental data was fitted to models developed in PREDICI in order to estimate termination ($k_t$) and depropagation ($k_{dep}$) rate coefficients for SHMeMB, assuming a constant $k_p$ value of 25 L·mol$^{-1}$·s$^{-1}$. The $k_t$ values were estimated to be ~$10^5$ L·mol$^{-1}$·s$^{-1}$, similar in magnitude to those reported for NaAA [38] and NaMAA [33], and a ceiling temperature of ~90 °C was estimated. While these estimates have considerable uncertainty, the modeling effort has provided some valuable insights into the polymerization behavior of the system. The finding that $k_{t,SHMeMB}$ is of similar magnitude to other ionic monomers indicates that electrostatic repulsion of charged radical species, rather than steric hindrance from bulky substituents, is the reason for slow termination. The addition of salt and increase in monomer concentration both increased $k_p$ but had a greater effect on the estimated values of $k_t$, showing that addition of salt had a similar effect to an increased concentration of ionized monomers on the polymerization rate. The knowledge gathered from parameter estimations were then implemented to estimate $k_t$ values for SHMeMB:AM copolymerizations, which were found to be much lower than the known value of $10^8$ L·mol$^{-1}$·s$^{-1}$ for $k_{t,AM}$ but similar in magnitude to the estimates of $k_{t,SHMeMB}$. This result suggests that $k_t$ in the SHMeMB:AM copolymerization system is largely dominated by $k_{t,SHMeMB}$ because of the large fraction of charged SHMeMB radicals.

**Supplementary Materials:** The table and additional figures discussed in the text are available online at www.mdpi.com/2227-9717/5/4/55/s1.

**Acknowledgments:** The authors would like to thank the Natural Science and Engineering Research Council for their financial support.

**Author Contributions:** SBL and RAH conceived and designed the experiments; SBL performed the experiments and analyzed the results; SBL and RAH developed the model; SBL performed the parameter estimation; and SBL and RAH wrote the paper.

**Conflicts of Interest:** The authors declare no conflict of interest.

## References

1.  Syed, N.; Habib, W.W.; Kuhajda, A.M. Water-Soluble Polymers in Hair Care. In *Water Soluble Polymers*; Springer: Boston, MA, USA, 2002; pp. 231–244. ISBN 978-0-306-46915-2.

2.  Hayashi, Y.; Lu, D.; Kobayashi, N. Application of Ultra-High Molecular Weight Amphoteric Acrylamide Copolymers to Detergents. In *Water Soluble Polymers*; Springer: Boston, MA, USA, 2002; pp. 245–250. ISBN 978-0-306-46915-2.

3.  Ahmed, E.M. Hydrogel: Preparation, characterization, and applications: A review. *J. Adv. Res.* **2015**, *6*, 105–121. [CrossRef] [PubMed]

4.   Kadajji, V.; Betageri, G. Water Soluble Polymers for Pharmaceutical Applications. *Polymers* **2011**, *3*, 1972–2009. [CrossRef]

5.   Vedoy, D.; Soares, J. Water-soluble polymers for oil sands tailing treatment: A Review. *Can. J. Chem. Eng.* **2015**, *93*, 888–904. [CrossRef]

6.   Rivas, B.L.; Pereira, E.; Palencia, M.; Sánchez, J. Water-soluble functional polymers in conjunction with membranes to remove pollutant ions from aqueous solutions. *Prog. Polm. Sci.* **2011**, *36*, 294–322. [CrossRef]

7.   Akkapeddi, M.K. Poly(α-methylene-γ-butyrolactone) Synthesis Configurational Structure, and Properties. *Macromolecules* **1979**, *12*, 546–551. [CrossRef]

8.   Suenaga, J.; Sutherlin, D.M.; Stille, J. Polymerization of (RS)-and (R)-a-Methylene-y-methyl-y-butyrolactone. *Macromolecules* **1984**, *17*, 2913–2916. [CrossRef]

9.   Ueda, M.; Takahashi, M. Radical-initiated homo- and copolymerization of α-methylene-γ-butyrolactone. *J. Polym. Sci.* **1982**, *20*, 2819–2828. [CrossRef]

10.  Cockburn, R.A.; McKenna, T.F.; Hutchinson, R.A. An Investigation of Free Radical Copolymerization Kinetics of the Bio-renewable Monomer γ-Methyl-α-methylene-γ-butyrolactone with Methyl methacrylate and Styrene. *Macromol. Chem. Phys.* **2010**, *211*, 501–509. [CrossRef]

11.  Kollár, J.; Mrlík, M.; Moravčíková, D.; Kroneková, Z.; Liptaj, T.; Lacík, I.; Mosnáček, J. Tulips: A Renewable Source of Monomer for Superabsorbent Hydrogels. *Macromolecules* **2016**, *49*, 4047–4056. [CrossRef]

12.  Luk, S.B.; Kollár, J.; Chovancová, A.; Mrlík, M.; Lacík, I.; Mosnáček, J.; Hutchinson, R.A. Superabsorbent hydrogels made from bio-derived butyrolactone monomers in aqueous solution. *Polym. Chem.* **2017**. [CrossRef]

13.  Buback, M.; Gilbert, R.G.; Russell, G.T.; Hill, D.J.T.; Moad, G.; O'Driscoll, K.F.; Shen, J.; Winnik, M.A. Consistent values of rate parameters in free radical polymerization systems. II. Outstanding dilemmas and recommendations. *J. Polym. Sci. Part A Polym. Chem.* **1992**, *30*, 851–863. [CrossRef]

14.  Beuermann, S.; Buback, M.; Hesse, P.; Lacík, I. Free-Radical Propagation Rate Coefficient of Nonionized Methacrylic Acid in Aqueous Solution from Low Monomer Concentrations to Bulk Polymerization. *Macromolecules* **2006**, *39*, 184–193. [CrossRef]

15.  Lacík, I.; Beuermann, S.; Buback, M. PLP-SEC Study into Free-Radical Propagation Rate of Nonionized Acrylic Acid in Aqueous Solution. *Macromolecules* **2003**, *36*, 9355–9363. [CrossRef]

16.  Lacík, I.; Beuermann, S.; Buback, M. PLP-SEC Study into the Free-Radical Propagation Rate Coefficients of Partially and Fully Ionized Acrylic Acid in Aqueous Solution. *Macromol. Chem. Phys.* **2004**, *205*, 1080–1087. [CrossRef]

17.  Lacík, I.; Učňová, L.; Kukučková, S.; Buback, M.; Hesse, P.; Beuermann, S. Propagation Rate Coefficient of Free-Radical Polymerization of Partially and Fully Ionized Methacrylic Acid in Aqueous Solution. *Macromolecules* **2009**, *42*, 7753–7761. [CrossRef]

18.  Lacík, I.; Chovancová, A.; Uhelska, L.; Preusser, C.; Hutchinson, R.A.; Buback, M. PLP-SEC Studies into the Propagation Rate Coefficient of Acrylamide Radical Polymerization in Aqueous Solution. *Macromolecules* **2016**, *49*, 3244–3253. [CrossRef]

19.  Kuchta, F.D.; van Herk, A.M.; German, A.L. Propagation Kinetics of Acrylic and Methacrylic Acid in Water and Organic Solvents Studied by Pulsed-Laser Polymerization. *Macromolecules* **2000**, *33*, 3641–3649. [CrossRef]

20.  Preusser, C.; Chovancová, A.; Lacík, I.; Hutchinson, R.A. Modeling the Radical Batch Homopolymerization of Acrylamide and Aqueous Solution. *Macromol. React. Eng.* **2016**, *10*, 490–501. [CrossRef]

21.  Preusser, C.; Hutchinson, R.A. An In Situ NMR Study of Radical Copolymerization Kinetics of Acrylamide and Non-Ionized Acrylic Acid in Aqueous Solution. *Macromol. Symp.* **2013**, *333*, 122–137. [CrossRef]

22.  Preusser, C.; Ezenwajiaku, I.H.; Hutchinson, R.A. The Combined Influence of Monomer Concentration and Ionization on Acrylamide/Acrylic acid Composition in Aqueous Solution Radical Batch Copolymerization. *Macromolecules* **2016**, *49*, 4746–4756. [CrossRef]

23.  Riahinezhad, M.; McManus, N.; Penlidis, A. Effect of Monomer Concentration and pH on Reaction Kinetics and Copolymer Microsctructure of Acrylamide/Acrylic Acid Copolymer. *Macromol. React. Eng.* **2015**, *9*, 100–113. [CrossRef]

24.  Schier, J.E.; Hutchinson, R.A. The influence of hydrogen bonding on radical chain-growth parameters for butyl methacrylate/2-hydroxyethyl acrylate solution copolymerization. *Polym. Chem.* **2016**, *7*, 4567–4574. [CrossRef]

25. Li, D.; Li, N.; Hutchinson, R.A. High-Temperature Free Radical Copolymerization of Styrene and Butyl Methacrylate with Depropagation and Penultimate Kinetics Effects. *Macromolecules* **2006**, *39*, 4366–4373. [CrossRef]
26. Szablan, Z.; Stenzel, M.H.; Davis, T.P.; Barner, L.; Barner-Kowollik, C. Depropagation Kinetics of Sterically Demanding Monomers: A Pulsed Laser Size Exclusion Chromatography Study. *Macromolecules* **2005**, *38*, 5944–5954. [CrossRef]
27. Penelle, J.; Collot, J.; Rufflard, G. Kinetic and thermodynamic analysis of methyl ethacrylate radical polymerization. *J. Polym. Sci.* **1993**, *31*, 2407–2412. [CrossRef]
28. Brandrup, J.; Immergut, E.; Grulke, E. *Polymer Handbook*, 4th ed.; John Wiley & Sons: New York, NY, USA, 1999; ISBN 978-0-471-47936-9.
29. Morris, L.; Davis, T.; Chaplin, R. Radical copolymerization propagation kinetics of methyl ethacrylate and styrene. *Polymer* **2001**, *42*, 941–952. [CrossRef]
30. Ueda, M.; Takahashi, M.; Imai, Y.; Pittman, C.U. Synthesis and homopolymerization kinetics of α-methylene-δ-valerolactone, an exo-methylene cyclic monomer with a nonplanar ring system spanning the radical center. *Macromolecules* **1983**, *16*, 1300–1305. [CrossRef]
31. Drawe, P.; Buback, M.; Lacík, I. Radical Polymerization of Alkali Acrylates in Aqueous Solution. *Macromol. Chem. Phys.* **2015**, *216*, 1333–1340. [CrossRef]
32. Drawe, P. Kinetic of the Radical Polymerization of Ionic Monomers in Aqueous Solution: Spectroscopic Analysis and Modelling. Ph.D. Thesis, University of Göttingen, Göttingen, Germany, 2016.
33. Barth, J.; Buback, M. SP-PLP-EPR Study into the Termination Kinetics of Methacrylic Acid Radical Polymerization in Aqueous Solution. *Macromolecules* **2011**, *44*, 1292–1297. [CrossRef]
34. Wulkow, M. Computer Aided Modeling of Polymer Reaction Engineering—The Status of Predici, 1—Simulation. *Macromol. React. Eng.* **2008**, *2*, 461–494. [CrossRef]
35. Wittenburg, N. Kinetics and Modeling of the Radical Polymerization of Acrylic Acid and of Methacrylic Acid in Aqueous Solution. Ph.D. Thesis, University of Göttingen, Göttingen, Germany, 2013.
36. Wako Pure Chemical Industries Ltd. "V-50". Available online: http://www.wako-chem.co.jp/kaseihin_en/waterazo/V-50.htm (accessed on 31 March 2017).
37. Wako Pure Chemical Industries Ltd. "VA-086". Available online: http://www.wako-chem.co.jp/kaseihin_en/waterazo/VA-086.htm (accessed on 31 March 2017).
38. Barth, J.; Buback, M. Termination and Transfer Kinetics of Sodium Acrylate Polymerization. *Macromolecules* **2012**, *45*, 4152–4157. [CrossRef]
39. Luk, S.B. Radical Polymerization Kinetics of Bio-Renewable Monomers in Aqueous Solution. Master's Thesis, Queen's University, Kingston, ON, Canada, 2017.
40. Kattner, H.; Drawe, P.; Buback, M. Chain-Length-Dependent Termination of Sodium Methacrylate Polymerization in Aqueous Solution Studied by SP-PLP-EPR. *Macromolecules* **2017**, *50*, 1386–1393. [CrossRef]
41. Kattner, H.; Buback, M. Termination and Transfer Kinetics of Acrylamide Homopolymerization in Aqueous Solution. *Macromolecules* **2015**, *48*, 7410–7419. [CrossRef]

*processes*

MDPI

*Article*

# Organic Polymers as Porogenic Structure Matrices for Mesoporous Alumina and Magnesia

**Zimei Chen [1,2], Christian Weinberger [1], Michael Tiemann [1,\*] and Dirk Kuckling [2,\*]** 🄳

[1]    Department of Chemistry—Inorganic Functional Materials, University of Paderborn, Warburger Str. 100, 33098 Paderborn, Germany; zimei.chen@uni-paderborn.de (Z.C.); christian.weinberger@uni-paderborn.de (C.W.)
[2]    Department of Chemistry—Organic and Macromolecular Chemistry, University of Paderborn, Warburger Str. 100, 33098 Paderborn, Germany
\*    Correspondence: michael.tiemann@uni-paderborn.de (M.T.); dirk.kuckling@uni-paderborn.de (D.K.)

Received: 10 October 2017; Accepted: 6 November 2017; Published: 8 November 2017

**Abstract:** Mesoporous alumina and magnesia were prepared using various polymers, poly(ethylene glycol) (PEG), poly(vinyl alcohol) (PVA), poly($N$-(2-hydroxypropyl) methacrylamide) (PHPMA), and poly(dimethylacrylamide) (PDMAAm), as porogenic structure matrices. Mesoporous alumina exhibits large Brunauer–Emmett–Teller (BET) surface areas up to 365 m$^2$ g$^{-1}$, while mesoporous magnesium oxide possesses BET surface areas around 111 m$^2$ g$^{-1}$. Variation of the polymers has little impact on the structural properties of the products. The calcination of the polymer/metal oxide composite materials benefits from the fact that the polymer decomposition is catalyzed by the freshly formed metal oxide.

**Keywords:** mesoporous alumina; mesoporous magnesia; poly(ethylene glycol); poly(vinyl alcohol); poly($N$-(2-hydroxypropyl) methacrylamide); poly(dimethylacrylamide)

---

## 1. Introduction

Mesoporous metal oxides with large specific surface areas and uniform pore sizes have recently attained great interest, particularly regarding potential applications in such areas as catalysis [1], energy conversion and storage [2], and gas sensing [3,4]. By definition, mesopore widths range from 2 to 50 nm [5]. For metal oxides with uniform and ordered mesopores, a variety of synthesis methods have been established, mostly by utilization of porogens; said porogens may be supramolecular entities of amphiphilic species dispersed in liquid media ('soft templates' [6]) or solid structure matrices such as porous silica ('hard templates') in the so-called 'nanocasting' process [6–9].

Alumina (aluminum oxide, Al$_2$O$_3$) and magnesia (magnesium oxide, MgO) with high surface-to volume ratios play an important role as catalyst/catalyst support materials [10–13] and as adsorbents [14–16]. Both materials can be prepared by nanocasting, which leads to ordered and uniform mesopores. However, unlike for most other metal oxides, mesoporous silica is not suitable as a structure matrix here, because its removal requires chemical etching under strongly basic (e.g., NaOH) or acidic (HF) conditions. Both Al$_2$O$_3$ and MgO are amphoteric oxides that cannot withstand these conditions. Instead, mesoporous carbon materials have been employed as structure matrices for amphoteric oxides such as mesoporous Al$_2$O$_3$ [17,18], MgO [19,20], and ZnO [21–24], since their removal can be accomplished under milder condition by thermal oxidation [25,26]. Likewise, organic hydrogels have also been shown to be versatile porogenic matrices for porous oxidic materials [27–31]. We have recently described the utilization of photo cross-linked poly(dimethylacrylamide)-based hydrogels [32,33] as matrices for mesoporous alumina [34,35]. Here we report on the utilization of non-cross-linked water-soluble polymers as porogenic species for mesoporous Al$_2$O$_3$ and MgO; the synthesis process is thus simplified.

Polymer chains begin to overlap and form entanglements when the polymer solution is above a critical concentration. Hence, a physical network is formed between different polymer chains in a concentrated solution [36,37]. Therefore, a concentrated polymer solution could also theoretically work as a structure matrix to prepare mesoporous metal oxides. In this paper, we describe the synthesis of mesoporous alumina and magnesium oxide using simple polymers, such as poly(ethylene glycol) (PEG), poly(vinyl alcohol) (PVA), poly(*N*-(2-hydroxypropyl) methacrylamide) (PHPMA), and poly(dimethylacrylamide) (PDMAAm), as matrices. The strategy proposed here to prepare porous alumina and magnesium oxide is based on a one-pot synthesis approach using saturated aluminum/magnesium nitrate as precursor solutions and introducing direct polymers.

## 2. Materials and Methods

### 2.1. Materials

Poly(ethylene glycol) (PEG, Fluka, Darmstadt, Germany, $M_n$ 6000 g mol$^{-1}$), poly(vinyl alcohol) (PVA, Acros, Geel, Belgium, ≥98%, $M_n$ 16,000 g mol$^{-1}$), aluminum nitrate nonahydrate (Sigma-Aldrich, Darmstadt, Germany, ≥98.0%), magnesium nitrate hexahydrate (Sigma-Aldrich, Darmstadt, Germany, ≥97%), 1-amino-2-propanol (TCI, Eschborn, Germany, >98%), methacryloyl chloride (Fluka, Darmstadt, Germany, >97%), and 1,2-diaminoethane (Acros, Geel, Belgium, >99%) were used as received. *N,N*-dimethylacrylamide (DMAAm, TCI, Eschborn, Germany, 99%) was distilled under low pressure. α,α′-Azobisisobutyronitrile (AIBN, Fluka, Darmstadt, Germany, >98%) was recrystallized from methanol. Ammonia solution (Stockmeier, Bielefeld, Germany, 25%), diethyl ether (Hanke + Seidel, Steinhagen, Germany), tetrahydrofuran (THF, BASF, Ludwigshafen, Germany), 1,4-dioxane (Carl Roth, Karsruhe, Germany, ≥99.5%) ethyl acetate (Stockmeier, Bielefeld, Germany,), methanol (Stockmeier, Bielefeld, Germany,), magnesium sulfate (Grüssing, Filsum, Germany, 99%), acetone (Stockmeier, Bielefeld, Germany,), and sodium sulfate (Grüssing, Filsum, Germany, 99%) were used as received.

### 2.2. Characterization

$^1$H and $^{13}$C NMR spectra were recorded on a Bruker (Billerica, Massachusetts, USA) AV 500 spectrometer at 500 MHz and 125 MHz, respectively. Reference solvent signals at 7.26 and 2.56 ppm were used for spectra in CDCl$_3$ (99.8 atom-% Deuterium) and DMSO-d$_6$ (O=S(CD$_3$)$_2$, 99.9%), respectively.

Gel permeation chromatography (GPC) was performed in chloroform for PEG and PDMAAm at 30 °C and at a flow rate of 0.75 mL min$^{-1}$ on a Jasco (Groß-Umstadt, Germany) 880-PU Liquid Chromatograph connected to a Shodex (Yokohama, Japan) RI-101 detector. The instrument was equipped with four consecutive columns (PSS-SDV columns filled with 5 μm gel particles with a defined porosity of 10$^6$ Å, 10$^5$ Å, 10$^3$ Å, and 10$^2$ Å, respectively), and both samples were calibrated by poly(methyl methacrylate) standards. GPC was performed in hexafluoroisopropanol for PVA at 0 °C and at a flow rate of 1 mL min$^{-1}$ on a Merck (Darmstadt, Germany) LC-6200 liquid chromatograph connected to a Shodex (Yokohama, Japan) RI-101 detector. The instrument was equipped with a PSS-PFG 10$^3$ Å and PSS-PFG 10$^2$ Å column, and the sample was calibrated by poly(methyl methacrylate) standards. GPC was performed in *N,N*-dimethylacetamide for PHPMA at 50 °C and at a flow rate of 0.5 mL min$^{-1}$ on a Merck (Darmstadt, Germany) LC 655A-11 liquid chromatograph connected to a Waters (Milford, Massachusetts, United States) RI 2410 detector. The instrument was equipped with PSS-GRAM 10$^4$ Å, PSS-GRAM 10$^3$ Å, and PSS-GRAM 10$^2$ Å columns, and the sample was calibrated by poly(methyl methacrylate) standards. Thermogravimetric analysis (TGA) was conducted under synthetic air at a heating rate of 10 °C min$^{-1}$ using a Mettler Toledo (Columbus, Ohio, USA) TGA/SDTA851. N$_2$ physisorption analysis was performed at 77 K on a Quantachrome (Boynton Beach, Florida, United States) Autosorb 6B instrument; samples were degassed at 120 °C for 12 h prior to measurement. Specific surface areas were assessed via multi-point Brunauer–Emmett–Teller

(BET) analysis [38] in the range of $0.1 \leq p/p_0 \leq 0.3$. Pore volumes were calculated at $p/p_0 = 0.99$. Pore size distributions were calculated via Barrett–Joyner–Halenda (BJH) analysis [39] from the desorption branches of the isotherms. Powder X-ray diffraction was performed with a Bruker (Billerica, Massachusetts, USA) AXS D8 Advance diffractometer with Cu K$\alpha$ radiation (40 kV, 40 mA) with a step size of $0.02°$ and a counting time of 3 s per step.

*2.3. Monomer Synthesis*

$N$-(2-hydroxypropyl) methacrylamide (HPMA) was synthesized as described in the literature [40]. 1-Amino-2-propanol (45.5 mL, 589 mmol) and ethyl acetate (450 mL) were added in a 1 L three-neck round-bottom flask equipped with addition funnel. The flask was cooled to 10 °C and purged with argon for 15 min. Methacryloyl chloride (28 mL, 287 mmol) and ethyl acetate (50 mL) were added to the addition funnel and purged with argon for 15 min and left under an argon atmosphere. The methacryloyl chloride/ethyl acetate mixture was then added dropwise to the 1-amino-2-propanol/ethyl acetate mixture. The mixture was reacted in an ice bath for 1 h. Afterwards, the mixture was washed three times with an aqueous sat. sodium sulfate solution (250 mL) in a separatory funnel to remove any excess of reactants and side products. The aqueous phase was discarded and the organic phase was dried over magnesium sulfate and concentrated in vacuo to approximately 50 mL. The concentrate was then allowed to age for 1 h at 10 °C. The product was collected as colorless solid by filtration, dried under vacuum, and stored in the freezer. (11.52 g, 28%) $^1$H NMR (500 MHz, CDCl$_3$): $\delta$ (ppm) = 1.21 (d, $J$ = 6.3 Hz, 3 H, =CCH$_3$), 1.97 (dd, $J$ = 1.5, 1.0 Hz, 3 H, (HO)CCH$_3$), 2.36 (s,1H, OH), 3.18 (ddd, $J$ = 14.0, 7.6, 5.2 Hz, 1 H, CH$_2$), 3.51 (ddd, $J$ = 14.0, 6.5, 3.0 Hz, 1 H, CH$_2$), 3.96 (ddd, $J$ = 7.6, 6.3, 3.0 Hz, 1 H, CH), 5.33–5.37 (m, 1 H, =CH$_2$), 5.69–5.74 (m, 1 H, =CH$_2$), 6.24 (br. s, 1 H, NH). $^{13}$C NMR (125 MHz, CDCl$_3$): $\delta$ (ppm) = 18.64 (CH$_3$), 21.04 (CH$_3$), 47.17 (NH–CH$_2$), 67.52 (CH–OH), 119.88 (=CH$_2$), 139.77 (=C), 169.39 (C=O).

*2.4. Polymer Synthesis*

Homopolymer PDMAAm was synthesized by free radical polymerization initiated with AIBN as described in the literature [41]. Monomer DMAAm (5.2 mL, 50.4 mmol) and AIBN (10 mg, 0.06 mmol) were dissolved in 1,4-dioxane (92 mL) and purged with argon for 20 min. The polymerization was carried out at 70 °C for 7 h under an argon atmosphere. Afterwards, the polymer was precipitated in diethyl ether and reprecipitated from THF into diethyl ether for the purification. Finally, the product was obtained by low pressure drying and characterized by NMR spectroscopy and GPC (see Table 1). (3.39 g, 68%) $^1$H NMR (500 MHz, CDCl$_3$): $\delta$ (ppm)= 1.51–1.83 (m, CH$_2$), 2.30–2.74 (m, CH), 2.75–3.22 (m, CH$_3$).

**Table 1.** Characterization of the used homopolymers.

| Polymer | $M_n$/(g mol$^{-1}$) | Đ | Yield/% |
|---|---|---|---|
| PEG [1] | 12,000 | 1.1 | - |
| PVA [1] | 23,000 | 2.4 | - |
| PDMAAm [1] | 26,000 | 2.8 | 68 |
| PHPMA [1] | 43,000 | 6.3 | 77 |

[1] Poly(ethylene glycol) (PEG) and poly(dimethylacrylamide) (PDMAAm) determined by gel permeation chromatography (GPC) in CHCl$_3$, poly(vinyl alcohol) (PVA) determined in hexafluoroisopropanol, and poly($N$-(2-hydroxypropyl) methacrylamide) (PHPMA) determined in $N,N$-dimethylacetamide, all of which were calibrated by poly(methyl methacrylate) (PMMA) standards.

Homopolymer PHPMA was synthesized by free radical polymerization initiated with AIBN as described in the literature [42]. Monomer HPMA (1.5 g, 10.5 mmol) and AIBN (1.7 mg, 0.010 mmol) were dissolved in 1,4-dioxane (20 mL) in a 50 mL nitrogen flask and was degassed three times by freeze/thaw cycles. The HPMA was polymerized at 65 °C for 8 h under an argon atmosphere. The mixture was poured into acetone to get a white solid, which was collected and washed with acetone repeatedly. Further purification was carried out by dissolving the polymer in methanol and precipitating into acetone. The product was collected and dried under vacuum to obtain the homopolymer as a white powder and characterized by NMR spectroscopy and GPC (see Table 1). (1.16 g, 77%) $^1$H NMR (500 MHz, DMSO): δ (ppm) = 0.70–1.13 (m, CH$_3$), 1.43–2.04 (m, CH$_2$), 2.92 (m, NH–CH$_2$, OH), 3.69 (m, NH–CH$_2$), 4.69 (m, CH), 7.14 (br, NH).

### 2.5. Preparation of Mesoporous Metal Oxides

One hundred ninety six milligrams of polymer were dissolved in 800 μL of a saturated aqueous solution of aluminum nitrate (1.9 mol L$^{-1}$) or magnesium nitrate (4.9 mol L$^{-1}$). The Al(NO$_3$)$_3$-containing solution was treated at 60 °C with a vapor of an aqueous ammonia solution (12.5%) for 3 h to convert Al(NO$_3$)$_3$ to Al(OH)$_3$/AlO(OH); the resulting material was dried overnight at 60 °C and then calcined in a tube furnace for 4 h at 500 °C (heating rate 1 °C min$^{-1}$) to form Al$_2$O$_3$ and to combust the polymer. The Mg(NO$_3$)$_2$-containing solution was dried overnight at 120 °C; the resulting material was calcined in a tube furnace for 2 h at 300 °C and for 2 h at 500 °C (heating rate 1 °C min$^{-1}$) to convert Mg(NO$_3$)$_2$ to MgO and to combust the polymer.

## 3. Results and Discussion

A variety of four simple water soluble polymers were used as porogenic structure directors for mesoporous Al$_2$O$_3$ and MgO. The polymers possess different hydrophilicity and distinct ability to coordinate to Al$^{3+}$ and Mg$^{2+}$ metal cations: (i) poly(ethylene glycol) (PEG; ether groups), (ii) poly(vinyl alcohol) (PVA; hydroxyl groups), (iii) poly(dimethylacrylamide) (PDMAAm; tertiary amido groups), and (iv) poly(N-(2-hydroxypropyl) methacrylamide) (PHPMA; secondary amido with hydroxyl groups). The latter two polymers were synthesized by free-radical polymerization, as shown in Figure 1b,c. Their properties are summarized in Table 2; molecular weights, dispersities, and yields are typical of free-radical polymerization synthesis.

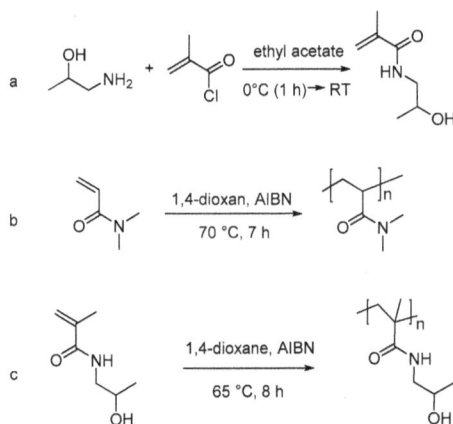

**Figure 1.** Synthesis of (**a**) monomer N-(2-hydroxypropyl) methacrylamide (HPMA) and homopolymers (**b**) PDMAAm and (**c**) PHPMA.

The aim of this study was to investigate the impact of the polymers on the porosity of the metal oxides. For this purpose, the respective polymer was dissolved in a saturated aqueous solution of aluminum nitrate, followed by treatment in ammonia vapor at 60 °C to convert Al(NO$_3$)$_3$ to Al(OH)$_3$/AlO(OH), as described in the experimental section. After evaporation of the water, the material was then calcined at 500 °C to turn Al(OH)$_3$/AlO(OH) into Al$_2$O$_3$ and simultaneously combust the polymer. For MgO, the same procedure was applied, but without the ammonia treatment step; magnesium nitrate was directly converted to magnesium oxide by calcination. By this procedure, a composite of the metal oxide precursor ($Al(OH)_3$/$AlO(OH)$ or $Mg(NO_3)_2$, respectively) and the polymer was formed first, with the polymer being entangled within the inorganic phase. Then, simultaneous conversion of the precursor into the metal oxide and thermal combustion of the polymer led to a mesoporous product.

To study the calcination/polymer combustion step in some detail, thermogravimetric analysis (TGA) was carried out. As an example, the TGA curves of the $Al(OH)_3$/PDMAAm composite and of the pure PDMAAm polymer were compared and are shown in Figure 2. A mass loss of ca. 72% can be observed for the composite material in the temperature range up to 230 °C, which can be attributed to both the dehydration of $Al(OH)_3$/$AlO(OH)$ (i.e., $Al_2O_3$ formation) and the combustion of the polymer. Further mass loss of ca. 14% can be observed between 230 and 570 °C. By comparison, the pure polymer shows an initial mass loss of 6% below 200 °C, probably due to loss of residual water, then a mass loss of about 74% between 300 to 400 °C, followed by another 18% up to ca. 600 °C. Obviously, the presence of the aluminum hydroxide/oxide led to a combustion of the polymer at lower temperature; this effect has already been observed for the combustion of amorphous carbon [18,26] and organic hydrogel matrices [35]. Very similar results were obtained for $Al_2O_3$ prepared using the other polymers (see Figures S1–S3).

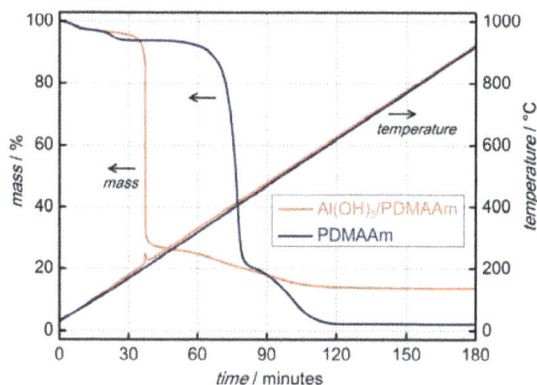

**Figure 2.** Thermo-gravimetric analysis (TGA) of the $Al(OH)_3$-polymer composite and of the pure polymer PDMAAm.

The porogenic impact of the polymers on the polymer-free metal oxides was confirmed by $N_2$ physisorption analysis. Figure 3 (left) shows the sorption isotherms of four $Al_2O_3$ materials prepared with different polymers. All isotherms exhibit a faint type-IV(a) behavior [43] with a more or less well-pronounced hysteresis. This indicates mesopores with an ill-defined shape, but with a fairly uniform size, as confirmed in the BJH pore size distribution curves [41] derived from the isotherms (Figure 3, right). Pore widths from 3 to 8 nm can be observed, with a clear peak occurring at 3.6 nm in all materials. The pore size distribution is somewhat narrower in the two samples prepared with PVA and PEG, respectively. The specific pore volumes and BET surface areas are shown in Table 2, confirming that a reproducible synthesis of porous alumina with a large surface area up to 365 $m^2$ $g^{-1}$ is possible by the utilization of these polymers as porogens. Comparison of all prepared $Al_2O_3$ materials reveals similar mesopore sizes, mesopore volumes, and specific BET areas. The choice of the porogenic polymer matrix has little impact on the porosity. Although polymers with different binding sites were used, the appearance of a polymer rich phase due to physical network formation in concentrated solution can be considered the sole reason for pore formation.

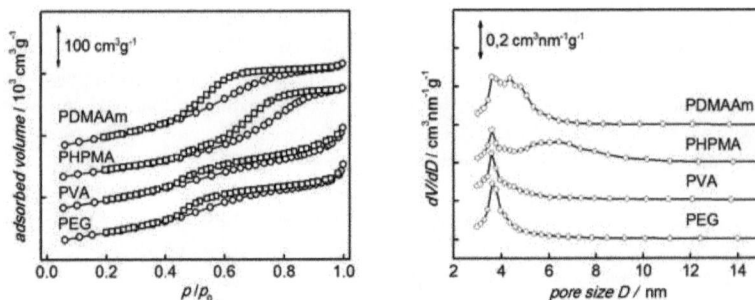

**Figure 3.** N$_2$ physisorption isotherm (**left**) and pore size distribution (**right**) of mesoporous γ-Al$_2$O$_3$ prepared using various polymers as the porogenic structure matrices as indicated. (Data are vertically shifted for clarity).

**Table 2.** Specific Brunauer–Emmett–Teller (BET) surface areas $A_{BET}$, pore volumes $V$, and mean pore widths $r$ obtained from N$_2$ physisorption of mesoporous alumina synthesized using various polymers.

| Polymer Used | $A_{BET}$/m$^2$ g$^{-1}$ | $V$/cm$^3$ g$^{-1}$ | $r$/nm |
|---|---|---|---|
| PDMAAm | 365 | 0.51 | 3.6 |
| PHPMA | 312 | 0.54 | 3.6 |
| PEG | 325 | 0.44 | 3.6 |
| PVA | 343 | 0.48 | 3.6 |

Figure 4 shows the powder X-ray diffraction patterns of the alumina materials. Again, the differences between the materials are rather low. All samples exhibit only a few broad reflections, two of which can be attributed to the cubic defect spinel structure of γ-Al$_2$O$_3$. (JCPDS card number 75-0921). The formation of this phase with low crystallinity is commonly observed for Al$_2$O$_3$ syntheses under these conditions [17,18]. The crystallite sizes calculated by the Scherrer method are between 5 and 6 nm.

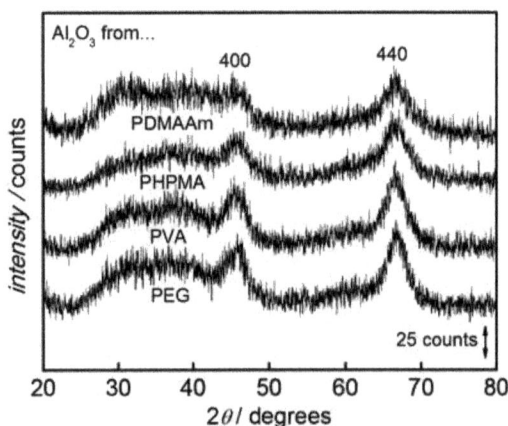

**Figure 4.** Powder XRD patterns of mesoporous γ-Al$_2$O$_3$ prepared using various polymers as the porogenic structure matrices as indicated. (Data are vertically shifted for clarity).

Since the choice of polymer turned out not to have any significant impact on the Al$_2$O$_3$ synthesis, only one polymer, PDMAAm, was chosen for the preparation of porous MgO. The TGA curves of the

Mg(NO$_3$)$_2$/PDMAAm composite and of the pure PDMAAm polymer are shown in Figure 5. For the composite, the mass loss occurs in two distinct steps: by ca. 54% up to a temperature of 265 °C and by another 32% between 265 and 500 °C. It is fair to assume that the first step is mainly attributable to the conversion of magnesium nitrate into magnesium oxide, while the second step corresponds mostly to the polymer decomposition. This seems likely because the pure polymer starts to combust only above ca. 300 °C (after some initial mass loss of 6% below 200 °C, presumably due to loss of residual water); a steep reduction in mass by ca. 74% occurs between 300 and 400 °C, followed by another 18% between 400 and 600 °C. Again, the presence of the magnesium species results in a polymer decomposition at a slightly lower temperature, although this effect is less pronounced than in the case of the aluminum species.

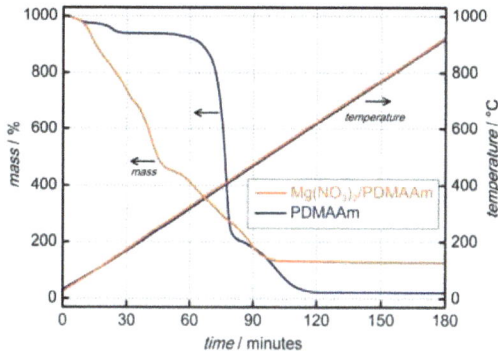

**Figure 5.** Thermo-gravimetric analysis (TGA) of the Mg(NO$_3$)$_2$-polymer composite and of the pure PDMAAm polymer.

Figure 6 shows the N$_2$ physisorption data of the porous MgO sample. The isotherm shape is mostly type II, with a slight type-IV character and little hysteresis, indicating a fairly low degree of porosity. Accordingly, the pore size distribution peak is very low in intensity. The specific BET surface area and pore volume are 111 m$^2$ g$^{-1}$ and 0.37 cm$^3$ g$^{-1}$, respectively. Obviously, the polymer failed to have a pronounced porogenic impact in the case of MgO, which may be explained by the sintering of MgO particles during calcination upon combustion of the polymer. During the Al$_2$O$_3$ synthesis, by contrast, a solid network of Al(OH)$_3$/AlO(OH) was formed before the combustion of the polymer.

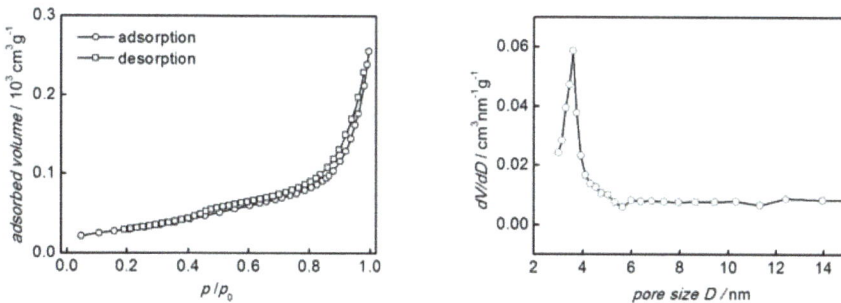

**Figure 6.** N$_2$ physisorption isotherm (**left**) and pore size distribution (**right**) of MgO prepared using PDMAAm polymer as the porogenic structure matrix.

The powder X-ray diffraction diagram of MgO is shown in Figure 7, confirming the cubic rock salt structure of MgO (JCPDS card number 77-2179) with a substantially higher degree of crystallinity than in case of $Al_2O_3$. This is consistent with the above-made assumption of strong sintering upon polymer combustion. The crystallite size calculated by the Scherrer method is ca. 12 nm.

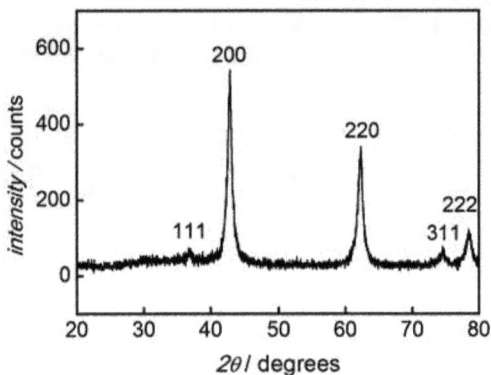

**Figure 7.** Powder XRD pattern of MgO prepared by using PDMAAm polymer as the porogenic structure matrix.

## 4. Conclusions

Mesoporous $\gamma$-$Al_2O_3$ and mesoporous MgO with large specific BET surface areas were successfully synthesized using simple polymers (PEG, PVA, PDMAAm, and PHPMA) as porogenic matrices under relatively mild conditions. The polymers were mixed with a metal nitrate solution. The polymer matrices were removed by thermal combustion, while the metal oxides were formed at the same time. The mesoporous alumina products exhibit mesopore sizes in the range from 3.6 to 6.4 nm, large specific BET surface areas up to 365 $m^2$ $g^{-1}$, and specific pore volumes up to 0.54 $cm^3$ $g^{-1}$. Variation of the polymer has little impact on the structural properties of the products. The mesoporous magnesium oxide product has a mesopore size of 3.6 nm, a specific BET surface area of 111 $m^2$ $g^{-1}$, and a specific pore volume of 0.37 $cm^3$ $g^{-1}$. The calcination of the polymer/metal oxides composite materials benefits from the fact that polymer decomposition is catalyzed by the freshly formed metal oxides.

**Supplementary Materials:** The following are available online at http://www.mdpi.com/2227-9717/5/4/70/s1. Figure S1: TGA of the $Al(OH)_3$-PVA composite and of the pure polymer PVA; Figure S2: TGA of the $Al(OH)_3$-PHPMA composite and of the pure polymer PHPMA; Figure S3: TGA of the $Al(OH)_3$-PEG composite and of the pure PEG.

**Acknowledgments:** The authors thank Manuel Traut for help with the TGA measurements.

**Author Contributions:** M.T. and D.K. conceived and designed the experiments; Z.C. performed the experiments; Z.C. and C.W. analyzed the data; Z.C., M.T. and D.K. wrote the paper.

**Conflicts of Interest:** The authors declare no conflict of interest.

## References

1. Tüysüz, H.; Schüth, F. Ordered Mesoporous Materials as Catalysts. *Adv. Catal.* **2012**, *55*, 127–239.
2. Li, W.; Liu, J.; Zhao, D. Mesoporous materials for energy conversion and storage devices. *Nat. Rev. Mater.* **2016**, *1*, 16023. [CrossRef]
3. Tiemann, M. Porous metal oxides as gas sensors. *Chem. Eur. J.* **2007**, *13*, 8376–8388. [CrossRef] [PubMed]
4. Wagner, T.; Haffer, S.; Weinberger, C.; Klaus, D.; Tiemann, M. Mesoporous materials as gas sensors. *Chem. Soc. Rev.* **2013**, *42*, 4036–4053. [CrossRef] [PubMed]

5.  Sing, K.S.W.; Everett, D.H.; Haul, R.A.W.; Moscou, L.; Pierotti, R.A.; Rouquérol, J.; Siemieniewska, T. Reporting physisorption data for gas/solid systems with special reference to the determination of surface area and porosity. *Pure Appl. Chem.* **1985**, *57*, 603–619. [CrossRef]
6.  Gu, D.; Schüth, F. Synthesis of non-siliceous mesoporous oxides. *Chem. Soc. Rev.* **2014**, *43*, 313–344. [PubMed]
7.  Tiemann, M. Repeated templating. *Chem. Mater.* **2007**, *20*, 961–971. [CrossRef]
8.  Ren, Y.; Ma, Z.; Bruce, P.G. Ordered mesoporous metal oxides: Synthesis and applications. *Chem. Soc. Rev.* **2012**, *41*, 4909–4927. [PubMed]
9.  Deng, X.; Chen, K.; Tüysüz, H. Protocol for the nanocasting method: Preparation of ordered mesoporous metal oxides. *Chem. Mater.* **2017**, *29*, 40–51.
10.  Čejka, J. Organized mesoporous alumina: Synthesis, structure and potential in catalysis. *Appl. Catal. A Gen.* **2003**, *254*, 327–338.
11.  Choudary, B.M.; Mulukutla, R.S.; Klabunde, K.J. Benzylation of aromatic compounds with different crystallites of MgO. *J. Am. Chem. Soc.* **2003**, *125*, 2020–2021. [CrossRef] [PubMed]
12.  Trueba, M.; Trasatti, S.P. γ-Alumina as a support for catalysts: A review of fundamental aspects. *Eur. J. Inorg. Chem.* **2005**, *17*, 3393–3403. [CrossRef]
13.  Morris, S.M.; Fulvio, P.F.; Jaroniec, M. Ordered mesoporous alumina-supported metal oxides. *J. Am. Chem. Soc.* **2008**, *130*, 15210–15216. [CrossRef] [PubMed]
14.  Rajagopalan, S.; Koper, O.; Decker, S.; Klabunde, K.J. Nanocrystalline metal oxides as destructive adsorbents for organophosphorus compounds at ambient temperatures. *Chem. Eur. J.* **2002**, *8*, 2602–2607. [CrossRef]
15.  Li, L.; Wen, X.; Fu, X.; Wang, F.; Zhao, N.; Xiao, F.; Wie, W.; Sun, Y. MgO/Al$_2$O$_3$ Sorbent for CO$_2$ Capture. *Energy Fuels* **2010**, *24*, 5773–5780. [CrossRef]
16.  Wie, J.; Ren, Y.; Luo, W.; Sun, Z.; Cheng, X.; Li, Y.; Deng, Y.; Elzatahry, A.A.; Al-Dahyan, D.; Zhao, D. Ordered mesoporous alumina with ultra-large pores as an efficient absorbent for selective bioenrichment. *Chem. Mater.* **2017**, *29*, 2211–2217.
17.  Liu, Q.; Wang, A.; Wang, X.; Zhang, T. Ordered crystalline alumina molecular sieves synthesized via a nanocasting route. *Chem. Mater.* **2006**, *18*, 5153–5155. [CrossRef]
18.  Haffer, S.; Weinberger, C.; Tiemann, M. Mesoporous Al$_2$O$_3$ by nanocasting: Relationship between crystallinity and mesoscopic order. *Eur. J. Inorg. Chem.* **2012**, *2012*, 3283–3288. [CrossRef]
19.  Roggenbuck, J.; Tiemann, M. Ordered mesoporous magnesium oxide with high thermal stability synthesized by exotemplating using CMK-3 carbon. *J. Am. Chem. Soc.* **2005**, *127*, 1096–1097. [CrossRef] [PubMed]
20.  Roggenbuck, J.; Koch, G.; Tiemann, M. Synthesis of mesoporous magnesium oxide by CMK-3 carbon structure replication. *Chem. Mater.* **2006**, *18*, 4151–4156. [CrossRef]
21.  Waitz, T.; Tiemann, M.; Klar, P.J.; Sann, J.; Stehr, J.; Meyer, B.K. Crystalline ZnO with an enhanced surface area obtained by nanocasting. *Appl. Phys. Lett.* **2007**, *90*, 123108. [CrossRef]
22.  Wagner, T.; Waitz, T.; Roggenbuck, J.; Fröba, M.; Kohl, C.-D.; Tiemann, M. Ordered mesoporous ZnO for gas sensing. *Thin Solid Films* **2007**, *515*, 8360–8363. [CrossRef]
23.  Chernikov, A.; Horst, S.; Waitz, T.; Tiemann, M.; Chatterjee, S. Photoluminescence properties of ordered mesoporous ZnO. *J. Phys. Chem. C* **2011**, *115*, 1375–1379. [CrossRef]
24.  Polarz, S.; Orlov, A.V.; Schüth, F.; Lu, A.-H. Preparation of High-Surface-Area Zinc Oxide with Ordered Porosity, Different Pore Sizes, and Nanocrystalline Walls. *Chem. Eur. J.* **2007**, *13*, 592–597. [CrossRef] [PubMed]
25.  Roggenbuck, J.; Waitz, T.; Tiemann, M. Synthesis of Mesoporous Metal Oxides by Structure Replication: Strategies of Impregnating Porous Matrices with Metal Salts. *Microporous Mesoporous Mater.* **2008**, *113*, 575–582. [CrossRef]
26.  Weinberger, C.; Roggenbuck, J.; Hanss, J.; Tiemann, M. Synthesis of Mesoporous Metal Oxides by Structure Replication: Thermal Analysis of Metal Nitrates in Porous Carbon Matrices. *Nanomaterials* **2015**, *5*, 1431–1441. [CrossRef] [PubMed]
27.  Llusar, M.; Pidol, L.; Roux, C.; Pozzo, J.L.; Sanchez, C. Templated Growth of Alumina-Based Fibers through the Use of Anthracenic Organogelators. *Chem. Mater.* **2002**, *14*, 5124–5133. [CrossRef]
28.  Jiu, J.; Kurumada, K.; Tanigaki, M. Preparation of oxide with nano-scaled pore diameters using gel template. *J. Non-Cryst. Solids* **2003**, *325*, 124–132. [CrossRef]

29. Kurumada, K.; Suzuki, A.; Baba, S.; Otsuka, E. Relationship between polarity of template hydrogel and nanoporous structure replicated in sol–gel-derived silica matrix. *J. Appl. Polym. Sci.* **2009**, *114*, 4085–4090. [CrossRef]

30. Cui, X.; Tang, S.; Zhou, H. Mesoporous alumina materials synthesized in different gel templates. *Mater. Lett.* **2013**, *98*, 116–119. [CrossRef]

31. Jiang, R.; Zhu, H.-Y.; Chen, H.-H.; Yao, J.; Fu, Y.-Q.; Zhang, Z.-Y.; Xu, Y.-M. Effect of calcination temperature on physical parameters and photocatalytic activity of mesoporous titania spheres using chitosan/poly(vinyl alcohol) hydrogel beads as a template. *Appl. Surf. Sci.* **2014**, *319*, 189–196. [CrossRef]

32. Kuckling, D.; Hoffmann, J.; Plötner, M.; Ferse, D.; Kretschmer, K.; Adler, H.-J.P.; Arndt, K.-F.; Reichelt, R. Photo cross-linkable poly(*N*-isopropylacrylamide) copolymers III: micro-fabricated temperature responsive hydrogels. *Polymer* **2003**, *44*, 4455–4462. [CrossRef]

33. Döring, A.; Birnbaum, W.; Kuckling, D. Responsive hydrogels—structurally and dimensionally optimized smart frameworks for applications in catalysis, micro-system technology and material science. *Chem. Soc. Rev.* **2013**, *40*, 7391–7420. [CrossRef] [PubMed]

34. Birnbaum, W.; Weinberger, C.; Schill, V.; Haffer, S.; Tiemann, M.; Kuckling, D. Synthesis of mesoporous alumina through photo cross-linked poly(dimethylacrylamide) hydrogels. *Colloid Polym. Sci.* **2014**, *292*, 3055–3060. [CrossRef]

35. Weinberger, C.; Chen, Z.; Birnbaum, W.; Kuckling, D.; Tiemann, M. Photo-cross-linked polydimethylacrylamide hydrogels as porogens for mesoporous alumina. *Eur. J. Inorg. Chem.* **2017**, *2017*, 1026–1031. [CrossRef]

36. Cottet, H.; Gareil, P. Electrophoretic behavior of fully sulfonated polystyrenes in capillaries filled with entangled polymer solutions. *J. Chromatogr. Coruña* **1997**, *772*, 369–384. [CrossRef]

37. Daoud, M.; Stanley, H.E.; Stauffer, D. Scaling, Exponents, and Fractal Dimensions. In *Physical Properties of Polymers Handbook*, 2nd ed.; Mark, J.E., Ed.; Springer: New York, NY, USA, 2007; pp. 83–92.

38. Brunauer, S.; Emmett, P.H.; Teller, E. Adsorption of gases in multimolecular layers. *J. Am. Chem. Soc.* **1938**, *60*, 309–319. [CrossRef]

39. Barrett, E.P.; Joyner, L.G.; Halenda, P.P. The determination of pore volume and area distributions in porous substances. I. Computations from nitrogen isotherms. *J. Am. Chem. Soc.* **1951**, *73*, 373–380. [CrossRef]

40. Rowe, M.D.; Chang, C.C.; Thamm, D.H.; Kraft, S.L.; Harmon, J.F.; Vogt, A.P.; Sumerlin, B.S.; Boyes, S.G. Tuning the magnetic resonance imaging properties of positive contrast agent nanoparticles by surface modification with RAFT polymers. *Langmuir* **2009**, *25*, 9487–9499. [CrossRef] [PubMed]

41. Kuckling, D.; Harmon, M.E.; Frank, C.W. Photo-cross-linkable PNIPAAm copolymers. 1. Synthesis and characterization of constrained temperature-responsive hydrogel layers. *Macromolecules* **2002**, *35*, 6377–6383. [CrossRef]

42. Javadi, A. Synthesis of Thin Hydrogel Layers Based on Photo-Cross-Linkable Polymers. Ph.D. Thesis, Tarbiat Moallem University, Tehran, Iran, May 2012.

43. Thommes, M.; Kaneko, K.; Neimark, A.V.; Olivier, J.P.; Rodriguez-Reinoso, F.; Rouquerol, J.; Sing, K.S.W. Physisorption of gases, with special reference to the evaluation of surface area and pore size distribution (IUPAC technical report). *Pure Appl. Chem.* **2015**, *87*, 1051–1069. [CrossRef]

# processes

MDPI

*Article*

# Development of Molecularly Imprinted Polymers to Target Polyphenols Present in Plant Extracts

Catarina Gomes [1] , Gayane Sadoyan [1], Rolando C. S. Dias [1,*] and Mário Rui P. F. N. Costa [2]

1   LSRE-Instituto Politécnico de Bragança, Quinta de Santa Apolónia, 5300 Bragança, Portugal;
    cpgomes@ipb.pt (C.G.); gay.sadoyan@mail.ru (G.S.)
2   LSRE-Faculdade de Engenharia da Universidade do Porto, Rua Roberto Frias s/n, 4200-465 Porto, Portugal;
    mrcosta@fe.up.pt
*   Correspondence: rdias@ipb.pt; Tel.: +351-273-303-088

Received: 29 September 2017; Accepted: 9 November 2017; Published: 14 November 2017

**Abstract:** The development of molecularly imprinted polymers (MIPs) to target polyphenols present in vegetable extracts was here addressed. Polydatin was selected as a template polyphenol due to its relatively high size and amphiphilic character. Different MIPs were synthesized to explore preferential interactions between the functional monomers and the template molecule. The effect of solvent polarity on the molecular imprinting efficiency, namely owing to hydrophobic interactions, was also assessed. Precipitation and suspension polymerization were examined as a possible way to change MIPs morphology and performance. Solid phase extraction and batch/continuous sorption processes were used to evaluate the polyphenols uptake/release in individual/competitive assays. Among the prepared MIPs, a suspension polymerization synthesized material, with 4-vinylpyridine as the functional monomer and water/methanol as solvent, showed a superior performance. The underlying cause of such a significant outcome is the likely surface imprinting process caused by the amphiphilic properties of polydatin. The uptake and subsequent selective release of polyphenols present in natural extracts was successfully demonstrated, considering a red wine solution as a case study. However, hydrophilic/hydrophobic interactions are inevitable (especially with complex natural extracts) and the tuning of the polarity of the solvents is an important issue for the isolation of the different polyphenols.

**Keywords:** molecular imprinting; crosslinking polymerization; polyphenols; adsorbents; amphiphilic materials; vegetable extracts; continuous processes

## 1. Introduction

Polyphenols are organic compounds produced by plants to provide them protection to different kinds of attacks such as UV radiation, parasites, insect pests and many other environmental threats. Close to 10,000 different polyphenols have already been identified. Their incidence changes among the different plant species, as well as the kind of vegetable parts (roots, stalks, barks, leaves, fruits, skin-fruits, shells, seeds, stones, etc.) inside the same species [1].

It is well known that polyphenols present important antioxidant and anti-inflammatory activities, thus providing beneficial effects on human health, such as the protection of the cardiovascular system, anti-cancer effects or anti-aging actions. Besides the preventive effects on human health, polyphenols are also being considered in the treatment of several diseases [1]. Owing to these useful properties, a growing interest on these natural compounds is being observed namely by the pharmaceutical, biomedical, biotechnological and cosmetics industries [1]. Hence, the development of techniques and processes to perform the efficient extraction, purification, fractionation, isolation and concentration of polyphenols is also a subject attracting wide interest from the scientific community and the above-mentioned industries. Extraction methods for the polyphenols present in the vegetable matrices can range from the use of conventional solvents (e.g., hydro-alcoholic solutions) at different operation temperatures (e.g., up to 100 °C) in refluxing extraction [2], microwave assisted and sonication extractions [3], Soxhlet processes [4] to the application of supercritical fluids [5] (usually allowing the efficient operation at lower temperatures, such as 40 °C, making it a "green" process). Due to their particular chemical structures, polyphenols tend to undergo sorption in different kinds of materials, namely activated carbons (by adsorption) and synthetic resins (mainly by dissolution), which are therefore often chosen for the purification, fractionation and isolation of such compounds (see e.g., chapter 16 in [1] and references therein and review works [6] and [7–10] for examples of application systems using different combinations of resins and extracts).

Indeed, a sorption process is a relatively simple option, namely concerning the operation requirements and scale-up, which can be considered for separating compounds present in dilute solutions. Moreover, this is a low-cost technology, allowing, in principle, the facile regeneration of active sorbents. It usually avoids the use of toxic solvents and prevents the degradation of the target compounds. Owing to these benefits, many important processes used in practice to recover and concentrate phenolic compounds are based upon sorption technology [6]. Recent studies also explore the coupled use of ultrafiltration/diafiltration and sorption or membrane technology to recover polyphenols from vegetable extracts [4,10].

Activated carbons, minerals (e.g., siliceous materials or clay), synthetic resins (e.g., styrene/divinylbenzene or acrylic polymer networks) and materials originated from industrial or agricultural wastes (e.g., fly ash, lignocellulosic materials or polysaccharides) are examples of different of adsorbents used to recover and concentrate/purify phenolic compounds [6]. Among these classes, synthetic polymeric adsorbents are inert and durable materials, presenting high stability, selectivity and sorption capability. Ease of regeneration and limited toxicity are also generally associated to adsorbents based on synthetic resins [6]. Furthermore, tailored adsorbents can in principle be designed when synthetic polymers are considered, namely through the control of the hydrophilic and hydrophobic moieties or the inclusion of specific functional groups to improve the selectivity and efficiency (e.g., anionic or cationic groups as in the ion-exchange resins, hydrogen bonding functionalities, etc.).

Molecularly imprinted polymers (MIPs) are polymer networks with tailor-made cavities, having high specificity and affinity with respect to a specific target molecule, thus playing the role of artificial antibodies [11–13]. Sorption and separation, among other important applications (e.g., biological and biomedical processes, development of sensors, catalysis, etc. [14]), are potential areas where MIPs features can yield important gains comparatively to alternative materials. Indeed, after network formation (often through a free radical polymerization using a high crosslinker content), the generated binding sites should preserve geometrical stability, allowing the uptake and release of target molecules in several cycles. Note, however, that performance of MIPs is affected by several factors, namely their preparation process (e.g., chemical composition, target/functional monomer interactions, final morphology, etc.), as discussed in the reference works [11–14] and evidenced with imprinted hydrogels [15] or precipitation/suspension products [16]. Furthermore, specific molecular recognition capabilities of MIPs depend also on their application conditions, particularly the extent of hydrophilic and hydrophobic interactions between the MIPs, the solvents considered and the different molecules present in the solutions. Currently, new strategies are being implemented by the scientific community to address some shortcomings of MIPs, namely through the design of materials combining molecular recognition capabilities and stimulation or by modification of their surface properties (e.g., tuning of the hydrophobic/hydrophilic balance). The grafting of functional polymer brushes on the surface of molecularly imprinted particles (e.g., using RAFT polymerization) is an example of such approaches [17–24]. Indeed, the improvement of water compatibility of MIPs for biological applications [17–19,21], the design of multifunctional materials allowing stimulation [20,22] and the development of drug delivery carriers [23,24] are some new possibilities being explored in this context.

Considering the aforementioned potential affinity of MIPs towards the selected template molecules, adsorbents in this class have been chosen for the selective uptake of phenolic compounds in past research works aiming at different application fields, as shown in works [25–43] and references therein, considering diverse retention materials and plant extracts. Indeed, the isolation and purification of polyphenols present in different vegetable sources [25–35] (e.g., resveratrol in *Polygonum Cuspidatum*, giant knotweed or grape skin) or the detection and quantification of these compounds in beverages [36–39] (e.g., in wine or fruit juices) are examples of such approaches. Resveratrol—due to its popular benefits for human heath—was targeted in many of these studies but the same principles can be applied to other kinds of polyphenols present in vegetable extracts, such as emodin [29], catechin, piceid or A-type procyanidins [31], quercetin and other flavonoids [34,35,37,38], etc. Furthermore, the synthesis of tailored adsorbents based on the molecular imprinting technique can be used for the retention of other kinds of phenolic compounds, namely bisphenol A [40–43], whose presence in beverages, water and food constitutes an important health damaging issue. It should also be emphasized that MIPs show promising capabilities for application in continuous processes, as reported in recent research involving ions separation [44], the extraction of melamine from milk [45], the treatment of wastewater membranes [46] and the valorization of agricultural wastes through the isolation of the biophenol oleuropein from olive leaves [47].

Here, we present results of an experimental research concerning the effect of the molecular imprinting synthesis conditions on the performance of the resulting materials for the retention of polyphenols, in particular when making use of continuous processes. Polydatin was selected as a template molecule and different kinds of MIPs were produced and characterized, namely concerning the targeting of polyphenols present in plant extracts, as described in the next sections.

## 2. Materials and Methods

### 2.1. Reagents

Acrylic acid (AA), methacrylic acid (MAA), acrylamide (AAm), 2-(dimethylamino)ethyl methacrylate (DMAEMA), N-vinylpyrrolidone (NVP), ethylene glycol dimethacrylate (EGDMA), trimethylolpropane triacrylate (TMPTA) were purchased from Sigma Aldrich and 4-vinylpyridine (4VP) from Alfa Aesar. Analytical reagent grade acetonitrile (ACN), dimethylformamide (DMF) and methanol (MeOH) were bought from Fisher Chemical (Loughborough, Leicestershire, UK). Azobisisobutyronitrile (AIBN), acetic acid and n-heptane were purchased from Sigma-Aldrich (Steinheim, Germany). The surfactant sorbitan mono-oleate (Span 80) was purchased from Panreac (Barcelona, Spain). Millipore water (Milli-Q quality) was used in all the experiments unless otherwise mentioned. Polydatin (purity $\geq$ 95%), *trans*-resveratrol (purity $\geq$ 99%), gallic acid (assay 97.5–102.5%, titration), tannic acid (Chinese natural gall nuts) and bisphenol A (purity $\geq$ 99%) were also purchased from Sigma Aldrich. Caffeine (98.5% purity) and quercetin (hydrate, 95% purity) were bought from Acros Organics (Geel, Belgium). Catechin (hydrate, purity $\geq$ 98%) was purchased from Cayman Chemical Company (Ann Arbor, MI, USA).

### 2.2. Synthesis of Molecularly Imprinted and Non-Imprinted Polymer Particles

Molecularly imprinted (MIP), as well as non-imprinted (NIP) polymer particles, have been synthesized following procedures similar to previous works [16]. Briefly, for MIP synthesis with precipitation polymerization, the template polyphenol (polydatin) was mixed with the selected functional monomer and the solvent (a binary mixture was used) up to the formation of a clear solution. Note that, owing to the limited solubility of polydatin (and of many polyphenols in general) in common pure solvents, an appropriate amount of MeOH was generally used to compose the final solvent, as will be discussed in the next section. Then, the template–functional monomer (T-FM) interaction was promoted in an ultrasounds bath during 30 min. A solution containing the required crosslinker and initiator was then added and the final homogeneous mixture was transferred to a glass vessel and purged with a flow of dry argon for 30 min. The vessel was then sealed and polymerization was started, using magnetic stirring at low speed (c.a. 100 rpm), in a thermostatic oil bath pre-set at 60 °C. The reaction was allowed to run for 24 h.

For MIP synthesis with suspension polymerization, Span 80 was first dissolved in n-heptane under vigorous stirring using a thermostatic oil bath pre-set at 70 °C. In parallel, a solution containing the polydatin, selected functional monomer, chosen crosslinker, AIBN and the selected solvent was prepared, following a procedure similar to that above described for precipitation polymerization, but at higher monomer concentrations, as will be discussed below. Then, this solution was drop wise fed to the reaction vessel containing the oil phase (n-heptane/Span 80), under magnetic vigorous stirring (c.a. 1000 rpm) and polymerization was also carried out during 24 h.

NIP particles were also synthesized by precipitation and suspension polymerization, using procedures similar to those above described, but in the absence of the template molecule (polydatin).

The yield of the preparation was estimated using gravimetric analysis and values in the range 90% to 100% were generally observed. Note, however, that some dispersion in these data occurs due to the small particle sizes of certain products and the number of cleaning steps performed.

### 2.3. Isolation and Purification of the Polymer Particles

The synthesized polymer particles were submitted to several cleaning steps in order to perform their isolation and purification. Centrifugation in a large excess of MeOH (c.a. $1/10$ $v/v$) was used in the first stages. The extraction and the cleaning with a methanol/acetic acid mixture (90:10) were afterwards carried out in order to remove impurities and the bonded polyphenol template. The washing processes were monitored through UV–vis spectroscopy measurements up to the time at which the level of the template molecule became lower than the detection limit. Finally, the purified MIP and NIP materials were dried in a vacuum oven at 40 °C.

### 2.4. SEM Analysis

The morphology of the different kinds of synthesized MIP and NIP particles was analyzed by SEM of the dried products, at the Materials Center of the University of Porto (CEMUP). A SEM FEI Quanta 400FEG instrument with the EDS (Energy Dispersive Spectroscopy) system Edax Genesis X4M was used in these analyses.

### 2.5. FTIR Analysis

Characterization of the dried MIP and NIP particles was also performed using IR spectroscopy with an ABB Bomem Fourier Transform Infra-Red instrument, model FTLA2000-104. Particles were mixed with KBr and pressed into pellets in order to collect the correspondent IR spectra.

### 2.6. Solid Phase Extraction Measurements

Solid phase extraction (SPE) measurements were carried out using packing extraction cartridges containing the MIP or NIP particles (e.g., 150 mg of dried products). The cartridges were first eluted with a methanol/acetic acid mixture (90:10), then with methanol and finally conditioned with the solvent selected for the polyphenols uptake assessment during 24 h. A final elution step with this same solvent was performed before testing. Afterwards, cartridges were loaded with the solution containing a specific phenolic compound at the desired concentration (or a mixture phenolic compounds in the competitive testing), using a constant percolation flow rate [16]. After loading, the washing step was performed by percolating the same solvent through the particles. At the end, the elution step was carried out using pure MeOH. The collected fractions correspondent to loading, washing and elution steps were monitored using batch UV–vis spectrometry or HPLC analysis also with UV detection (this analysis was mainly used with SPE competitive polyphenols sorption). From these data, the retention capabilities ($1 - C_{out}/C_{in}$) of the different materials were calculated. These SPE measurements were performed at least in duplicate.

## 2.7. Batch Sorption Measurements

Fixed amounts of the purified MIPs and NIPs in the dried state (e.g., 20 mg) were replicated and first subjected to a conditioning process in SPE tubes (similar to above described the processes with SPE measurements). Tubing ends were after sealed and the particles were mixed with the required amount of the solution (e.g., 2 mL) containing the polyphenol at different concentrations. A blank solution was also included in this procedure. The sorption systems were then equilibrated at room temperature for 24 h under agitation using an orbital shaker. At the end, the supernatant was collected by centrifugation and the concentration measured by UV–vis monitoring. From these absorbance data, the adsorbed amount of polyphenol was calculated using the previously reported calculation method [16,23].

## 2.8. Experiments with Continuous Sorption Processes

Different kinds of continuous sorption processes have also been considered in this research in order to assess the performance of the synthesized materials for the retention and release of different kinds of polyphenols. Generically, the polymer particles were packed in HPLC columns (different column sizes were considered, ranging from 10 to 300 mm bed lengths and 4.6 to 8 mm of internal diameter) and then submitted to saturation and cleaning steps with selected polyphenols, mixtures of these molecules or natural extracts (red wine was used). A frontal analysis procedure was used in these studies, considering an approach similar to that before described [16], but continuous sorption with recycling was also here tried, as discussed below. Continuous sorption with the particles placed in transparent syringe tubes was also carried out in order to visually observe the possible retention of anthocyanins in the adsorbents. In-line and off-line UV measurements were both used for putting into evidence the dynamics of retention and release of the polyphenols during these continuous processes.

## 2.9. HPLC Analysis for Polyphenols Identification and Quantification

HPLC analysis needed for polyphenols identification and quantification (e.g., with the competitive polyphenols sorption/desorption runs) were performed using an Ascentis® C18 column, 5 μ particle size and with dimensions L × ID = 25 cm × 4.6 mm. This column was assembled in a Viscotek GPC pumping system—the module Viscotek TDA 305 and UV detection was used for these purposes. HPLC analysis were performed in isocratic conditions, with $H_2O$/ACN 70/30 (*v/v*) at pH = 3 (adjusted with acetic acid), T = 45 °C and at a constant flow rate Q = 1 mL/min. Note that a straightforward HPLC analysis of the solutions containing polyphenols was here considered. Improved separation and identification of similar polyphenol molecules demand much more complex analysis conditions, as reported in the literature; see e.g., [48,49] and references therein for the discussion of the complexities involved in such kinds of protocols.

## 3. Results and Discussion

In this research, the polydatin molecule (also known as piceid) was selected as a template polyphenol for the molecular imprinting studies. This choice was driven by the hybrid hydrophilic/hydrophobic character of this molecule, namely in comparison with resveratrol, owing to the presence of a glucoside group. Additionally, polydatin is a molecule with a relatively high size. Besides the hydrophilic and hydrophobic interactions, molecular size is a parameter with an important effect on molecular imprinting efficiency through the formation of specific cavities inside the polymer network. Moreover, polydatin is a natural precursor of resveratrol, which is present in many vegetable extracts (e.g., vineyard products, *Polygonum cuspidatum* extracts, etc.), with possible important beneficial effects on human health, namely the protection against hemorrhagic shock, ischemia/reperfusion, heart failure, endometriosis and cancer (e.g., see [50] and references therein).

Different kinds of MIPs were synthesized by changing the functional monomers (acrylic acid, methacrylic acid, 4-vinylpyridine, acrylamide, 2-(dimethylamino)ethyl methacrylate and

*N*-vinylpyrrolidone were alternatively used), the crosslinkers considered for the polymer network generation (ethylene glycol dimethacrylate and trimethylolpropane triacrylate were chosen for that purpose) and also the solvents (toluene, acetonitrile, methanol, dimethylformamide and water, or their mixtures, were used). Diverse types of interactions between the template molecule and the functional monomers were thence tried. The effect of hydrophilic/hydrophobic interactions on molecular imprinting efficiency was also assessed. Precipitation and suspension polymerization were both considered for MIP particles production using AIBN as free radical initiator.

The morphology of the products was studied by SEM and the chemical composition was determined by FTIR. The assessment of the synthesized MIPs with respect to polyphenols uptake was carried out using different techniques, namely batch sorption, individual/competitive solid phase extraction and individual/competitive sorption/desorption in continuous processes (e.g., by packing the MIP particles in HPLC columns). In these assays, solvents with different polarities were considered in order to highlight the effect of the hydrophilic and hydrophobic interactions on the MIPs capability to perform the selective uptake and release of different kinds of polyphenols. Finally, the usefulness of the MIPs for the recovering, isolation and concentration of polyphenols present in natural extracts was assessed through the analysis of sorption and desorption processes with red wine.

The above described strategy was aimed at elucidating possible links between the molecular imprinting conditions (template, functional monomers, solvent, polymerization mechanisms), the structure of the products (morphology, chemical composition) and their performance on polyphenols retention, as thoroughly discussed in the next sections.

### 3.1. Rationale for MIP Synthesis

The molecular imprinting process is based on the promotion of specific interactions (e.g., hydrogen bonding) between the template molecule and a functional monomer. Through the polymerization with a crosslinker, stereo-specific three-dimensional cavities could form in the polymer network. An example of such kinds of interactions in the framework of the present MIP research is presented in Figure 1, with polydatin as the template (T) molecule and 4VP as a possible functional monomer (FM), highlighting the plausible role of hydrogen bonding in this imprinting system.

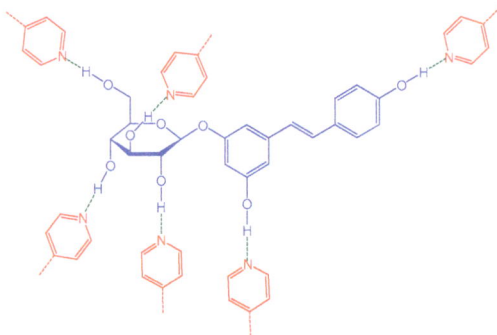

**Figure 1.** Depiction of expected interactions between the template molecule (polydatin) and functional monomers (interaction with pyridyl functional groups was here considered for illustration purposes).

However, the imprinting efficiency (effective formation of selective binding sites towards the target molecule) strongly depends on many factors such as the strength of the interactions between the template and the functional monomer (different functional monomers lead to MIPs with dissimilar performance) or the solvent being used in the synthesis process [11–13,16]. For instance, many solvents (e.g., water) can break the preferential assembly T/FM by hydrogen bonding, which justifies the use solvents with poor H-bond capacity (e.g., ACN) in many imprinting systems [11–13,16].

Additional kinds of intermolecular forces such as hydrophobic interactions, π-π stacking and others, can also enhance or damage the imprinting process, depending on the particular chemical system addressed [11–13]. Moreover, the common solubility of all the components within the reaction locus for molecular imprinting is another issue with important impact on the performance of the produced MIPs.

The behavior of MIPs in molecular recognition is also affected by the morphology of the obtained materials due to possible limitation of the involved mass transfer mechanisms (the size of the template molecule plays here an important role) and the available surface areas, such as in other kinds of adsorbents (e.g., activated carbons, etc.). The morphology of the MIPs can also be tailored (in principle) by the choice of the polymerization conditions: nanoparticles, microparticles or macroscopic monoliths can alternatively be synthesized. Finally, it should also be stressed that the performance of the MIPs in selective uptake and release process is also strongly dependent on the working conditions and the same material can show different capabilities at different environments. One of the most important parameters in this context is the polarity of the solvent considered in the sorption/desorption processes. Non-specific retention mechanisms are often inevitable due to hydrophilic/hydrophobic effects. Thus, the design of amphiphilic MIPs and the change of their surface properties (e.g., inclusion of hydrophilic grafted chains) are also important issues to tailor these materials.

Considering the above discussed effects of the synthesis conditions on MIPs performance, different kinds of materials were produced in our research with polydatin as template, as described in Table 1. MAA, 4VP, AAm, AA, DMAEMA and NVP were alternatively used as functional monomers in order to explore possible differences in T/FM assembly and a concomitant effect on MIPs performance. EGDMA and TMPTA were tried as crosslinkers leading to the formation of networks with dissimilar crosslinking topology (bi and tri-functional, respectively). Assessment of the change of solvent polarity on molecular imprinting was studied considering different mixtures, namely ACN/MeOH, TOL/MeOH, MeOH/H$_2$O and DMF. Note that a small amount of MeOH was generally needed in order to obtain total solubility of polydatin in the reaction medium. Modification in the morphology of the MIPs was tried considering precipitation (usually at a very low monomer concentration) and suspension polymerization (higher concentrations can therefore be specified).

**Table 1.** Polymerization conditions used in the preparation of different MIPs.

| MIP | Polym. Technique | Funct. Monomer | Crosslinker | Solvent | $Y_M$ | $Y_I$ | $Y_{CL}$ | $Y_{FM/T}$ | $Y_{C/D}$ |
|-----|------------------|----------------|-------------|---------|-------|-------|----------|------------|-----------|
| MIP1 | Precipitation | MAA | EGDMA | ACN/MeOH [1] | 5.0 | 3.5 | 79.7 | 6.1 | - |
| MIP2 | Precipitation | 4VP | EGDMA | ACN/MeOH [1] | 5.1 | 3.5 | 80.0 | 6.0 | - |
| MIP3 | Precipitation | AAm | EGDMA | ACN/MeOH [1] | 4.9 | 3.5 | 79.6 | 6.2 | - |
| MIP4 | Precipitation | 4VP | EGDMA | TOL/MeOH [2] | 5.1 | 3.5 | 80.0 | 6.0 | - |
| MIP5 | Precipitation | 4VP | EGDMA | MeOH/H$_2$O [3] | 5.1 | 3.5 | 80.0 | 6.0 | - |
| MIP6 | Suspension | 4VP | EGDMA | DMF | 58.1 | 1.8 | 55.7 | 6.0 | 1.4 [4] |
| MIP7 | Suspension | 4VP | EGDMA | MeOH/H$_2$O [3] | 62.4 | 1.8 | 55.7 | 6.0 | 3.1 [4] |
| MIP8 | Precipitation | AA | TMPTA | ACN/MeOH [1] | 4.6 | 3.8 | 65.6 | 8.2 | - |
| MIP9 | Precipitation | DMAEMA | TMPTA | ACN/MeOH [1] | 4.9 | 3.8 | 65.3 | 8.0 | - |
| MIP10 | Precipitation | NVP | TMPTA | ACN/MeOH [1] | 4.1 | 3.8 | 65.0 | 8.1 | - |

[1] ACN/MeOH = 10/1 (*v/v*). [2] TOL/MeOH = 10/3.4 (*v/v*). [3] MeOH/H$_2$O = 10/2.5 (*v/v*). [4] n-heptane containing 1% (*w/w*) of Span80 was used as continuous phase. NIPs were prepared in parallel, following the same experimental procedure as for MIPs, eliminating the presence of polydatin.

Definitions for the composition parameters described in Table 1 are the following:

- $Y_M$—mass fraction of FM + crosslinker in the solution (%).
- $Y_I$—mole fraction of initiator comparatively to FM + crosslinker (%).
- $Y_{CL}$—mole fraction of crosslinker in the mixture FM + crosslinker (%).
- $Y_{FM/T}$—mole ratio between the FM and the template molecule.
- $Y_{C/D}$—mass ratio between the oil phase and the monomer phase in suspension polymerization.

Differences in the specified monomer concentrations for precipitation and suspension polymerization become evident when the values for $Y_M$ are compared (around 5% with precipitation polymerization and up to c.a. 60% with suspension polymerization). Besides their possible impact on the imprinting process (e.g., due to the effect on products morphology), these differences are also important when the large-scale production of MIPs is intended [16]. Note, however, that thermodynamic solubility effects involving polymers, monomers and solvents present a critical impact on the precipitation and suspension reaction mechanisms, as also below discussed in the framework of the entropic and enthalpic phase segregation.

*3.2. SEM Analysis*

In Figure 2 are presented the SEM micrographs for some products obtained in this research, highlighting the effect of the synthesis conditions on MIPs morphology (see also the Supplementary Materials for additional SEM images). Particles with diameter smaller than 1 μm were formed with MIP1 and MIP2 (MAA and 4VP as FM, respectively, both with ACN/MeOH as solvent) but a higher agglomeration seems to have occurred in the latter case. Aggregates consisting of particles with particle diameter ~2 μm were formed with MIP3 (AAm as FM in ACN/MeOH) and a moderate agglomeration phenomena was observed. MIP4 shows a very different morphology (almost plain surface at the μm scale) due to the effect of the solvent used (TOL/MeOH) with 4VP as FM. Conversely, individual particles with particle diameter ~1 μm, without agglomeration effects, were formed with 4VP as FM and MeOH/H$_2$O as solvent (MIP5). Formation of similar particles was also possible with suspension polymerization with 4VP as FM and DMF as solvent in the monomer phase (MIP6). Stronger agglomeration effects should have occurred for MIP7, resulting also from suspension polymerization with 4VP as FM but MeOH/H$_2$O as solvent in the monomer phase. Aggregates consisting of particles with diameter <1 μm, thus also exhibiting some degree of product agglomeration, have also been observed with other imprinting systems, such as MIP8 (AA as FM in ACN/MeOH and TMPTA as crosslinker). Note that, besides the concentration effect on particles agglomeration, other issues such as the thermodynamics of phase segregation or even the stirring conditions can regulate the formation of such aggregates.

Interesting features concerning the impact of polymerization conditions on products morphologies were also observed with other reaction systems (see also the Supplementary Materials). A much higher agglomeration extent is observed for MIP9 (DMAEMA as FM), namely comparatively to MIP8, highlighting the important effect of the FM on products morphology, even at a low content ($Y_{CL} \sim$ 65%). The SEM micrograph for NIP1, obtained with conditions similar to MIP1, but in the absence of polydatin, shows the formation of individual particles with lower diameter and without agglomeration effects. A possible impact of the template molecule on particles formation is thus evidenced. The morphologies correspondent to EGDMA particles and MAA particles, respectively, obtained with conditions similar to MIP1/NIP1, highlight the entropic and enthalpic precipitation processes involved in the formation of such kind of polymer structures [51,52]. Indeed, in these kinds of polymerization processes, the growing polymer chains can phase-separate from the continuous medium by precipitation due to adverse enthalpic polymer-solvent interactions, such as in the case of MAA polymerization in ACN/MeOH. Additionally, precipitation can also be a consequence of an entropic effect due to the crosslinking action that prevents the mixing of solvent and polymer, such in the case of EGDMA polymerization. Note, however, that in many cases, such as MIPs/NIPs formation, both entropic and enthalpic precipitation mechanisms could be involved with concomitant effects on products morphology. Important insights on these precipitation/dispersion processes can also be obtained through polymer reaction engineering modeling studies (see e.g., [53–55] and references therein) but these issues are beyond the scope of the present paper.

**Figure 2.** SEM images for different MIPs synthesized in this research: (**a**) MIP1 (prec. MAA/ACN/MeOH); (**b**) MIP3 (prec. AAm/ACN/MeOH); (**c**) MIP5 (prec. 4VP/MeOH/H$_2$O); (**d**) MIP6 (susp. 4VP/DMF); (**e**) MIP7 (susp. 4VP/MeOH/H$_2$O); (**f**) MIP8 (prec. AA/ACN/MeOH).

### 3.3. FTIR Analysis

Important information concerning MIPs formation can be obtained through FTIR analysis, namely the extent of carbon-carbon double bonds depletion, functionalization and the incorporation of both the functional monomer and crosslinker in the final polymer network. Examples of such kinds of analysis are presented in Figure 3 for two imprinting systems explored in this research (MIP1 and MIP7 were selected for illustration purposes). Besides the FTIR spectra for the isolated and purified MIPs, the spectra for the constitutive functional monomers (MAA or 4VP) and crosslinker (EGDMA) are also included in the plots.

Observation of Figure 3a,b allows to conclude that a very high conversion of carbon-carbon double bonds must have occurred since only a very small peak is observed for the final materials at around 1630 cm$^{-1}$ (a well-known C=C assignment). Moreover, the incorporation of the functional monomer in the final MIP (an important issue in view of the molecular imprinting efficiency) is also clearly put into evidence in these analyses. Indeed, peak assignments correspondent to the FM (e.g., aromatic C=C in 4VP at ~1550 cm$^{-1}$ or ~1600 cm$^{-1}$), which are not observed in the crosslinker, can be indubitably identified in the synthesized MIPs.

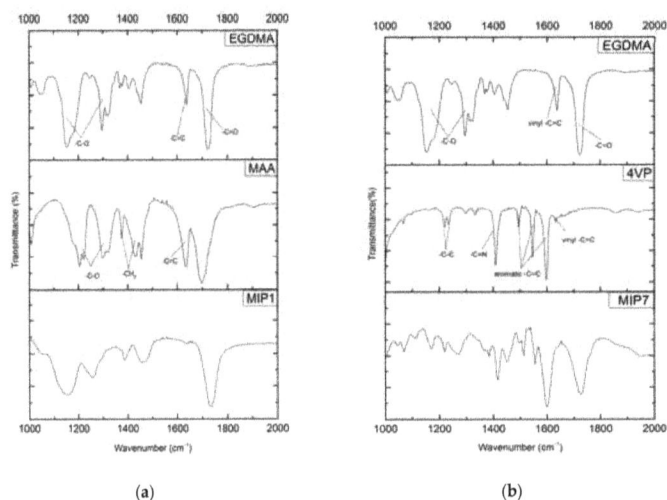

(a)                                    (b)

**Figure 3.** (a) FTIR spectra collected for the MIP1 product (synthesized with MAA as functional monomer and EGDMA as crosslinker); (b) FTIR spectra collected for the MIP7 product (synthesized with 4VP as functional monomer and EGDMA as crosslinker).

### 3.4. SPE Testing

Due to its simplicity and the fast analysis procedure involved, solid phase extraction (SPE) was considered as a starting point in screening studies regarding the performance of the different materials synthesized (MIPs and NIPs) for different polyphenols uptake and release. In view of the expectable influence of the work conditions in the retention and liberation of phenolic compounds, assays with change on solvent polarity were also performed. Below are presented the most relevant results obtained in the framework of these SPE assays.

#### 3.4.1. Study of the MIPs Performance in Low Polarity Solvent

A common strategy for MIPs evaluation consists in using the same solvent of the molecular imprinting step to measure their retention capability. Additionally, in order to avoid breaking possible specific hydrogen bonding effects between the MIP and the target molecules, solvents lacking H-bond capacity, such as ACN, are often used. Thus, in Figure 4 are presented results for the SPE analysis of different materials with ACN/MeOH as solvent together with different polyphenols (note that, as above stated, the use of a small fraction of MeOH (e.g.,) is necessary due to solubility limitations of many compounds). The obtained results show a clearly higher capacity of MIP4, MIP6 and MIP7 to retain polyphenols such as polydatin, resveratrol, quercetin or gallic acid (see Figure 4a–d). However, MIP7 show a superior performance for the uptake of these polyphenols. Also, the relatively high retained amount of these molecules also after the washing process puts into evidence the plausible benefits of the molecular imprinting mechanism. Nonetheless, these MIPs were produced with polydatin as the template and these results show that non-specific retention of other polyphenols in the solid sorbent is unavoidable, even with a favorable solvent. The fitting of small molecules in cavities generated by molecular imprinting of polydatin is a possible explanation for such observations. Additional molecular-imprinting experiments with other template molecules (e.g., gallic acid) and considering the same polymerization conditions used in this work are needed to clarify these issues. Moreover, NIP7 also presents a very high retention capability for these polyphenols, which seems to indicate an enhancement of uptake properties when the polymerization process considered with MIP7/NIP7 is used (suspension polymerization with 4VP as FM and MeOH/$H_2O$ as solvent), as will be further discussed below.

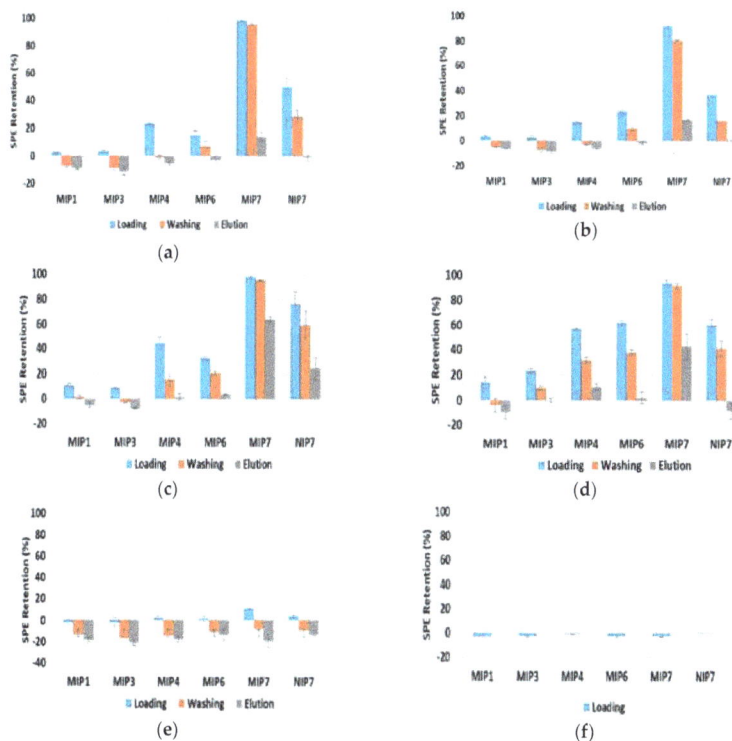

**Figure 4.** Comparison of the SPE performance for different materials with different phenolic compounds: (**a**) polydatin (**b**) resveratrol (**c**) quercetin (**d**) gallic acid (**e**) bisphenol A (**f**) caffeine. In all experiments, loading step was performed with a 0.02 mM polyphenol solution in ACN/MeOH (10/1), washing with the mixture ACN/MeOH (10/1) and elution with pure MeOH. Measurements were performed at least in duplicate (*n* = 2).

On the other hand, results presented in Figure 4e,f, concerning the retention of bisphenol A and caffeine in these same materials, show that almost no affinity with these molecules is observed when ACN/MeOH is used as solvent ; only a very small holding capacity is observed in these conditions for these species. Therefore, some of the synthesized materials, especially MIP4, MIP6 and MIP7, present a good selective ability for polyphenols uptake with working conditions leading to the total elution of other kind of species (e.g., bisphenol A or caffeine).

### 3.4.2. Effect of the Solvent Polarity on MIPs Sorption Capabilities

As observed with many other sorbents, the retention capability of MIPs and their selectivity are strongly dependent on the chosen operating conditions, namely the polarity of the solvent; the temperature is also often used to tune the sorption capabilities. Indeed, due to hydrophilic and/or hydrophobic interactions critical changes on the uptake of a target molecule by a MIP can be observed, leading to possible impairing of the designed selectivity of the material (see e.g., the high retention of caffeine in a 5-fluouracil MIP reported in [16] when water was used as solvent).

These issues are highlighted in Figure 5, where the SPE performance of different MIPs for polydatin retention is compared considering two solvents with different polarities, namely $H_2O/MeOH$ (9/1) and $MeOH/H_2O$ (4/1). A very high retention capability of polydatin in all materials is observed with $H_2O/MeOH$ (9/1), namely when measurements presented in Figure 5a are compared with the results shown in Figure 4a. This outcome is a consequence of the strong hydrophobic interactions between the MIPs and the polydatin molecule arising when $H_2O/MeOH$ (9/1) is used as solvent. Note that very high retention is observed in conditions similar to other molecules (e.g., bisphenol A, namely in comparison with Figure 4e), due to hydrophobic interactions, thus breaking a possible selectivity of the adsorbent.

Still, results presented in Figure 5b show that a strong elution effect is observed when $MeOH/H_2O$ (4/1) is used as solvent (see also Figure 4a). However, even with these unfavorable conditions for sorption, a significant uptake capability is observed with MIP7 (c.a. 50% retention of polydatin), indicating an indisputable ability of this material for polyphenols retention.

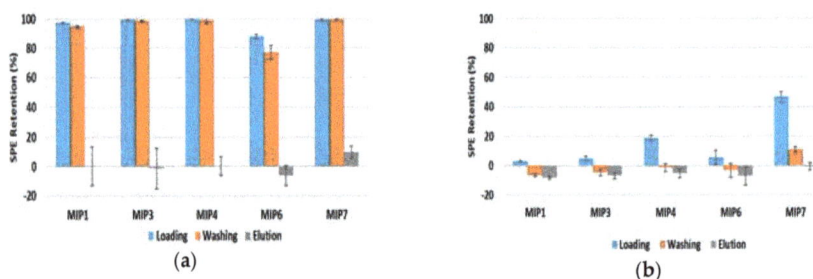

**Figure 5.** Comparison of the SPE performance for different materials in polydatin sorption with different solvents: (**a**) $H_2O/MeOH$ (9/1) with $n = 5$; (**b**) $MeOH/H_2O$ (4/1) with $n = 3$. Loading steps were performed with a 0.02 mM polydatin solution in the selected solvent, washing with the same solvent and elution with pure MeOH.

### *3.5. Batch Sorption*

Batch sorption is a technique regularly considered for MIPs evaluation and some correspondent results for polydatin retention in different materials with ACN/MeOH 10/1 as solvent are presented in Figures 6 and 7. Globally, these results confirm the superior ability of MIP7 for the retention of polydatin (and other polyphenols, as presented in Figure 4), namely in comparison with remaining prepared materials. With the working conditions used, a maximum retention capability $q_{max} \sim 300$ µmol/g is estimated for MIP7, which is much higher than the sorption capabilities reported in the bibliography for related systems. Indeed, $q_{max} \sim 116$ μmol/g [26], $q_{max} \sim 20$ μmol/g [27], $q_{max} \sim 83$ μmol/g [33] and $q_{max} \sim 30$ μmol/g [39] are examples of reported values for the maximum binding capacity of resveratrol in different MIPs, considering ACN as solvent (contribution of non-specific sites is included in these values). As expected, a smaller value $q_{max} \sim 3$ μmol/g was measured for the retention of resveratrol in a MIP choosing ethanol as solvent [32].

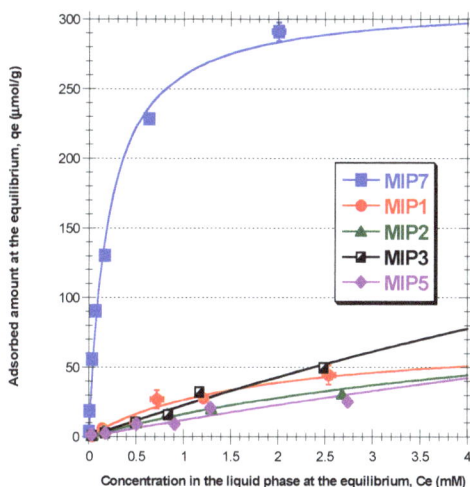

**Figure 6.** Measured equilibrium binding isotherms of polydatin in different MIPs synthesized in this work. Measurements were performed with ACN/MeOH (10/1 *v/v*) as solvent.

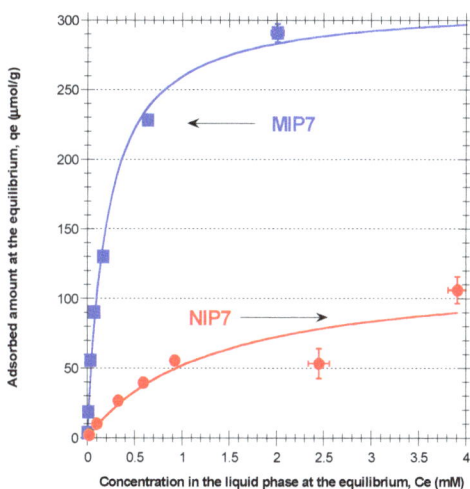

**Figure 7.** Measured equilibrium binding isotherms of polydatin in the MIP7 and NIP7 materials. Measurements were performed with ACN/MeOH (10/1 *v/v*) as solvent.

In Figure 8 it is presented a possible explanation for the very high sorption capability for polyphenols measured with MIP7. Indeed, the hybrid hydrophilic/hydrophobic (amphiphilic) character of polydatin (a glucoside moiety is present in this molecule in contrast with resveratrol), can lead to its simultaneous interaction with the hydrophobic and the hydrophilic phases in a suspension polymerization process, as depicted in Figure 8. The hydrophilic part of polydatin should interact preferentially with the aqueous phase containing water, MeOH, monomer 4VP (the FM) and EGDMA. Note that these three later species have affinity with the hydrophobic phase and should also be distributed along the interface layer. Conversely, the hydrophobic moieties of polydatin tend to shun water, thus migrating to the hydrophobic phase containing n-heptane. Moreover, Span 80 should also be located at the interface layer between the two phases (see Figure 8), with the correspondent water-liking group interacting with hydrophilic moieties (e.g., water, glucoside group of polydatin, MeOH). After polymerization of this assembly and removal of the template (and also Span 80), the resulting polymer network particles should bear surface cavities with high affinity for polydatin and related polyphenols. Mass transfer mechanisms should also be enhanced owing to these surface binding sites, contributing to the higher performance observed with MIP7.

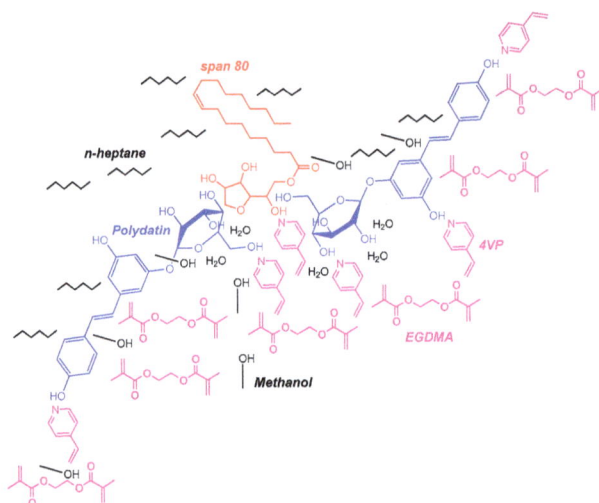

**Figure 8.** Depiction of a plausible surface-molecular imprinting process taking place when polydatin is used as template in a suspension polymerization with n-heptane in the oil phase, Span 80 as surfactant, 4VP as FM, EGDMA as crosslinker and MeOH/$H_2O$ as solvent in the monomer phase.

Note that a close association between polydatin and functional monomer should occur with chosen MIP7 synthesis conditions, due to the higher mass fraction used in the suspended phase, leading to an expected increase of molecular imprinting efficiency. Moreover, polydatin is c.a. 50 times more soluble in DMF than in water/methanol, which is likely to avoid the above depicted surface molecular imprinting process in MIP6.

The relatively high capability of NIP7 (non-imprinted material) for retention of polyphenols (see Figures 4 and 7), should also be a consequence of an interfacial phenomena similar to the above depicted surface-molecular imprinting process. Actually, NIP7 was produced in the absence of polydatin but in the presence of Span 80. Therefore, it is also likely the formation of surface cavities owing to the hydrophilic moiety of Span 80 (containing OH groups) that should be effective (at a lower extent comparatively to MIP7) for polyphenols retention.

### 3.6. Competitive Sorption with SPE

Considering the superior performance of MIP7 concerning polyphenols retention, this material was selected for the main further characterization studies, having in view final applications with natural extracts. At a first stage, the competitive sorption in SPE of a mixture containing gallic acid +

polydatin + resveratrol was chosen to provide information concerning the selectivity of this material and correspondent non-imprinted analogue, with different polyphenols. Results presented in Figure 9 confirm the previous findings with individual sorption assays, namely the high ability of MIP7 and NIP7 to uptake different kinds of polyphenols (eventually present in natural extracts). However, these results also highlight the benefits of molecular imprinting for polyphenols retention because a much higher and stronger uptake ability is shown for MIP7 comparatively to NIP7 (see the results for the washing step).

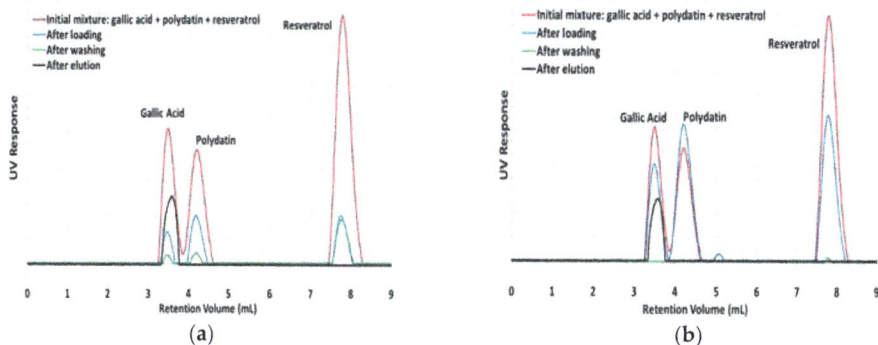

**Figure 9.** HPLC analysis for different steps involved in the SPE testing of MIP7 and NIP7 with the competitive sorption of gallic acid, polydatin and resveratrol. Chromatograms correspondent to the initial mixture (0.02 mM in each component) and to the liquid phases collected after loading, washing and elution are presented. Loading was performed with the polyphenols mixture in $H_2O$/ACN 50/50 (*v/v*), washing with the same solvent and elution with MeOH. HPLC analysis were performed in $H_2O$/ACN 70/30 (*v/v*) at pH = 3 with UV detection at 273 nm. (**a**) Results with MIP7; (**b**) Results with NIP7.

In Figure 10a are presented additional results concerning the competitive sorption of different polyphenols in MIP7 and tannic acid, a plausible compound in many natural extracts, which was now included in the mixture. These results show again the likely ability of this material to uptake polyphenols eventually present in more complex natural matrices. Note that the peak appearing at about 11 mL should be due to some component of tannic acid, which is not a well-defined compound. Elution of other impurities present in the used polyphenols (e.g., in gallic acid, polydatin, resveratrol) is another possible cause for those unexpected peaks.

Previously (see Figures 4a and 5), it was put into evidence the influence of solvent polarity on MIPs performance, namely owing to the interference of hydrophilic and hydrophobic effects. Results presented in Figure 10b–d show similar outcomes with competitive sorption in three solvents with different polarities ($H_2O$/ACN 50/50, $H_2O$/ACN 90/10 and ACN/MeOH 10/1). Note, again, the superior retention in MIP7 comparatively to NIP7 in all the solvents and the noticeable influence of the hydrophobic effects when a polar solvent in used (total retention for the three polyphenols in MIP7 when $H_2O$/ACN 90/10 is considered). These results also show that the tuning of the polarity of the solvent is a critical issue if these materials are intended to separate different kinds of polyphenols because the non-specific retention is inevitable owing to the hydrophobic interactions. The use of some gradient of solvent composition is a possible way to address this issue, as discussed below.

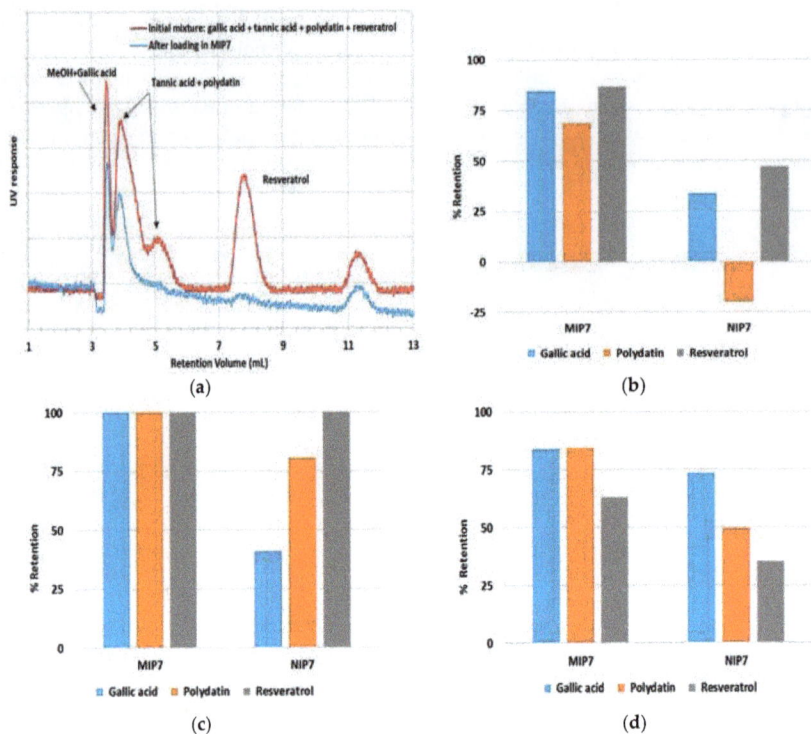

**Figure 10.** (**a**) HPLC analysis for the competitive retention of gallic acid + tannic acid + polydatin + resveratrol in MIP7; (**b**) Comparative competitive retention of gallic acid, polydatin and resveratrol in MIP7 and NIP7 with loading in $H_2O$/ACN 50/50; (**c**) Similar analysis with loading in $H_2O$/ACN 90/10; (**d**) Similar analysis with loading in ACN/MeOH 10/1.

*3.7. Continuous Sorption Processes*

Continuous sorption processes are important if industrial applications are sought. They also make easier the study of many fundamental mechanisms, such as mass transfer phenomena and equilibrium conditions. This approach was also considered in the present research and two different kinds of operating conditions we used (with/without recycling) are depicted in the Supplementary Materials file. In these experimental systems, the dynamics of polyphenols retention/release was measured using in-line, off-line UV detection or HPLC analysis of collected samples. Note that much more elaborated continuous processes for the isolation of polyphenols from plant extracts were recently reported in the literature, namely concerning the recovery of oleuropein from olive leaf extracts with MIPs [47]. In that work, besides a sophisticated experimental set-up, contour plots were used for the selection of the solvent and adsorbent (ethyl acetate and a MIP based on 1-(4-vinylphenyl)-3-(3,5-bis(trifluoromethyl)phenyl)urea/styrene with q ~ 228 µmol/g and IF ~ 12.2 were chosen) [47]. Here, we only present first assessment studies concerning the performance of MIP7 in continuous processes.

Typical profiles measured for the dynamics of sorption in MIP7 of individual polydatin or a mixture of gallic acid + polydatin + resveratrol are presented in Figure 11a,b, respectively. Note the saturation of the MIP adsorbent that it is attained in both situations. For the individual polydatin sorption, equilibrium is reached with q ~ 19 µmol/g (with $H_2O$/MeOH 50/50 as solvent), while the global UV absorbance is used to show the column saturation with the polyphenols mixture.

The potential use of MIP7 as adsorbent in a continuous process for a natural extract containing polyphenols is showed in Figure 12. Red wine (Portuguese red wine from the Douro region) was directly diluted in $H_2O/ACN$ 70/30 at 6% (v/v) and the solution was percolated through MIP7. Results for the dynamics of sorption are presented in Figure 12 with UV detection at 273 nm and also at 515 nm. The ability of the material to retain compounds with response at both wavelengths is shown.

Gallic acid, tannic acid, polydatin or resveratrol are examples of compounds probably present in the red wine with response at 273 nm, while anthocyanins are the likely molecules in red wine with absorbance at 515 nm. Indeed, the high ability of MIP7 to retain anthocyanins (compounds with important applications in bioprocesses) was visually confirmed through the elution of the red wine solution with the adsorbent placed in a transparent syringe tube. An obvious coloration of the polymer particles and discoloration of the solution was thus observed.

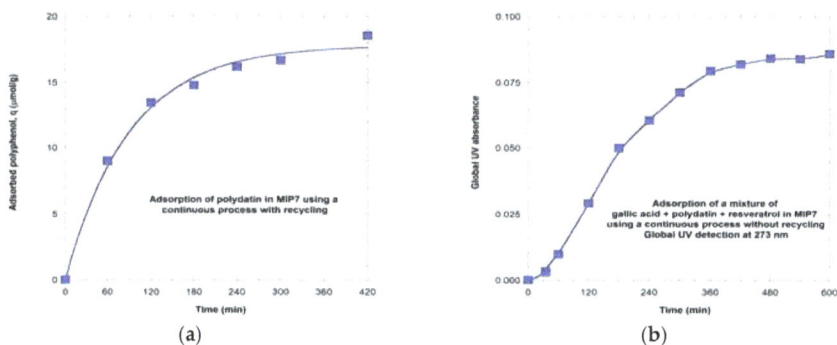

(a)                                        (b)

**Figure 11.** (a) Dynamics of sorption of polydatin in MIP7 considering a continuous process with recycling (the adsorbed amount of polyphenol in μmol per gram of adsorbent is here showed). Process performed with a flow rate Q = 0.25 mL/min, $H_2O/MeOH$ 50/50 (v/v) as solvent, initial polydatin concentration $C_0$ = 0.05 mM and solution volume V = 200 mL; (b) Dynamics of sorption of a mixture of gallic acid + polydatin + resveratrol in MIP7 considering a continuous process without recycling (the UV absorption of the column eluent is here presented). Process performed with a flow rate Q = 0.35 mL/min, $H_2O/ACN$ 70/30 (v/v) as solvent and concentration $C_0$ = 0.02 mM in each component.

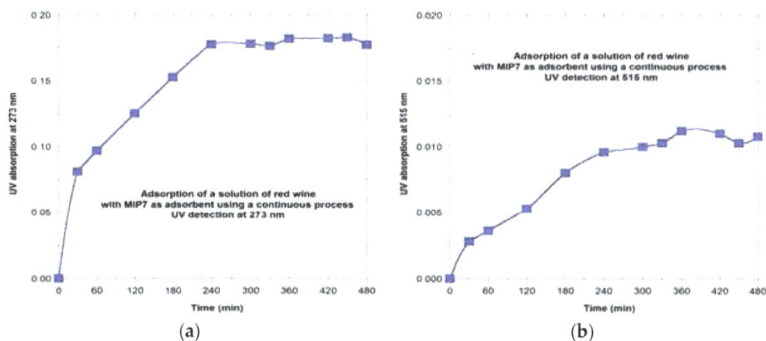

(a)                                        (b)

**Figure 12.** Dynamics of sorption of a solution of red wine in MIP7 considering a continuous process performed with a flow rate Q = 0.35 mL/min and $H_2O/ACN$ 70/30 (v/v) as solvent. Red wine was directly diluted in $H_2O/ACN$ at 6% (v/v). (a) Global UV absorption of the solution at column outlet measured at 273 nm; (b) Similar measurement at 515 nm.

Having demonstrated the good performance of MIP7 on the global retention of polyphenols present in composed mixtures, or eventually in natural extracts, it is important to get insights on the selectivity of the uptake process and the subsequent release from the adsorbent. Actually, these issues are critical if the application of the materials is also aimed at the identification, separation and purification of the individual molecules.

Some results obtained in this context for the composed mixture gallic acid + polydatin + resveratrol are presented in Figure 13. Information concerning the individual amounts of each molecule in the collected samples was obtained by HPLC analysis. The dynamics for the composition of the eluent stream in a continuous process, with $H_2O/ACN$ 70/30 as solvent (see also Figure 11b), is shown in Figure 13a. A preferential initial retention of resveratrol (probably due to hydrophobic interactions) and gallic acid seems to occur but close to saturation the retention order appears to be polydatin > resveratrol > gallic acid and some selectivity of the material through the three components is observed.

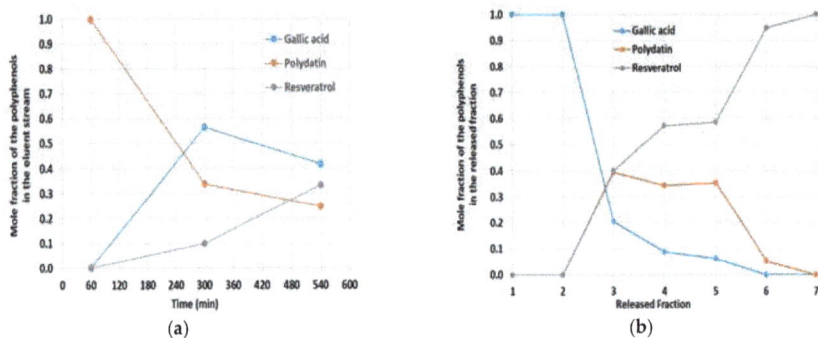

**Figure 13.** (**a**) Composition of the eluent stream in a continuous sorption process for the mixture of gallic acid + polydatin + resveratrol in MIP7 (conditions specified in Figure 11b); (**b**) Composition of different fractions released from the pre-saturated column described in (**a**). $H_2O$ was used as release eluent in fractions 1/2, $H_2O/ACN$ 50/50 in fractions 3/4/5 and ACN/MeOH 10/1 in fractions 6/7.

After saturation, the selectivity on the release of the three components was assessed and a kind of gradient of solvents with different polarities was considered, owing to the expected effect of the hydrophobic interactions on the release of the individual species. Measurements for the composition of the successive fractions obtained are presented in Figure 13b. Fractions 1/2 were obtained with water as eluent and only gallic acid was recovered. With fractions 3/4/5, $H_2O/ACN$ 50/50 was used as eluent and products with mixed compositions were recovered, but with prevalence to resveratrol + polydatin. Finally, with fractions 6/7, ACN/MeOH 10/1 was considered and almost only resveratrol was recovered.

The results above presented confirm the usefulness of the MIP7 adsorbent in the uptake and subsequent release of polyphenols, having in mind the concentration, separation and purification of such kinds of compounds. However, the effect of concomitant hydrophilic and hydrophobic interactions is inevitable in these processes and the tuning of the solvents polarity is a critical issue to achieve good separation of the individual species when complex mixtures are targeted.

### 3.8. Application to Polyphenols Identification and Separation in Natural Extracts

The above described strategy for the synthetic mixtures of polyphenols was also tried with natural extracts, namely a diluted solution of red wine (6% *v/v*) in $H_2O/ACN$ 70/30 (see also Figure 12). The dynamics of sorption of such solution of red wine in MIP7 is further put into evidence in Figure 14, through HPLC analysis of samples collected at the column outlet at different operation times. Note the clear change of the UV absorbance for the samples collected at the column end, namely when compared with the feeding solution, showing the sorption of species with UV response at 273 nm (e.g., gallic acid, tannic acid, polydatin, resveratrol, etc.) and also at 306 nm (resveratrol and polydatin have stronger absorbance at this wavelength).

After column saturation, the release of the adsorbed species was studied by collecting fractions with solvents of different polarities, as above described for the mixture gallic acid + polydatin + resveratrol. Some of the results thus obtained are presented in Figure 15, considering again the HPLC analysis of the fractions collected. In Figure 15a are compared the chromatograms correspondent to the initial red wine solution and to a fraction collected with $H_2O/ACN$ 50/50. Differences observed for the two chromatograms are indicative of composition changes achieved with the sorption process. Among other differences, a new peak at around 6.5 mL retention volume is clearly identified in the chromatogram of the eluted fraction but not in the original extract. This outcome is a plausible consequence of the concentration process achieved with the MIP for some specie present in the original red wine. Indeed, a third chromatogram presented in Figure 15a shows that the concentrated species should be resveratrol. This third chromatogram concerns the same eluted fraction spiked with resveratrol. The matching of the new peak identified in the collected fraction with the one correspondent to the spike of resveratrol is indicative of the likely concentration process achieved for this polyphenol. Similar results are presented in Figure 15b, considering detection at 306 nm and also a sample spiked with polydatin + resveratrol. Besides the clear peak arising at 6.5 mL retention volume, consistent with resveratrol concentration, comparison of the three chromatograms is also indicative of polydatin (or similar species) concentration in consequence of the sorption process performed in the MIP (see change in the chromatogram of the original red wine extract at around 4 mL, that is the region correspondent to polydatin).

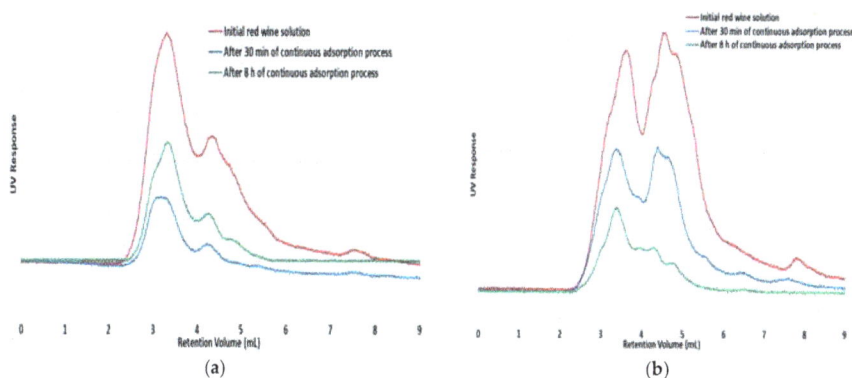

**Figure 14.** HPLC analysis of the liquid phase, correspondent to different sorption time instants, in a continuous process. MIP7 was used as adsorbent with a solution of red wine in $H_2O/ACN$ 70/30 (red wine was directly diluted in $H_2O/ACN$ at 6% (*v/v*)). (**a**) UV measurements at 273 nm; (**b**) UV measurements at 306 nm.

As described in Section 2.9, a simple isocratic HPLC analysis was here considered and some changes in peak positions are possible due to eluent composition variations (see e.g., resveratrol in Figures 10a and 15). Injection of the pure components and spiking studies were used to identify individual molecules in more complex mixtures. However, improved results for the identification and quantification of such compounds can be achieved using more elaborated HPLC gradient strategies, as reported in the literature (see e.g., [48,49] and references therein).

Results here presented are indisputable evidence for the usefulness of the tested MIP for the uptake, identification and separation of polyphenols present in natural extracts. However, many process upgrades are possible in order to improve the results obtained with complex natural extracts where a huge number of different species can be present. A possible strategy is the designing of different MIPs to target diverse polyphenols and consider a train of columns in a continuous sorption process (see also [31]). Indeed, a MIP such as MIP9 can undoubtedly be used to selectively retain acidic species present in natural extracts, such as gallic acid and tannic acid, due to ionic interactions with the functional monomer DMAEMA (see also [56], where additional characterization data for some materials here addressed are also available). Other tailored materials can in principle be synthesized with expected outcomes in the simplification of these complex mixtures. The long-term stability and reusability of these MIPs is also an important issue to assure the reliability and economic viability of the continuous sorption processes [57].

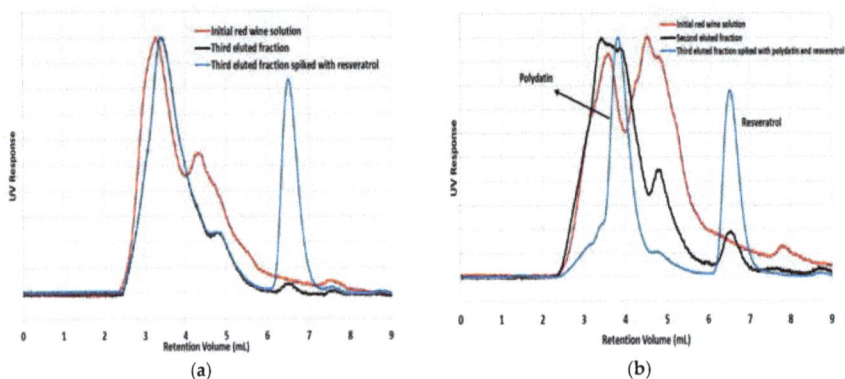

**Figure 15.** (**a**) HPLC analysis of an eluted fraction collected after the sorption of a red wine solution in MIP7 (the third eluted fraction with $H_2O/ACN$ 50/50 is here considered). The chromatograms for initial red wine solution and the eluted fraction spiked with resveratrol are also presented. The plausible concentration enhancement of resveratrol in the MIP is here evidenced with UV measurements at 273 nm; (**b**) Similar analysis with spiking of polydatin and resveratrol and UV measurements at 306 nm.

## 4. Conclusions

This work was devoted to the development of molecularly imprinted polymers to target polyphenols present in vegetable extracts. Polydatin was selected as a template polyphenol due to the relatively high size of the molecule and its hybrid hydrophilic/hydrophobic character. Different kinds of MIPs were synthesized in order to explore preferential interactions between the functional monomers (methacrylic acid, 4-vinylpyridine or 2-(dimethylamino)ethyl methacrylate, among others, were alternatively used) and the template molecule. The role of the polarity of the selected solvent on the molecular imprinting efficiency, namely owing to hydrophobic interactions, was also assessed. Precipitation and suspension polymerization were also considered in order to change the products morphology with potential effects on MIPs performance. Different techniques (SPE, batch sorption, continuous sorption processes) were used to measure the ability of the prepared MIPs to uptake and release polyphenols. Individual and competitive assays (e.g., with composed mixtures of gallic acid + tannic acid + polydatin + resveratrol) were both considered.

Among all prepared MIPs, a material resulting from a suspension polymerization process, with 4VP as functional monomer, EGDMA as crosslinker, water/methanol as solvent, n-heptane as oil and Span 80 as surfactant showed a superior performance on polyphenols uptake. A maximum retention capability $q_{max} \sim 300$ µmol/g is estimated for this MIP, which is much higher than sorption capabilities reported in the literature for related systems. The origin of such significant outcome is the likely surface imprinting process caused by the amphiphilic properties of polydatin. The better performances observed with this suspension product comparatively to the precipitation particles should also rely on these interfacial phenomena. The uptake and release of polyphenols present in natural extracts, considering a red wine solution as a case study, was also assessed using this material as sorbent. The ability of the MIP to uptake polyphenols present in the red wine was put into evidence and the subsequent selective release of the adsorbed species was also demonstrated. However, hydrophilic/hydrophobic interactions are inevitable in such processes, especially with complex natural extracts and the tuning of the polarity of the solvents used (eventually in a gradient process) is an important issue to achieve the improved isolation of the different polyphenols.

Despite the promising results obtained with the above-described MIP, further studies are needed to confirm the likely surface imprinting process. New synthesis tasks with similar suspension processes, but involving other template molecules (changing the amphiphilic interactions) and the measurement of the surface area and pore size distribution (e.g., using BET) for such products and analogues (e.g., correspondent NIPs) are some issues to be addressed in the future. Additionally, many other developments are possible to tailor materials and processes devoted to polyphenols targeting. Tuning of the hydrophilic/hydrophobic balance of the MIPs by changing the polymerization conditions (e.g., composition) or designing more complex architectures (e.g., MIP particles with RAFT surface—grafted hydrophilic chains) are strategies being explored. Use of a train of different MIPs in continuous sorption processes in order to address complex natural extracts or the designing of solvent gradient formulations are other ongoing studies in this context. Valorization of different kinds of agriculture and forest residues is the final goal of this research line.

**Supplementary Materials:** The following are available online at http://www.mdpi.com/2227-9717/5/4/72/s1, Figure S1: Supplementary SEM images for different kinds of products synthesized in this research, Figure S2: Depiction of the continuous processes considered for the sorption of polyphenols and natural extracts in the prepared MIPs.

**Acknowledgments:** This work is a result of project "AIProcMat@N2020—Advanced Industrial Processes and Materials for a Sustainable Northern Region of Portugal 2020," with the reference NORTE-01-0145-FEDER-000006, supported by Norte Portugal Regional Operational Programme (NORTE 2020), under the Portugal 2020 Partnership Agreement, through the European Regional Development Fund (ERDF) and of Project POCI-01-0145-FEDER-006984—Associate Laboratory LSRE-LCM funded by ERDF through COMPETE2020—Programa Operacional Competitividade e Internacionalização (POCI)—and by national funds through FCT—Fundação para a Ciência e a Tecnologia.

**Author Contributions:** Rolando Dias and Mário Rui Costa conceived the idea of design molecularly imprinted polymers to target polyphenols and lead the research team. Catarina Gomes carried out most of the experimental work of this research that was supervised by Rolando Dias. Catarina Gomes also produced the first manuscript draft that was revised by Rolando Dias and Mário Rui Costa. Gayane Sadoyan contribution was performed in the framework of her master thesis at Instituto Politécnico de Bragança, namely concerning the synthesis, characterization and assessment of MIP8, MIP9 and MIP10.

**Conflicts of Interest:** The authors declare no conflict of interest.

## References

1. Watson, R. *Polyphenols in Plants: Isolation, Purification and Extract Preparation*, 1st ed.; Academic Press: London, UK, 2014; ISBN 978-0-12-397934-6.
2. Wang, D.-G.; Liu, W.-Y.; Chen, G.-T. A simple method for the isolation and purification of resveratrol from Polygonum cuspidatum. *J. Pharm. Anal.* **2013**, *3*, 241–247. [CrossRef]
3. Hofmann, T.; Tálos-Nebehaj, E.; Albert, L.; Németh, L. Antioxidant efficiency of Beech (*Fagus sylvatica* L.) bark polyphenols assessed by chemometric methods. *Ind. Crops Prod.* **2017**, *108*, 26–35. [CrossRef]
4. Syed, U.T.; Brazinha, C.; Crespo, J.G.; Ricardo-da-Silva, J.M. Valorisation of grape pomace: Fractionation of bioactive flavan-3-ols by membrane processing. *Sep. Purif. Technol.* **2017**, *172*, 404–414. [CrossRef]
5. Natolino, A.; Porto, C.D.; Rodríguez-Rojo, S.; Moreno, T.; Cocero, M.J. Supercritical antisolvent precipitation of polyphenols from grape marc extract. *J. Supercrit. Fluids* **2016**, *118*, 54–63. [CrossRef]
6. Soto, M.L.; Moure, A.; Domíngues, H.; Parajó, J.C. Recovery, concentration and purification of phenolic compounds by adsorption: A review. *J. Food Eng.* **2011**, *105*, 1–27. [CrossRef]
7. Kühn, S.; Wollseifen, H.R.; Galensa, R.; Schulze-Kaysers, N.; Kunz, B. Sorption of flavonols from onion (Allium cepa L.) processing residues on a macroporous acrylic resin. *Food Res. Int.* **2014**, *65*, 103–108. [CrossRef]

8.  Sun, C.; Xiong, B.; Pan, Y.; Cui, H. Adsorption removal of tannic acid from aqueous solution by polyaniline: Analysis of operating parameters and mechanism. *J. Colloid Interface Sci.* **2017**, *487*, 175–181. [CrossRef] [PubMed]

9.  Buran, T.J.; Sandhu, A.K.; Li, Z.; Rock, C.R.; Yang, W.W.; Gu, L. Adsorption/desorption characteristics and separation of anthocyanins and polyphenols from blueberries using macroporous adsorbent resins. *J. Food Eng.* **2014**, *128*, 167–173. [CrossRef]

10. Pinto, P.R.; Mota, I.F.; Pereira, C.M.; Ribeiro, A.M.; Loureiro, J.M.; Rodrigues, A.E. Separation and recovery of polyphenols and carbohydrates from *Eucalyptus* bark extract by ultrafiltration/diafiltration and adsorption processes. *Sep. Purif. Technol.* **2017**, *183*, 96–105. [CrossRef]

11. Sellergren, B. *Molecularly Imprinted Polymers Man-Made, Mimics of Antibodies and Their Applications in Analytical Chemistry*; Elsevier: Amsterdam, The Netherlands, 2001; ISBN 0-444-82837-0.

12. Ye, L.; Mosbach, K. Molecularly imprinted microspheres as antibody binding mimics. *React. Funct. Polym.* **2001**, *48*, 149–157. [CrossRef]

13. Ye, L.; Mattiasson, B. *Molecularly Imprinted Polymers in Biotechnology*; Springer International Publishing: Cham, Switzerland, 2015; ISBN 978-3-319-20728-5.

14. Whitcombe, M.J.; Kirsch, N.; Nicholls, I.A. Molecular imprinting science and technology: A survey of the literature for the years 2004–2011. *J. Mol. Recognit.* **2014**, *27*, 297–401. [CrossRef] [PubMed]

15. Kadhirvel, P.; Machado, C.; Freitas, A.; Oliveira, T.; Dias, R.C.S.; Costa, M.R.P.F.N. Molecular imprinting in hydrogels using reversible addition-fragmentation chain transfer polymerization and continuous flow micro-reactor. *J. Chem. Technol. Biotechnol.* **2015**, *90*, 1552–1564. [CrossRef]

16. Oliveira, D.; Freitas, A.; Kadhirvel, P.; Dias, R.C.S.; Costa, M.R.P.F.N. Development of high performance and facile to pack molecularly imprinted particles for aqueous applications. *Biochem. Eng. J.* **2016**, *111*, 87–99. [CrossRef]

17. Pan, G.; Zhang, Y.; Guo, X.; Li, C.; Zhang, H. An efficient approach to obtaining water compatible and stimuli-responsive molecularly imprinted polymers by the facile surface-grafting of functional polymer brushes via RAFT polymerization. *Biosens. Bioelectron.* **2010**, *26*, 976–982. [CrossRef] [PubMed]

18. Pan, G.; Ma, Y.; Zhang, Y.; Guo, X.; Li, C.; Zhang, H. Controlled synthesis of water-compatible molecularly imprinted polymer microspheres with ultrathin hydrophilic polymer shells via surface-initiated reversible addition-fragmentation chain transfer polymerization. *Soft Matter* **2011**, *7*, 8428–8439. [CrossRef]

19. Ma, Y.; Zhang, Y.; Zhao, M.; Guo, X.; Zhang, H. Efficient synthesis of narrowly dispersed molecularly imprinted polymer microspheres with multiple stimuli-responsive template binding properties in aqueous media. *Chem. Commun.* **2012**, *48*, 6217–6219. [CrossRef] [PubMed]

20. Zhang, H. Controlled/'living' radical precipitation polymerization: A versatile polymerization technique for advanced functional polymers. *Eur. Polym. J.* **2013**, *49*, 579–600. [CrossRef]

21. Zhao, M.; Chen, X.; Zhang, H.; Yan, H.; Zhang, H. Well-defined hydrophilic molecularly imprinted polymer microspheres for efficient molecular recognition in real biological samples by facile RAFT coupling chemistry. *Biomacromolecules* **2014**, *15*, 1663–1675. [CrossRef] [PubMed]

22. Zhou, T.; Jørgensen, L.; Mattebjerg, M.A.; Chronakis, I.S.; Ye, L. Molecularly imprinted polymer beads for nicotine recognition prepared by RAFT precipitation polymerization: A step forward towards multi-functionalities. *RSC Adv.* **2014**, *4*, 30292–30299. [CrossRef]

23. Oliveira, D.; Gomes, C.P.; Dias, R.C.S.; Costa, M.R.P.F.N. Molecular imprinting of 5-fluorouracil in particles with surface RAFT grafted functional brushes. *React. Funct. Polym.* **2016**, *107*, 35–45. [CrossRef]

24. Oliveira, D.; Dias, R.C.S.; Costa, M.R.P.F.N. Modeling RAFT Gelation and Grafting of Polymer Brushes for the Production of Molecularly Imprinted Functional Particles. *Macromol. Symp.* **2016**, *370*, 52–65. [CrossRef]

25. Schwarz, L.J.; Danylec, B.; Yang, Y.; Harris, S.J.; Boysen, R.I.; Hearn, M.T.W. Enrichment of (*E*)-Resveratrol from Peanut Byproduct with Molecularly Imprinted Polymers. *J. Agric. Food Chem.* **2011**, *59*, 3539–3543. [CrossRef] [PubMed]

26. Cao, H.; Xiao, J.B.; Xu, M. Evaluation of New Selective Molecularly Imprinted Polymers for the Extraction of Resveratrol from *Polygonum Cuspidatum. Macromol. Res.* **2006**, *14*, 324–330. [CrossRef]

27. Ma, S.; Zhuang, X.; Wang, H.; Liu, H.; Li, J.; Dong, X. Preparation and Characterization of Trans-Resveratrol Imprinted Polymers. *Anal. Lett.* **2007**, *40*, 321–333. [CrossRef]

28.  Schwarz, L.J.; Danylec, B.; Harris, S.J.; Boysen, R.I.; Hearn, M.T.W. Preparation of molecularly imprinted polymers for the selective recognition of the bioactive polyphenol, (*E*)-resveratrol. *J. Chromatogr. A* **2011**, *1218*, 2189–2195. [CrossRef] [PubMed]

29.  Zhuang, X.; Dong, X.; Ma, S.; Zhang, T. Selective On-Line Extraction of *Trans*-Resveratrol and Emodin from *Polygonum cuspidatum* Using Molecularly Imprinted Polymer. *J. Chromatogr. Sci.* **2008**, *46*, 739–742. [CrossRef] [PubMed]

30.  Hashim, S.N.N.S.; Schwarz, L.J.; Danylec, B.; Potdar, M.K.; Boysen, R.I.; Hearn, M.T.W. Selectivity mapping of the binding sites of (*E*)-resveratrol imprinted polymers using structurally diverse polyphenolic compounds present in Pinot noir grape skins. *Talanta* **2016**, *161*, 425–436. [CrossRef] [PubMed]

31.  Schwarz, L.J.; Danylec, B.; Harris, S.J.; Boysen, R.I.; Hearn, M.T.W. Sequential molecularly imprinted solid-phase extraction methods for the analysis of resveratrol and other polyphenols. *J. Chromatogr. A* **2016**, *1438*, 22–30. [CrossRef] [PubMed]

32.  Zhang, Z.; Liu, L.; Li, H.; Yao, S. Synthesis, characterization and evaluation of uniformly sized core–shell imprinted microspheres for the separation trans-resveratrol from giant knotweed. *Appl. Surf. Sci.* **2009**, *255*, 9327–9332. [CrossRef]

33.  Schwarz, L.J.; Potdar, M.K.; Danylec, B.; Boysen, R.I.; Hearn, M.T.W. Microwave-assisted synthesis of resveratrol imprinted polymers with enhanced selectivity. *Anal. Methods* **2015**, *7*, 150–154. [CrossRef]

34.  Song, X.; Li, J.; Wang, J.; Chen, L. Quercetin molecularly imprinted polymers: Preparation, recognition characteristics and properties as sorbent for solid-phase extraction. *Talanta* **2009**, *80*, 694–702. [CrossRef] [PubMed]

35.  Xia, Y.-Q.; Guo, T.-Y.; Song, M.-D.; Zhang, B.-H.; Zhang, B.-L. Selective separation of quercetin by molecular imprinting using chitosan beads as functional matrix. *React. Funct. Polym.* **2006**, *66*, 1734–1740. [CrossRef]

36.  Mugo, S.M.; Edmunds, B.J.; Berg, D.J.; Gill, N.K. An integrated carbon entrapped molecularly imprinted polymer (MIP) electrode for voltammetric detection of resveratrol in wine. *Anal. Methods* **2015**, *7*, 9092–9099. [CrossRef]

37.  Euterpio, M.A.; Pagano, I.; Piccinelli, A.L.; Rastrelli, L.; Crescenzi, C. Development and Validation of a Method for the Determination of (*E*)-Resveratrol and Related Phenolic Compounds in Beverages Using Molecularly Imprinted Solid Phase Extraction. *J. Agric. Food Chem.* **2013**, *61*, 1640–1645. [CrossRef] [PubMed]

38.  Hashim, S.N.N.S.; Schwarz, L.J.; Boysen, R.I.; Yang, Y.; Danylec, B.; Hearn, M.T.W. Rapid solid-phase extraction and analysis of resveratrol and other polyphenols in red wine. *J. Chromatogr. A* **2013**, *1313*, 284–290. [CrossRef] [PubMed]

39.  Chen, F.-F.; Xie, X.-Y.; Shi, Y.-P. Preparation of magnetic molecularly imprinted polymer for selective recognition of resveratrol in wine. *J. Chromatogr. A* **2013**, *1300*, 112–118. [CrossRef] [PubMed]

40.  Bayramoglu, G.; Arica, M.Y.; Liman, G.; Celikbicak, O.; Salih, B. Removal of bisphenol A from aqueous medium using molecularly surface imprinted microbeads. *Chemosphere* **2016**, *150*, 275–284. [CrossRef] [PubMed]

41.  Zu, B.; Pan, G.; Guo, X.; Zhang, Y.; Zhang, H. Preparation of Molecularly Imprinted Polymer Microspheres via Atom Transfer Radical Precipitation Polymerization. *J. Polym. Sci. Part A* **2009**, *47*, 3257–3270. [CrossRef]

42.  Yang, Y.; Yu, J.; Yin, J.; Shao, B.; Zhang, J. Molecularly Imprinted Solid Phase Extraction for Selective Extraction of Bisphenol Analogues in Beverages and Canned Food. *J. Agric. Food Chem.* **2014**, *62*, 11130–11137. [CrossRef] [PubMed]

43.  Feng, Q.-Z.; Zhao, L.-X.; Yan, W.; Lin, J.-M.; Zheng, Z.-X. Molecularly imprinted solid-phase extraction combined with high performance liquid chromatography for analysis of phenolic compounds from environmental water samples. *J. Hazard. Mater.* **2009**, *167*, 282–288. [CrossRef] [PubMed]

44.  Jo, S.-H.; Park, C.; Yi, S.C.; Kim, D.; Mun, S. Development of a four-zone carousel process packed with metal ion-imprinted polymer for continuous separation of copper ions from manganese ions, cobalto ions, and the constituent metal ions of the buffer solution used as eluent. *J. Chromatogr. A* **2011**, *1218*, 5664–5674. [CrossRef] [PubMed]

45.  Dursun, E.M.; Üzek, R.; Bereli, N.; Şenel, S.; Denizli, A. Synthesis of novel monolithic cartridges with specific recognition sites for extraction of melamine. *React. Funct. Polym.* **2016**, *109*, 33–41. [CrossRef]

46.  Razali, M.; Kim, J.F.; Attfield, M.; Budd, P.M.; Drioli, E.; Lee, Y.M.; Szekely, G. Sustainable wastewater treatment and recycling in membrane manufacturing. *Green Chem.* **2015**, *17*, 5196–5205. [CrossRef]

47. Didaskalou, C.; Buyuktiryaki, S.; Kecili, R.; Fonte, C.P.; Szekely, G. Valorisation of agricultural waste with na adsorption/nanofiltration hybrid process: From materials to sustainable process design. *Green Chem.* **2017**, *19*, 3116–3125. [CrossRef]
48. Zhang, W.; Jia, Y.; Huang, Q.; Li, Q.; Bi, K. Simultaneous Determination of Five Major Compounds in *Polygonum cuspidatum* by HPLC. *Chromatographia* **2007**, *66*, 685–689. [CrossRef]
49. Fibigr, J.; Satinsky, D.; Solich, P. A study of retention characteristics and quality control of nutraceuticals containing resveratrol and polydatin using fused-core column chromatography. *J. Pharm. Biomed. Anal.* **2016**, *120*, 112–119. [CrossRef] [PubMed]
50. Wang, C.; Luo, Y.; Lu, J.; Wang, Y.; Sheng, G. Polydatin Induces Apoptosis and Inhibits Growth of Acute Monocytic Leukemia Cells. *J. Biochem. Mol. Toxicol.* **2016**, *30*, 200–205. [CrossRef] [PubMed]
51. Downey, J.S.; Frank, R.S.; Li, W.-H.; Stover, D.H. Growth Mechanism of Poly(divinylbenzene) Microspheres in Precipitation Polymerization. *Macromolecules* **1999**, *32*, 2838–2844. [CrossRef]
52. Yasuda, M.; Seki, H.; Yokoyama, H.; Ogino, H.; Ishimi, K.; Ishikawa, H. Simulation of a Particle Formation Stage in the Dispersion Polymerization of Styrene. *Macromolecules* **2001**, *34*, 3261–3270. [CrossRef]
53. Li, L.; Wu, L.; Bu, Z.; Gong, C.; Li, B.-G.; Hungenberg, K.-D. Graft Copolymerization of Styrene and Acrylonitrile in the Presence of Poly(propylene glycol): Kinetics and Modeling. *Macromol. React. Eng.* **2012**, *6*, 365–383. [CrossRef]
54. Sáens, J.M.; Asua, J.M. Mathematical modeling of dispersion copolymerization. *Colloids Surf. A* **1999**, *153*, 61–74. [CrossRef]
55. Lu, T.; Shan, G. Modeling of Two-Phase Polymerization of Acrylamide in Aqueous Poly(ethylene glycol) Solution. *AIChE J.* **2011**, *57*, 2493–2504. [CrossRef]
56. Sadoyan, G. Development of Amphiphilic Adsorbents for the Stimulated Uptake and Release of Polyphenols. Master's Thesis, Chemical Engineering, Instituto Politécnico de Bragança, Bragança, Portugal, July 2017.
57. Kupai, J.; Razali, M.; Buyuktiryaki, S.; Kecili, R.; Szekely, G. Long-term stability and reusability of molecularly imprinted polymers. *Polym. Chem.* **2017**, *8*, 666–673. [CrossRef]

MDPI AG

St. Alban-Anlage 66

4052 Basel, Switzerland

Tel. +41 61 683 77 34

Fax +41 61 302 89 18

http://www.mdpi.com

*Processes* Editorial Office

E-mail: processes@mdpi.com

http://www.mdpi.com/journal/processes

www.ingramcontent.com/pod-product-compliance
Lightning Source LLC
Chambersburg PA
CBHW051722210326
41597CB00032B/5572